AF361834

Machine Health Monitoring and Fault Diagnosis Techniques (Volume II)

Machine Health Monitoring and Fault Diagnosis Techniques (Volume II)

Guest Editors

Shilong Sun
Changqing Shen
Dong Wang

Basel • Beijing • Wuhan • Barcelona • Belgrade • Novi Sad • Cluj • Manchester

Guest Editors

Shilong Sun
Harbin Institute of
Technology
Shenzhen
China

Changqing Shen
Soochow University
Suzhou
China

Dong Wang
Shanghai Jiao Tong
University
Shanghai
China

Editorial Office
MDPI AG
Grosspeteranlage 5
4052 Basel, Switzerland

This is a reprint of the Special Issue, published open access by the journal *Sensors* (ISSN 1424-8220), freely accessible at: https://www.mdpi.com/journal/sensors/special_issues/JB1YIV154U.

For citation purposes, cite each article independently as indicated on the article page online and as indicated below:

Lastname, A.A.; Lastname, B.B. Article Title. *Journal Name* **Year**, *Volume Number*, Page Range.

ISBN 978-3-7258-2723-7 (Hbk)
ISBN 978-3-7258-2724-4 (PDF)
https://doi.org/10.3390/books978-3-7258-2724-4

Contents

Preface

We are pleased to present this Special Issue, a comprehensive collection of research that delves into the forefront of fault diagnosis and condition monitoring in mechanical and electrical systems. The motivation behind this Special Issue lies in addressing the increasing demand for innovative and reliable methods to maintain the operational safety and efficiency of critical systems. This compilation was curated to bridge the gap between theoretical advancements and practical implementations in sensor-based diagnostics, making it a valuable reference for researchers, engineers, and industry professionals.

The featured articles cover a wide range of approaches, including deep learning techniques for transformer and rolling bearing fault diagnosis, innovative use of vibration signal analysis, and time–frequency transformation methods. Notable contributions include the use of Bayesian-optimized machine learning and vision transformer models to enhance the diagnostic precision and robustness across varied operational scenarios. This Special Issue is particularly focused on cross-domain diagnostics and real-time monitoring solutions, emphasizing the adaptive technologies that uphold diagnostic accuracy despite changing conditions.

We extend our gratitude to the authors whose groundbreaking work has made this Special Issue possible, and we acknowledge the invaluable contributions of the peer reviewers who provided critical insights to refine and elevate the research presented. We also wish to thank the editorial team at Sensors for their steadfast support and assistance throughout this endeavor. We hope this Special Issue inspires continued innovation in the field of predictive maintenance and condition monitoring, supporting more resilient and sustainable industrial practices.

Shilong Sun, Changqing Shen, and Dong Wang
Guest Editors

 *sensors*

Editorial

Machine Health Monitoring and Fault Diagnosis Techniques (Volume II)

Shilong Sun [1,2,*], Changqing Shen [3] and Dong Wang [4]

1. School of Mechanical Engineering and Automation, Harbin Institute of Technology, Shenzhen 518055, China
2. Guangdong Provincial Key Laboratory of Intelligent Morphing Mechanisms and Adaptive Robotics, Shenzhen 518055, China
3. School of Rail Transportation, Soochow University, Suzhou 215131, China; cqshen@suda.edu.cn
4. The State Key Laboratory of Mechanical System and Vibration, Shanghai Jiao Tong University, Shanghai 200240, China; dongwang4-c@sjtu.edu.cn
* Correspondence: sunshilong@hit.edu.cn

Citation: Sun, S.; Shen, C.; Wang, D. Machine Health Monitoring and Fault Diagnosis Techniques (Volume II). *Sensors* **2024**, 24, 7177. https://doi.org/10.3390/s24227177

Received: 5 November 2024

Accepted: 6 November 2024

Published: 8 November 2024

This Special Issue highlights a diverse range of pioneering research dedicated to fault diagnosis, condition monitoring, and defect detection in various engineering systems. Focusing on leveraging state-of-the-art sensor technologies and intelligent diagnostic methodologies, the featured articles collectively aim to enhance the robustness, precision, and efficiency of diagnostic practices.

The collection presents significant strides in the domain, including innovative applications of deep learning algorithms for diagnosing faults in transformers and rolling bearings using vibration signals and time–frequency analyses. For instance, the study titled "*Convolutional Neural Network-Based Transformer Fault Diagnosis Using Vibration Signals*" showcases a convolutional neural network approach that significantly improves fault detection in transformers. Another valuable contribution, "*Bayesian-Optimized Hybrid Kernel SVM for Rolling Bearing Fault Diagnosis*", presents an enhanced machine learning framework that boosts the accuracy of rolling bearing diagnostics through Bayesian optimization.

Additionally, the research titled "*Research on Diesel Engine Fault Status Identification Method Based on Synchro Squeezing S-Transform and Vision Transformer*" integrates vision transformers with advanced signal transformation techniques, enabling precise fault identification in complex machinery. Another standout paper, "*A New Method for Bearing Fault Diagnosis across Machines Based on Envelope Spectrum and Conditional Metric Learning*", offers insights into cross-domain diagnostics and demonstrates the adaptability of machine learning models in varied operational environments.

Infrared thermographic (IRT) imaging features prominently within this Special Issue, especially for defect detection in the renewable energy and electronic industries. The review article "*Progress in Active Infrared Imaging for Defect Detection in the Renewable and Electronic Industries*" delves into the use of IRT for high-resolution, non-contact defect detection. This paper examines the integration of IRT with machine learning to identify structural anomalies in photovoltaic panels and electronic components, highlighting practical applications that bolster product quality and reliability.

Furthermore, the paper "*Evaluation of the Diagnostic Sensitivity of Digital Vibration Sensors Based on Capacitive MEMS Accelerometers*" explores digital sensors' effectiveness in continuous condition monitoring, emphasizing their role in detecting early-stage bearing faults. Complementing this, the study "*Evaluation of Hand-Crafted Feature Extraction for Fault Diagnosis in Rotating Machinery: A Survey*" offers an in-depth analysis of feature extraction methods, balancing computational efficiency and diagnostic accuracy.

In the context of startup conditions, "*Localized Bearing Fault Analysis for Different Induction Machine Start-Up Modes via Vibration Time–Frequency Envelope Spectrum*" investigates fault detection in varying operational states, expanding the understanding of time–frequency signal processing techniques. The article "*A Deep Learning Method for Bearing*

Cross-Domain Fault Diagnostics Based on the Standard Envelope Spectrum" highlights the adaptability of machine learning for reliable diagnostics across machine types.

The research *"Prediction of Pre-Loading Relaxation of Bolt Structure of Complex Equipment under Tangential Cyclic Load"* provides a predictive approach to understand and mitigate structural degradation in engineering applications. Additionally, the work *"Preventing Forklift Front-End Failures: Predicting the Weight Centers of Heavy Objects, Remaining Useful Life Prediction under Abnormal Conditions, and Failure Diagnosis Based on Alarm Rules"* offers practical solutions for real-time equipment monitoring and predictive maintenance.

The thorough exploration of these cutting-edge methodologies and applications makes this Special Issue an invaluable resource for professionals, researchers, and engineers dedicated to developing resilient and efficient solutions for equipment reliability and predictive maintenance. The collective expertise of the contributing authors and peer reviewers, alongside the editorial team's guidance, has enriched this compilation, advancing the pursuit of safer, more reliable, and sustainable industrial practices.

Funding: This research received no external funding.

Conflicts of Interest: The authors declare no conflicts of interest.

List of Contributions:

1. Li, C.; Chen, J.; Yang, C.; Yang, J.; Liu, Z.; Davari, P. Convolutional Neural Network-Based Transformer Fault Diagnosis Using Vibration Signals. *Sensors* **2023**, *23*, 4781.
2. Song, X.; Wei, W.; Zhou, J.; Ji, G.; Hussain, G.; Xiao, M.; Geng, G. Bayesian-Optimized Hybrid Kernel SVM for Rolling Bearing Fault Diagnosis. *Sensors* **2023**, *23*, 5137.
3. Li, S.; Liu, Z.; Yan, Y.; Wang, R.; Dong, E.; Cheng, Z. Research on Diesel Engine Fault Status Identification Method Based on Synchro Squeezing S-Transform and Vision Transformer. *Sensors* **2023**, *23*, 6447.
4. Yang, X.; Yang, J.; Jin, Y.; Liu, Z. A New Method for Bearing Fault Diagnosis across Machines Based on Envelope Spectrum and Conditional Metric Learning. *Sensors* **2024**, *24*, 2674.
5. Zhao, X.; Zhao, Y.; Hu, S.; Wang, H.; Zhang, Y.; Ming, W. Progress in Active Infrared Imaging for Defect Detection in the Renewable and Electronic Industries. *Sensors* **2023**, *23*, 8780.
6. Fidali, M.; Augustyn, D.; Ochmann, J.; Uchman, W. Evaluation of the Diagnostic Sensitivity of Digital Vibration Sensors Based on Capacitive MEMS Accelerometers. *Sensors* **2024**, *24*, 4463.
7. Sánchez, R.-V.; Macancela, J.C.; Ortega, L.-R.; Cabrera, D.; García Márquez, F.P.; Cerrada, M. Evaluation of Hand-Crafted Feature Extraction for Fault Diagnosis in Rotating Machinery: A Survey. *Sensors* **2024**, *24*, 5400.
8. Ruiz-Sarrio, J.E.; Antonino-Daviu, J.A.; Martis, C. Localized Bearing Fault Analysis for Different Induction Machine Start-Up Modes via Vibration Time–Frequency Envelope Spectrum. *Sensors* **2024**, *24*, 6935.
9. Zhai, L.; Wang, X.; Si, Z.; Wang, Z. A Deep Learning Method for Bearing Cross-Domain Fault Diagnostics Based on the Standard Envelope Spectrum. *Sensors* **2024**, *24*, 3500.
10. Lu, X.; Zhu, M.; Li, C.; Li, S.; Wang, S.; Li, Z. Prediction of Pre-Loading Relaxation of Bolt Structure of Complex Equipment under Tangential Cyclic Load. *Sensors* **2024**, *24*, 3306.
11. Lee, J.-G.; Kim, Y.-S.; Lee, J.H. Preventing Forklift Front-End Failures: Predicting the Weight Centers of Heavy Objects, Remaining Useful Life Prediction under Abnormal Conditions, and Failure Diagnosis Based on Alarm Rules. *Sensors* **2023**, *23*, 7706.

Article

Localized Bearing Fault Analysis for Different Induction Machine Start-Up Modes via Vibration Time–Frequency Envelope Spectrum

Jose E. Ruiz-Sarrio [1], Jose A. Antonino-Daviu [1,*] and Claudia Martis [2]

[1] Instituto Tecnológico de la Energía (ITE), Universitat Politècnica de València (UPV), 46022 Valencia, Spain; joruisar@die.upv.es

[2] Department of Electrical Machines and Drives, Technical University of Cluj-Napoca, 400114 Cluj-Napoca, Romania; claudia.martis@emd.utcluj.ro

* Correspondence: joanda@die.upv.es

Abstract: Bearings are the most vulnerable component in low-voltage induction motors from a maintenance standpoint. Vibration monitoring is the benchmark technique for identifying mechanical faults in rotating machinery, including the diagnosis of bearing defects. The study of different bearing fault phenomena under induction motor transient conditions offers interesting capabilities to enhance classic fault detection techniques. This study analyzes the low-frequency localized bearing fault signatures in both the inner and outer races during the start-up and steady-state operation of inverter-fed and line-started induction motors. For this aim, the classic vibration envelope spectrum technique is explored in the time–frequency domain by using a simple, resampling-free, Short Time Fourier Transform (STFT) and a band-pass filtering stage. The vibration data are acquired in the motor housing in the radial direction for different load points. In addition, two different localized defect sizes are considered to explore the influence of the defect width. The analysis of extracted low-frequency characteristic frequencies conducted in this study demonstrates the feasibility of detecting early-stage localized bearing defects in induction motors across various operating conditions and actuation modes.

Keywords: AC machines; vibration; bearing; fault diagnosis

Citation: Ruiz-Sarrio, J.E.; Antonino-Daviu, J.A.; Martis, C. Localized Bearing Fault Analysis for Different Induction Machine Start-Up Modes via Vibration Time–Frequency Envelope Spectrum. *Sensors* **2024**, *24*, 6935. https://doi.org/10.3390/s24216935

Academic Editors: Dong Wang, Shilong Sun and Changqing Shen

Received: 4 October 2024
Revised: 17 October 2024
Accepted: 22 October 2024
Published: 29 October 2024

1. Introduction

Induction motors are widely utilized in the industry due to their well-known advantages and their outstanding economic trade off. The maintenance of such equipment represents an economic burden, since only in North America, millions of electrical machines must undergo repairs every year [1]. In particular, the most critical constructive elements in low-voltage induction motors are the rolling bearings placed between the housing and the rotating motor shaft [2]. The early diagnosis of such components is crucial for reducing plant maintenance costs and preventing hazardous scenarios. Bearing fault diagnosis represents a prominent research field and has attracted a vast amount of research interest over the last years [3]. Additionally, the advent of machine learning and the availability of numerous open-source bearing fault datasets (e.g., Case Western Reserve University (CWRU) [4], Paderborn University dataset [5], etc.) have greatly increased the volume of research in this area [6,7]. Nevertheless, the sole data-based identification of bearing defects lacks the physical understanding of the failure mechanics, which hinders cross-domain failure identification. Therefore, research exploring the effects of various bearing defects across different applications and scenarios remains crucial in the field.

Vibration monitoring continues to be the most widespread methodology for rotating machinery diagnosis, including electrical machines. One of the main reasons is the high number of standards based on this physical magnitude [8]. On the other hand, vibration monitoring in electrical machines offers insights for both mechanical and electromagnetic fault signatures [9]. The main drawback of vibration monitoring is the need for external sensors (i.e., normally accelerometers) attached to accessible non-rotating parts and the influence of different mechanical transfer paths accentuating or attenuating potential fault signatures [10]. Other authors have explored alternative techniques for bearing diagnosis in the field of electrical machines by utilizing less-invasive methods exploiting electromagnetic signatures. Radial vibrations caused by bearing defects induce an air-gap variation that can be sensed in the stator current [11]. Some works exploiting current monitoring for bearing diagnosis can be found in [12,13]. In addition to the current monitoring, other authors have explored the utilization of stray-flux to diagnose bearings. The literature emphasizes the utilization of statistical indicators to identify localized bearing faults, with no evident fault signature in induction machines for the acquired stray-flux signals [14,15]. Despite its disadvantages, vibration monitoring remains the preferred methodology for diagnosing bearing faults in induction machines over electromagnetic-nature techniques due to the direct relation between the defect and the acquired signal [16].

Vibration signal processing in steady-state conditions is a broadly studied discipline [17]. The literature identifies various techniques for effectively diagnosing bearing faults, particularly in scenarios where the signals exhibit periodic and time-invariant characteristics [18]. Envelope spectrum analysis is considered as the baseline frequency–domain technique for identifying signal-modulating components. This is a simple and historically effective technique that successfully identifies bearing localized fault signatures in the low frequency range [19]. Other popular techniques for steady-state signal processing are the Discrete Wavelet Transform (DWT) [20], Empirical Mode Decomposition (EMD) [21], and cyclostationary tools [22,23], among others.

Operation at a constant speed for prolonged time is not common in all motoring scenarios, which hinders fault detection by using the classic signal processing approaches. For this reason, signal processing techniques under variable speed conditions have been gaining research attention over the last years [24]. Among the most widely utilized methods are those based on order tracking. These methods leverage the inherent periodicity of fault components relative to the rotating frequency [18]. Consequently, all components can be represented in both the frequency and angular domains, ensuring compliance under variable speed conditions. These resampling techniques provide successful results for localized bearing fault diagnosis, as demonstrated in works such as [25,26]. Nevertheless, the sampling or estimation of the instantaneous frequency is necessary to perform the transformation into the angular domain [27], which imposes some limitations in terms of hardware and computational burden. Resampling-free techniques overcome the limitations of order tracking but involve a necessary post-processing step [24]. Some of the traditional methods to analyze non-stationary signals include linear methods such as the Short Time Fourier Transform (STFT) [28,29] the Continuous Wavelet Transform (CWT) [30], and quadratic bi-linear methods such as Winger–Ville Distribution (WVD) [31]. Moreover, bearing fault diagnosis triggers the application of advanced time–frequency signal processing techniques such as various types of synchrosqueezing transformations [32–34], or Multiple Signal Classification (MUSIC) [35,36]. These advanced techniques offer increased resolution and enhanced energy concentration at a computational cost. The complexity and computational requirements of such transforms represent a limitation in many diagnosis scenarios where simpler methodologies provide an adequate solution, as demonstrated in [37].

In the field of induction machines, bearing diagnosis during different operating conditions, including start-up, is identified as a research gap by many authors [38,39]. In [40], the authors identify defect early detection and severity quantification under non-stationary regimes and the utilization of low-computational-burden processing techniques to ease technology implementation as future research directions. In a recent review [41], the authors also highlight a research gap related to transient analysis under different operating regimes. Moreover, induction machines can be line-started, soft-started, or inverter-fed, which impose different mechanical and electromagnetic conditions that may affect bearing fault identification. Very recently, several authors focused on the vibration transient start-up signature of induction motors to diagnose rotor dynamic defects such as misalignment and mass unbalances [42,43]. Few authors have explored the start-up signature to detect bearing faults. In [44], the authors utilize the CWT and feature extraction algorithms to identify different bearing defects. Other authors have explored the transient start-up current signal for the same purpose [45]. In [37], the authors detect characteristic bearing fault signatures for inverter-fed machines via vibrations acquired in the bearing housing. Thus, the detection of bearing defects under different start-up modes in induction motors represents a research gap in the field.

This work analyzes localized bearing race defects in induction motors under various starting modes using a resampling-free, straightforward time–frequency transformation. The utilized signal processing tool extends classic vibration signal envelope analysis through the STFT. This paper aims to generalize the detection of incipient bearing faults via vibration signals across different defect widths, excitation modes, and operating regimes in low-voltage induction motors. Moreover, the transient results are analyzed along with their steady-state counterparts. The vibration data are generated in a custom test bench where vibrations are acquired in the Drive End (DE) of the machine housing for different constant load points and two bearing defect widths. Two different induction motor starting modes (i.e., line-started and scalar-controlled inverter-fed) are implemented to elucidate the main differences between them. In addition, the obtained signature is compared with an existing open-source dataset including inverter-fed transient vibration signals acquired in the bearing surroundings. This paper is structured as follows. Section 2 presents the mechanics of localized bearing defects and the employed time–frequency transformation tool. Section 3 describes the test bench in which the vibration data are generated and acquired. Section 4 presents the main results of the analysis and compares the obtained signature with existing datasets. Finally, Section 5 discusses the main outcomes and limitations, and Section 6 concludes this work and defines future research steps on the topic.

2. Theoretical Background

This section aims to present the fault mechanics of localized bearing defects, thereby enhancing the reader's understanding of the analysis. Moreover, the basics of the utilized signal processing pipeline are described in a comprehensive manner.

2.1. Localized Bearing Fault Mechanics

Single-row deep-groove ball bearings are extensively utilized in low-voltage electrical machines. These are formed by inner and outer races, rolling elements (in this case, spherical), and a cage equally spacing the rolling elements. Figure 1 shows an expanded view of a spherical rolling element bearing and the identifications of its main components. Other fundamental parts are the lubricants and the seals containing the lubricant.

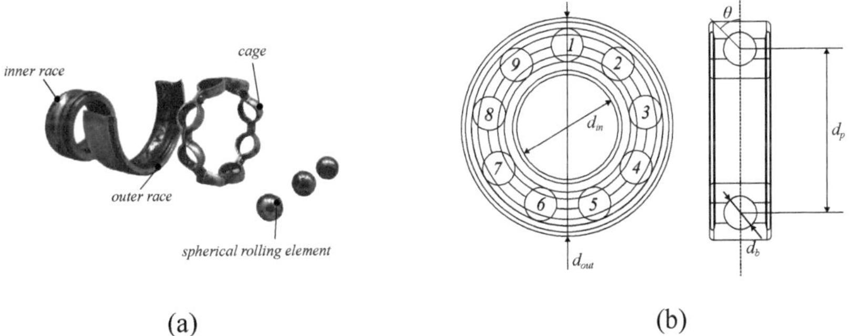

(a)

(b)

Figure 1. (**a**) Expanded deep-groove ball bearings view, (**b**) bearing geometry including numbering of rolling elements (i.e., 1 to 9 numbers) and main dimensions.

Rolling bearings may suffer from a wide variety of faults of a different nature. These faults are typically categorized into three primary groups: localized, extended, and distributed defects. Distributed bearing defects are equally spaced over the bearing circumferential space. One classic example of distributed bearing fault is race fluting due to bearing currents [46]. On the other hand, localized and extended bearing defects are confined in space within the different elements of the rolling bearing (i.e., races or rolling elements). The main difference among these two defect types is the extension of the defect. The extended type may span a larger space than the localized counterpart. Some examples of localized defects are pits and cracks, while extended defects are commonly found in the form of fatigue spalling [47]. The present work focuses only on the localized race defect type, which is often recognized as the most incipient type of bearing fault.

Localized bearing faults are ideally understood as repetitive shocks caused by the contact between rolling elements and localized defects. These periodic signals are characterized by characteristic frequencies depending on the defect location and bearing geometry. In the ideal case of a fixed outer ring and a rotating inner race, these frequencies are defined as follows [48]:

$$f_{BPO} = \frac{n_b}{2} \left[1 - \frac{d_b cos(\theta)}{d_p} \right] f_r \tag{1}$$

$$f_{BPI} = \frac{n_b}{2} \left[1 + \frac{d_b cos(\theta)}{d_p} \right] f_r \tag{2}$$

$$f_C = \frac{1}{2} \left[1 - \frac{d_b cos(\theta)}{d_p} \right] f_r \tag{3}$$

$$f_{BS} = \frac{d_p}{2d_b} \left[1 - \frac{d_b^2 cos^2(\theta)}{d_p^2} \right] f_r \tag{4}$$

where frequencies f_{BPO}, f_{BPI}, f_C, and f_{BS} correspond to the ball pass frequency in the outer and inner raceways, the cage frequency, and the ball spin frequency, respectively. These are determined by the number of rolling elements (n_b), and geometric parameters, including the pitch diameter (d_p), ball diameter (d_b), and contact angle (θ), as shown in Figure 1b.

A more realistic scenario considers the existence of slippage between rolling elements and the races, which causes the defect impulses to adopt a quasi-periodic behavior [49]. In this case, the periodicity of the pulses experiences some random fluctuations, slightly affecting the characteristic fault frequency locus. The prediction of the exact amplitude of the overall vibration signal represents a complex mechanical problem including nonlinear multi-body dynamics that requires specific and computationally expensive finite element models [47]. Nevertheless, even if the vibration signature depends on the specific topography and tribology of the defect, some studies relate the amplitude increase with the defect size and the shaft speed. According to [50], the increment in vibration is related to the defect size depending on the location and the defect size ratio (D_r), which is defined as follows:

$$D_r = \frac{\theta_d}{\Delta\theta} = \frac{n_b \theta_d}{2\pi r} \tag{5}$$

where θ_d represents the defect size, r is the bearing radius corresponding to the race containing the defect arc, and $\Delta\theta$ is the circumferential space between two rolling elements. Figure 2 presents a graphical explanation of the defect ratio. For small defect ratio values (i.e., $D_r < 1$) both inner and outer race defects present a linear relationship between the vibration rms value and the defect size. Thus, for increased speed and defect width, an increase in the vibration amplitude is expected.

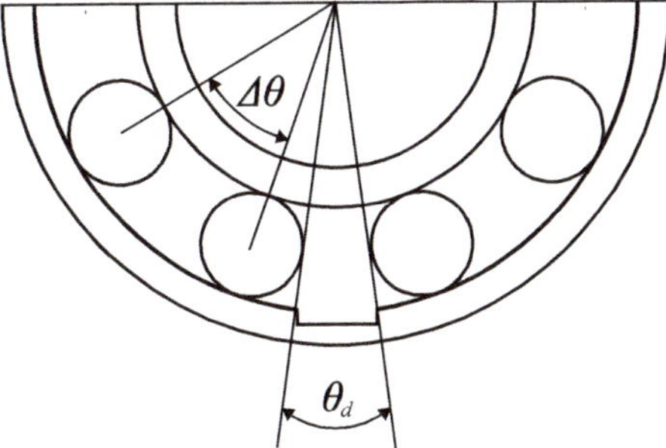

Figure 2. Defect ratio graphic description.

2.2. Time–Frequency Envelope Spectrum

Vibration signals caused by localized defects in bearings are characterized as amplitude-modulated, with a high-frequency carrier signal being modulated by a lower-frequency component. Classic signal processing tools exploit the demodulation of the vibration signal to extract information from the quasi-periodic vibration pulses. The benchmark processing strategy for localized bearing fault diagnosis is the envelope spectrum analysis [19]. Vibration signals are first demodulated and then transformed in the frequency domain by utilizing the Fast Fourier Transform (FFT). Demodulation is performed using the Hilbert Transform (HT), which generates the complex analytic signal. Given a discrete, time-dependent signal $x(t) = A\cos(\omega t)$, the corresponding analytic signal is defined as follows:

$$HT[x(t)] = \hat{x}(t) = A[\cos(\omega t) + j\sin(\omega t)] \tag{6}$$

Note that the analytic signal is defined as a complex-valued function in which the imaginary part is $90°$ shifted with respect to the real function. The envelope of the signal $x(t)$ is provided by the module of the analytical signal $\hat{x}(t)$. Figure 3 shows an example of a quasi-periodic bearing vibration signal and its envelope.

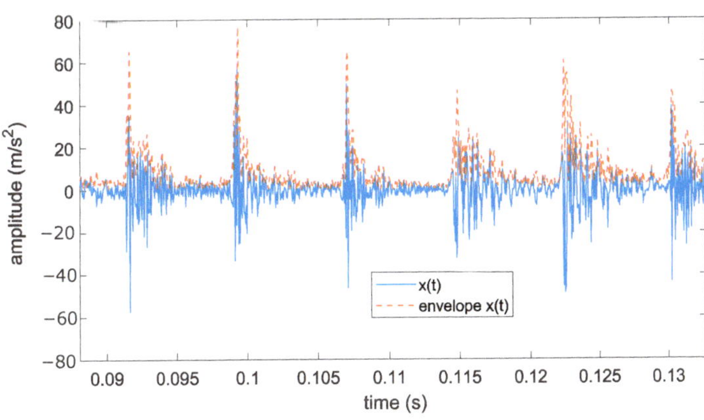

Figure 3. Bearing defect vibration signal $x(t)$ and its envelope.

Variable speed vibration signals are not periodic by nature. Thus, the direct application of the FFT to non-periodic signals does not provide physically meaningful results. Time–frequency transformations such as the STFT, DWT, or WVD overcome these limitations and provide qualitative and quantitative information regarding signal evolutions. The utilized signal processing method comprises several steps. First, a low-pass filter tuned at the defined upper frequency limit (i.e., 1000 Hz in the present study) is applied to the raw signal. Then, the envelope of the signal is obtained by utilizing the above-described

HT. To effectively track significant bearing fault frequencies, the signal should be band-pass filtered before applying the STFT. This process primarily serves to filter out high-frequency components and to eliminate the DC offset from the envelope signal, which would otherwise dominate the signal if not removed. The band-pass filter lower cut-off frequency is defined at 5 Hz, while the upper limit is set at the highest frequency of interest. Finally, the signal is down-sampled to match the Nyquist frequency with the highest frequency of interest to further prevent aliasing phenomena. The present analysis utilizes the STFT to interpret the signature of the envelope spectrum during the machine start-up, which is a windowed version of the classic FFT. This tool is mainly selected due to its simplicity and low computational burden. The STFT is defined as follows:

$$STFT[x(t)] = X(\tau, f) = \int_{-\infty}^{\infty} x(t)\, w(t - \tau) e^{-j2\pi ft} dt \tag{7}$$

where τ is the window length, and the function $w(t - \tau)$ represents a window of length τ centered at instant t. In addition, a Hanning window is applied to each signal section to minimize spectral leakage, and a high window overlap level is applied to improve the time–frequency map resolution. Note that the STFT offers an adequate time–frequency resolution for the application at hand, but it presents a limited frequency resolution depending on the window length. Figure 4 depicts the signal processing pipeline used to achieve adequate time–frequency representation of the transient vibration signature.

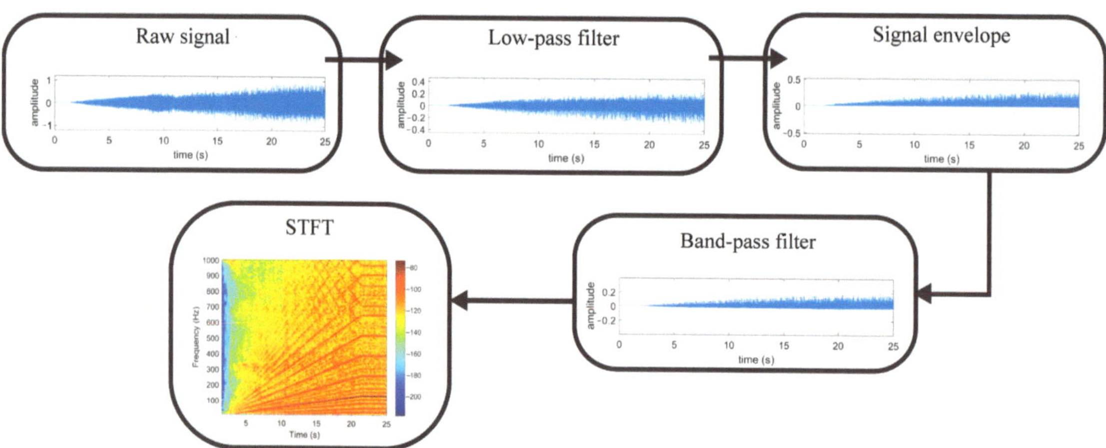

Figure 4. Signal processing pipeline graphic description with an inner race defect example.

3. Experiment Description

The proposed analysis is implemented by utilizing an induction motor test bench internally hosting the different bearings being tested. The utilized motor is a four-pole squirrel cage induction machine with 36 stator slots and 28 rotor bars. Figure 5 shows the machine cross-section, and Table 1 presents its main characteristics.

The bearings under test are internally allocated within the DE plate of the induction machine. This represents a realistic diagnosis scenario including the structural response of the motor and not only the bearing structural elements. The induction motor is coupled to a DC generator via flexible mechanical coupling that imposes a constant resistant torque. The load torque is manually controlled by utilizing a variable autotrasformer connected to the field winding. The armature winding of the DC generator is connected to a dissipation resistance. The induction machine is either directly supplied by a 50 Hz three-phase network or driven via a Variable Frequency Drive (VFD). The machine's alignment is precisely adjusted using a commercial tool that quantifies misalignment, ensuring that the levels remain within standard limits. Figure 6 shows a schematic of the test bench.

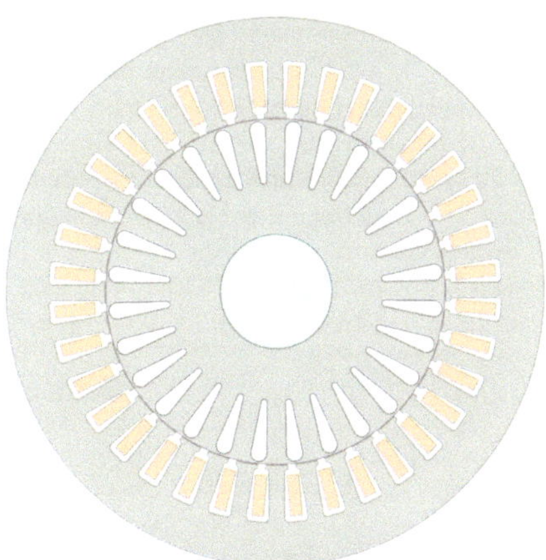

Figure 5. Induction motor specimen cross-section.

Table 1. Induction motor characteristics.

Number of poles	4
Rated power	1.1 kW
Rated speed	1440 rpm
Power factor	0.78
Rated voltage	230/400 V
Number of rotor bars	28
Number of stator slots	36

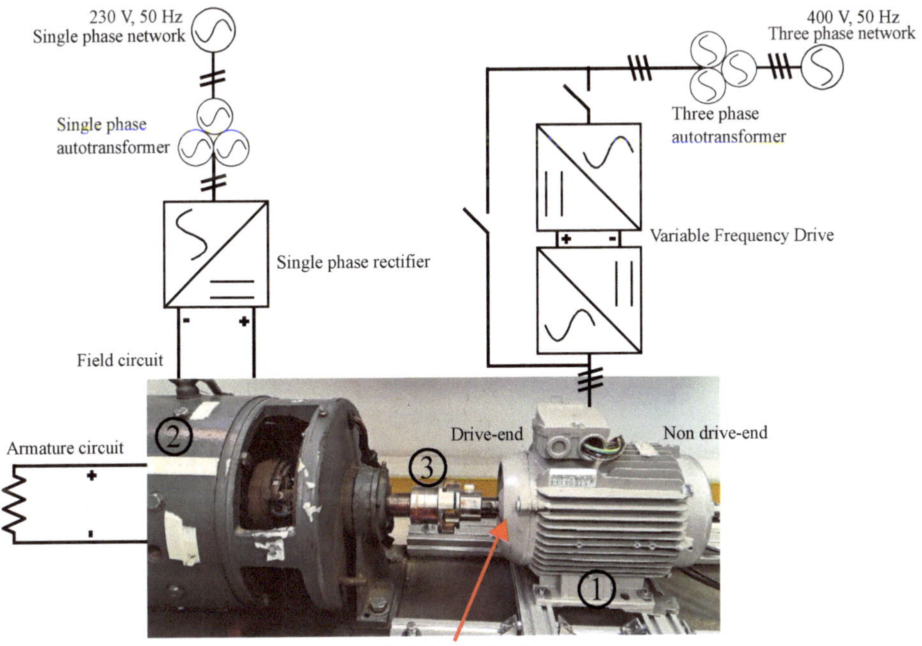

Figure 6. Test bench graphic description. (1) Induction machine including faulty bearing, (2) DC generator imposing constant resistant torque, (3) flexible coupling.

The signal acquisition is performed by utilizing commercial piezoelectric unidirectional accelerometers (PCB352C33). These are placed in two orthogonal circumferential directions in the DE plane (i.e., located at 12 o'clock and 3 o'clock). In this way, the reliability of the measurement is improved by considering slightly different structural responses. The sensor is attached using a well-known adhesive polymer (UHU® Patafix), which offers proper stability and attachment flexibility [51]. The sensors are connected to a signal conditioning unit, and the acquisition is performed using a wave recorder (Yokogawa DL350, Tokyo, Japan) at a sampling frequency of 20 kHz. Figure 7 shows the location description of the accelerometers in the induction machine specimen.

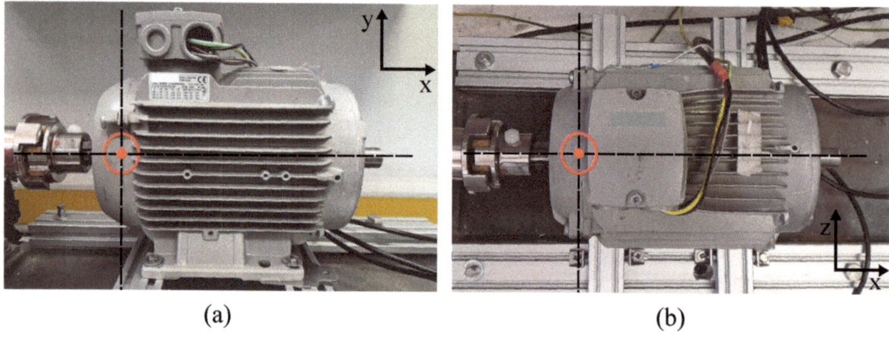

Figure 7. Accelerometer locus description. (**a**) Vertical xy-plane, (**b**) horizontal xz-plane.

This study includes different spherical rolling element bearings with different localized defects in the races. The defects are artificially induced via electric discharge machining in both the inner and outer races. Moreover, two different defect widths are implemented within the range of $D_r < 1$, corresponding to widths of 0.5 and 1 mm, respectively. Figure 8 depicts the different bearings under test with the implemented race faults. Table 2 shows the main bearing dimensions according to Figure 1b, together with the expected fault signature in the frequency domain following Equations (1)–(4). The characteristic fault signature is provided in terms of a coefficient k, only including geometry characteristics, which multiplies the rotating frequency to determine the characteristic fault locus in the frequency domain. The faulty bearings are allocated within the DE end plate, as shown in Figure 6. Outer race defects are placed in the maximum radial load area at the 6 o'clock circumferential position.

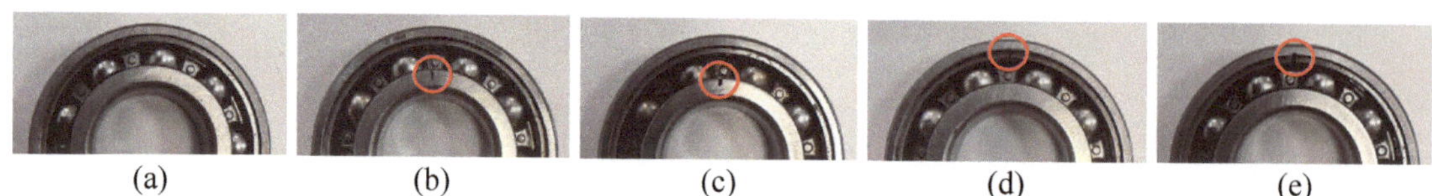

(a)   (b)   (c)   (d)   (e)

Figure 8. Bearing defect description. (**a**) Healthy, (**a**) 0.5 mm inner race defect, (**c**) 1 mm inner race defect, (**d**) 0.5 mm outer race defect, (**e**) 1 mm outer race defect.

Table 2. Bearing dimensions with $n_b = 9$ and characteristic frequency coefficients k assuming $\theta = 0$.

d_{in} [mm]	d_{out} [mm]	b_d [mm]	d_p [mm]
25	52.00	7.94	38.5
BFO	*BFI*	*BSF*	*CF*
3.59	5.41	2.37	0.40

The study of the bearing fault signature is performed during the machine's transient start-up. This strongly varies depending on the electrical actuation of the induction machine.

The present analysis examines two of the most commonly encountered actuation scenarios, specifically those involving VFD actuation and direct line-fed operation. The direct line-fed induction machine at the rated line-to-line voltage produces an extremely fast start due to the low inertia generally present in low-voltage machines. The abrupt start-up of the machine may mask the useful transient vibration response due to a high-amplitude initial shock. To mitigate this, a reduced voltage start-up at 50% of the rated line-to-line voltage is additionally implemented. In this way, the motor start-up is elongated and surpasses the duration of the acquired initial mechanical shock. Figure 9 exemplifies the differences between the rated and 50% line-to-line voltage start-ups at the rated steady-state slip and healthy bearing. The VFD-fed actuation is performed by utilizing a commercial inverter featuring an open-loop scalar control. The control system progressively varies the frequency and the supply voltage to keep a constant V/Hz ratio. This is performed at different variation rates, which are defined by a ramp length parameter. Thus, different transient start-up lengths are imposed to verify the behavior of faulty bearing components at different acceleration rates. The examined VFD-fed start-up ramps are defined with durations of 5 and 20 s, respectively.

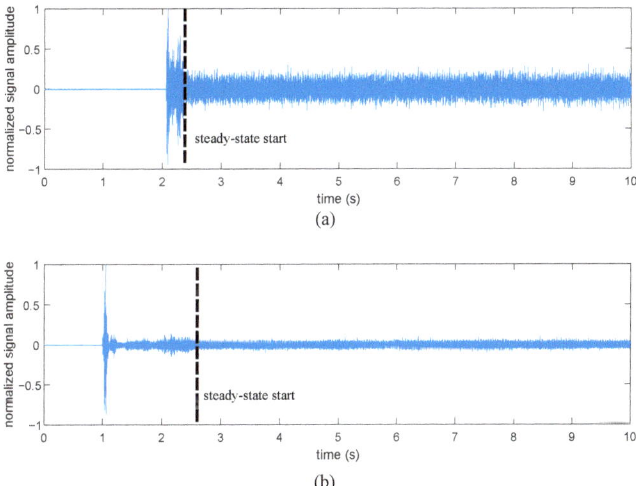

Figure 9. Line-fed induction machine startup vibration signal at 12 o'clock for (**a**) rated line-to-line voltage, (**b**) 50% rated line-to-line voltage.

The experimental campaign includes the vibration signal acquisition at different constant load points imposed by the DC generator. The machine load level is defined by the slip at steady-state (i.e., relative difference between synchronous and shaft rotating speeds). Note that the slip heavily swings from 1 to nearly 0 in the line-started case, while it is kept constant during the VFD-fed startup [52]. Five load points are defined, corresponding to the inherent slip at no-load and four points in steps of 25% of the rated slip (i.e., 1485 rpm, 1470 rpm, 1455 rpm, and 1440 rpm). Under conditions of reduced line-to-line voltage excitation, only four load points are defined, as the available torque scales with the square of the voltage. Additionally, no-load excitation already accounts for 25% of the rated slip. A total of 10 startups and steady-state signals of a 30 s duration are acquired per load point, startup type, and bearing fault topology.

4. Analysis of Results

This section presents the acquired results both in steady-state and different transient startup signals. Moreover, the results are compared with an open-source dataset, where faulty bearing vibration signals are directly acquired in the bearing surroundings during a VFD-fed startup.

4.1. Steady-State Analysis

The steady-state analysis at the rated slip of the vibration envelope spectrum elucidates the different fault signatures in the acquired signals. The spectrum is obtained by enveloping the vibration signal and further transforming it in the frequency domain though the well-known FFT. A Hanning window is implemented to reduce spectral leakage.

Figure 10 shows the envelope spectrum for different bearing specimens, including healthy, inner race, and outer race defects. The spectrum includes signals acquired during line starting at rated line-to-line voltage and VFD-fed excitation. The displayed signals are acquired at the 12 o'clock circumferential position at the rated slip (i.e., 1440 rpm shaft rotation). An immediate observation is the clear identification of the characteristic fault frequencies corresponding to inner and outer race-localized defects in both the line-started and VFD-fed cases. The envelope spectrum in healthy conditions does not show relevant information regarding bearing defects and only highlights the increased noise for VFD-acquired signals. Note that the envelope spectrum components observed in the healthy case corresponding to Figure 10a are not present in the faulty cases. This fact emphasizes the absence of significant electromagnetic vibration components influencing the bearing fault detection within the frequency range of interest when demodulation tools are utilized. Table 3 shows quantitative data regarding the amplitude and frequency location of characteristic faulty components for both defect sizes and actuation modes. A first conclusion drawn from the steady-state analysis is that the fault frequencies are independent of the actuation mode, as evidenced by the nearly identical amplitude values observed for both the line-started and VFD-fed modes. The light changes observed in the frequency location are derived from the manual tuning of the shaft speed by utilizing a manual tachometer and from the expected rolling element slippage. The amplitude comparison between 0.5 mm and 1 mm defects shows a slight decrease for the higher defect level. This fact contrasts with the expected rms increase in the vibration signal for higher defect sizes [50]. Nevertheless, the amplitude of the mechanical shocks follows the expected rms vibration increase, as shown in Figure 11. This increase is better observed for inner race defects in the present study. In addition, rotating frequency-modulated components around fault characteristic frequencies (i.e., $kf_{BPI} \pm nf_r \ \forall \ k = 1,2,3,\ldots \ n = 1,2,3,\ldots$; $kf_{BPO} \pm nf_r \ \forall \ k = 1,2,3,\ldots \ n = 1,2,3,\ldots$) are more prominent for reduced defect sizes.

Table 3. Characteristic frequency location and amplitudes for different faulty bearings.

	Line-Started		VFD-Fed	
	Frequency [Hz]	**Amplitude [dB]**	**Frequency [Hz]**	**Amplitude [dB]**
	Outer race, 0.5 mm			
f_{BFO}	86.67	−51.34	86.67	−60.21
$2f_{BFO}$	173.31	−52.82	173.31	−59.53
$3f_{BFO}$	260	−54.58	259.96	−61.24
	Outer race, 1 mm			
f_{BFO}	86.44	−62.4	86.6	−60.47
$2f_{BFO}$	172.9	−63.32	173.23	−60.32
$3f_{BFO}$	259.32	−63.33	259.83	−60.32
	Inner race, 0.5 mm			
f_{BFI}	129.37	−49.95	129.4	−49.81
$2f_{BFI}$	258.74	−51.12	258.8	−51.63
$3f_{BFI}$	388.1	−52.88	388.17	−52.98
	Inner race, 1 mm			
f_{BFI}	129.27	−58.39	129.33	−56.15
$2f_{BFI}$	258.53	−59.52	258.67	−56.63
$3f_{BFI}$	387.8	−61.19	388	−57.96

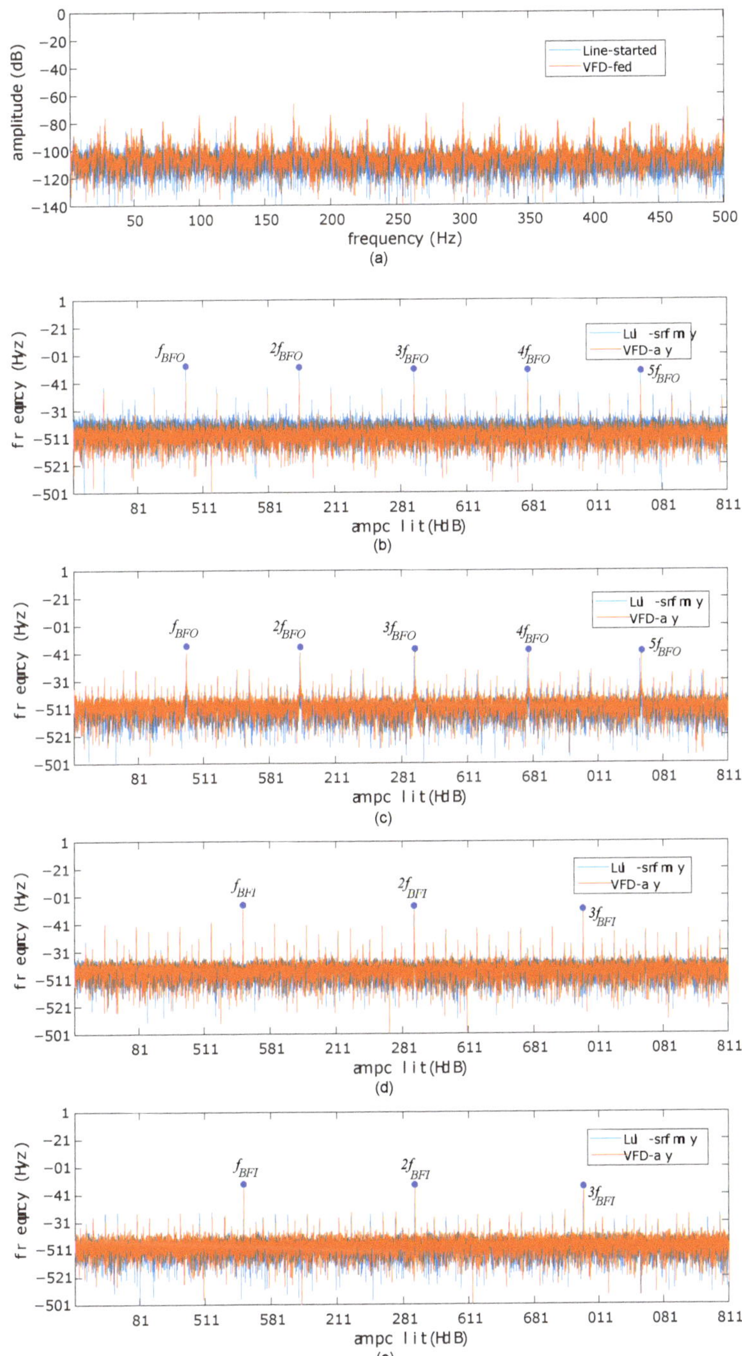

Figure 10. Vibration envelope spectrum analysis acquired at 12 o'clock position at rated slip, (**a**) healthy, (**b**) 0.5 mm outer race defect, (**c**) 1 mm outer race defect, (**d**) 0.5 mm inner race defect, (**e**) 1 mm inner race defect.

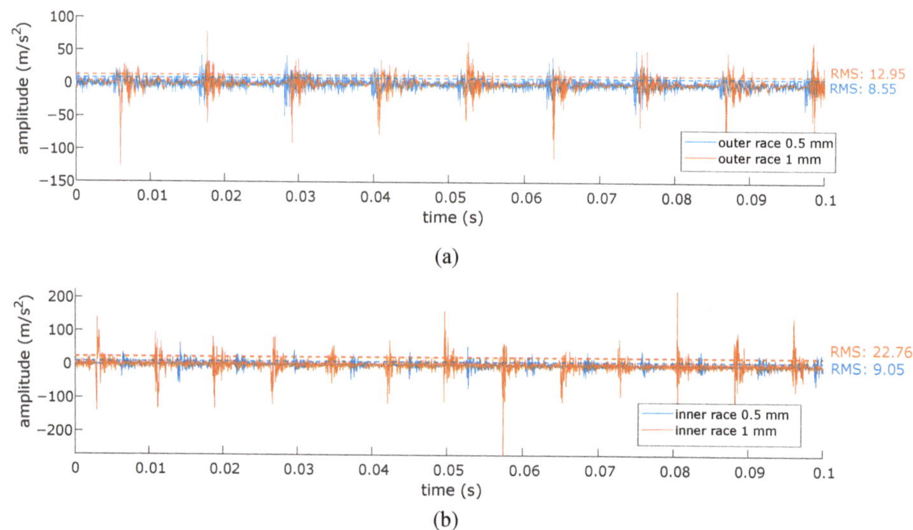

Figure 11. Vibration amplitude comparison among two defect widths. Signals acquired at 12 o'clock at rated slip. (**a**) Outer race defects, (**b**) inner race defects.

4.2. Transient Analysis

The present subsection describes the analysis of the induction motor vibration signals during the studied start-up excitation modes at the rated slip. Four distinct start-up modes were implemented, including line-started at 100% and 50% of the rated line-to-line voltage, along with VFD-fed modes featuring ramp-up times of 20 s and 5 s. Figures 12–16 show the results of the start-up analysis by utilizing the demodulated STFT together with the time-domain vibration signals at both the 12 o'clock and 3 o'clock circumferential positions. The dashed red line marks the start point of the time–frequency transformation, which is applied to exclude the processing of blank signal intervals. The STFT time window (i.e., $w(t - \tau)$) is kept at 0.3 s across the different start-up scenarios to allow for the quantitative analysis of time–frequency amplitudes.

Figure 12 shows the vibration signature when a brand new healthy bearing is located in the end plate. The effects of the VFD actuation are clearly identified by observing the time–domain transient signals. The VFD-induced harmonic content in the vibration signals is mainly located at higher frequencies, as evidenced by the steady-state comparison in Figure 10 and Table 3 even if the noise floor in the frequency range of interest is clearly increased. By observing the line-started signals, the overall vibration amplitude is lower for the reduced line-to-line voltage case. This observation is aligned with the expected electromagnetic vibration for reduced excitation levels, even if the shaft rotates at the rated speed. Moreover, in the case of the healthy bearing in line-started conditions, a high-amplitude starting shock is observed for both voltage levels. Sensors located at 12 o'clock and 3 o'clock provide similar information, with some amplitude differences, mainly due to the structural transfer function between acting forces and vibration acquisition points.

Figures 13 and 14 show the transient vibration analysis of the two studied bearings containing outer race defects. The outer race defect signature kf_{BPO} $\forall$ $k = 1, 2, 3, \ldots$ is clearly observed in all cases. However, the outer-race 0.5 mm defect provides less-evident signatures in both the line-started and VFD-fed excitation modes. The main reason for the signature masking in the case of the line-started 0.5 mm defect is the presence of a high-amplitude initial shock. In addition, the signature of this defect is not evident in the case of the 3 o'clock position, even in the case of VFD-fed excitation. For the 1 mm outer race defect case, the fault signature during transient evolution is much more observable. In this case, no initial shock is observed in the case of line-started excitation, which contributes to the clear identification of f_{BPO} frequencies. The fault signature is not evident at low rotating speeds. This is clearly observed in the transient evolutions of 50% line-started and VFD-fed excitation modes, even in the time domain signals. Moreover, the signature becomes observable at both the 12 o'clock and 3 o'clock positions, with only slight differences observed during long ramp excitation.

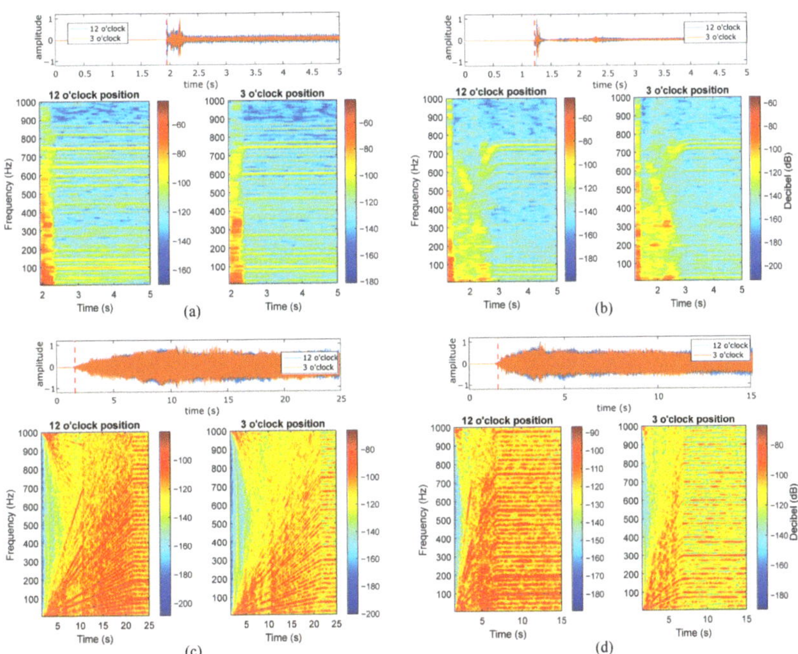

Figure 12. Healthy bearing at rated slip, (**a**) line-started 100% rated voltage, (**b**) line-started 50% rated voltage, (**c**) VFD-fed 20 s ramp, (**d**) VFD-fed 5 s ramp.

Figures 15 and 16 present the transient vibration analysis for the two bearings with inner race defects under study. Both 0.5 mm and 1 mm defects clearly provide observable fault signatures at the characteristic frequencies $kf_{BPI} \ \forall \ k = 1, 2, 3, \ldots$. Moreover, no initial shock masking is observed in any inner race case, which allows for the clear identification of characteristic frequencies in the line-started cases. All signatures are clearly evidenced in both the acquisition circumferential positions. In addition, the characteristic components are observed independently of the rotation speed for the VFD case. Under faster acceleration at 50% of the rated line-to-line voltage, the initial portion of the signal shows no contact shocks, making the characteristic frequencies unobservable at low speeds.

The mechanical impulses induced by race defects dominate the signal envelope spectrogram during the start-up for all types of actuation. The inner race is the most observable race defect for both defect widths, while the outer race 0.5 mm defect provides a less-dominant signature, even if the characteristic frequencies are clearly observed in the envelope spectrogram. The acquisition position in the circumferential direction does not heavily affect the bearing signature identification, which is only influenced under weaker excitation levels (e.g., outer race 0.5 mm defect or healthy specimen). On the other hand, the presence of initial mechanical shocks heavily hinders bearing signature identification during machine start-up. Another observation is the low levels of the characteristic fault frequencies for low speeds during the initial start. This initial attenuation is better observed for 1 mm defects in outer and inner races, while smaller defect widths are evident even for initial rotation phases.

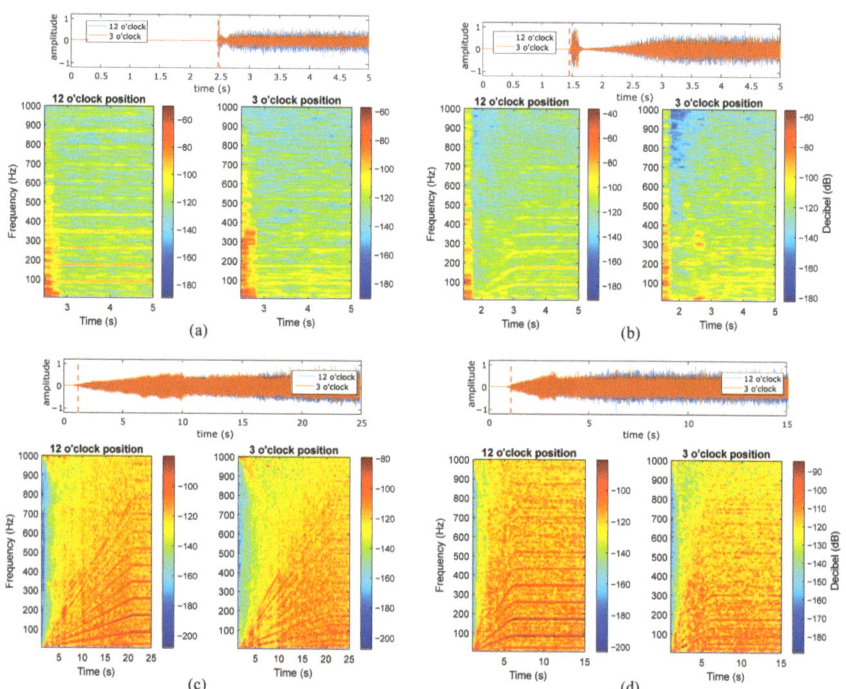

Figure 13. Outer race 0.5 mm defect at rated slip, (**a**) line-started 100% rated voltage, (**b**) line-started 50% rated voltage, (**c**) VFD-fed 20 s ramp, (**d**) VFD-fed 5 s ramp.

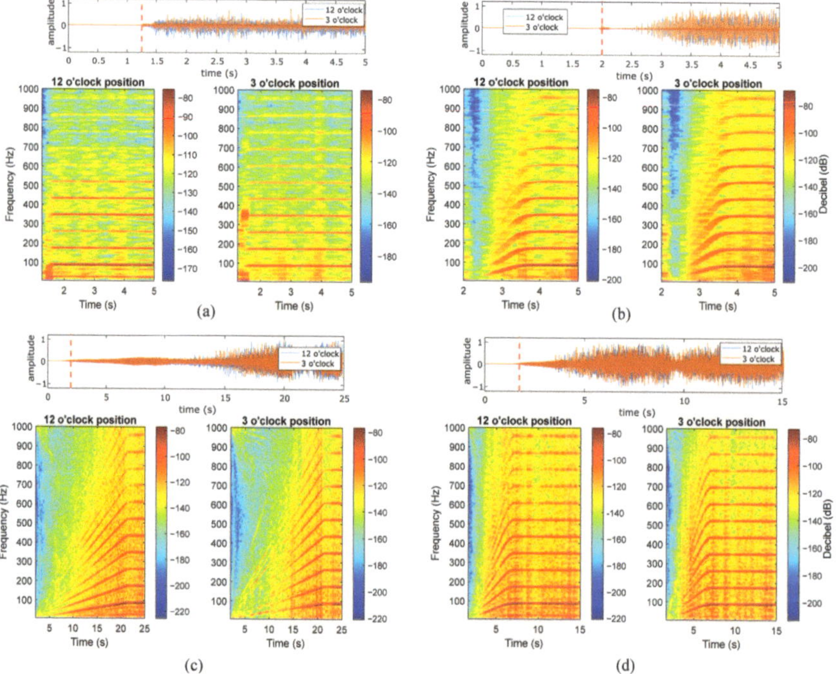

Figure 14. Outer race 1 mm defect at rated slip, (**a**) line-started 100% rated voltage, (**b**) line-started 50% rated voltage, (**c**) VFD-fed 20 s ramp, (**d**) VFD-fed 5 s ramp.

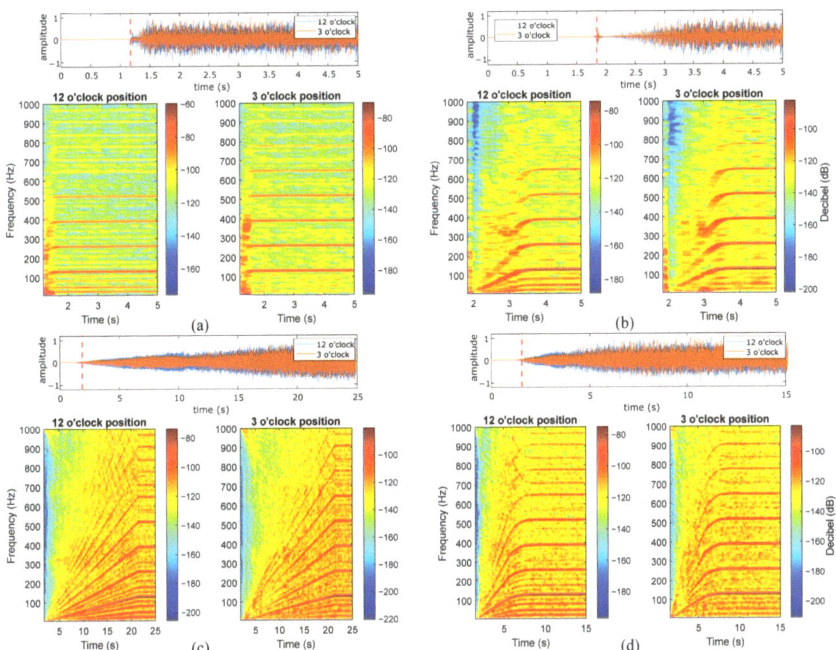

Figure 15. Inner race 0.5 mm defect at rated slip, (**a**) line-started 100% rated voltage, (**b**) line-started 50% rated voltage, (**c**) VFD-fed 20 s ramp, (**d**) VFD-fed 5 s ramp.

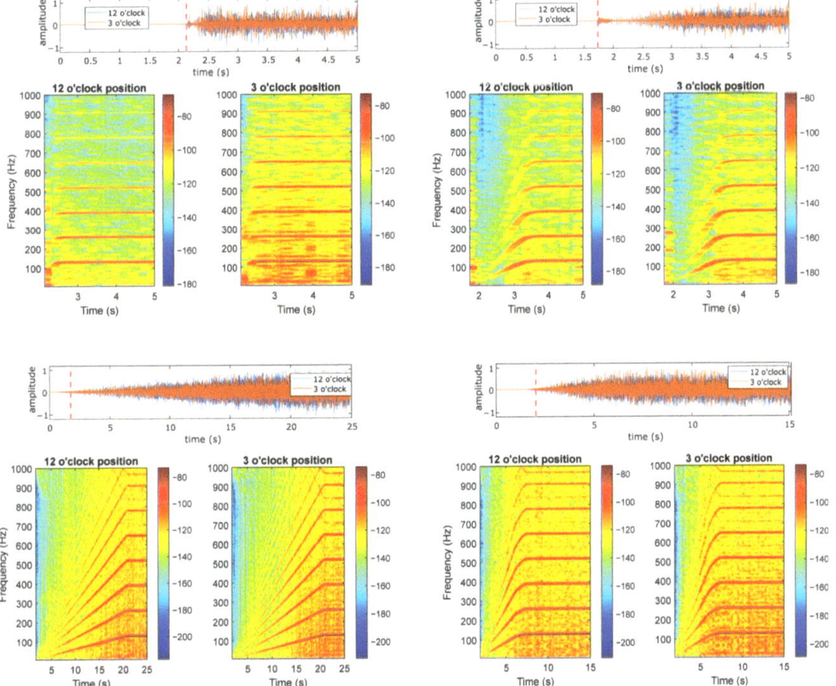

Figure 16. Inner race 1 mm defect at rated slip, (**a**) line-started 100% rated voltage, (**b**) line-started 50% rated voltage, (**c**) VFD-fed 20 s ramp, (**d**) VFD-fed 5 s ramp.

4.3. Effects of Load on the Characteristic Defect Signature

The present subsection analyzes the effects of machine load on the analytically estimated characteristic frequencies for both steady-state and transient conditions. This analysis is performed to verify the presence of the bearing fault components under different electromagnetic and speed conditions. For increased load levels, the slip slightly increases and the induction motor absorbs higher phase current levels, which may influence the level of electromagnetic vibration. The load variation analysis is performed for both line-started and VFD-fed cases as well as for different bearing defect widths. Figure 17 shows the steady-state analysis of the first two faulty components (i.e., f_{BFO}, $2f_{BFO}$, f_{BFI}, $2f_{BFI}$) for different load points. The left side of the figure shows the amplitude trends for different load points, while the right side shows the frequency locus of the different fault signatures for different load points. These right-side graphs are plotted for the VFD-fed case.

The faulty bearing components are identified independently of the load. A notable observation is the shift in frequency locus as a function of the load point. This is consistent with the expected variation in slip due to load changes in induction machines. Consequently, the characteristic faulty frequencies dynamically change with machine speed and the applied load. Nevertheless, the fault frequency does not exhibit significant variations across different load points, remaining within a maximum relative 4% deviation. By observing the amplitude trends of the outer race fault frequencies (i.e., Figure 17a,b), the amplitude generally decreases for increased load levels and lower speeds. On the other hand, the inner race signature does not exhibit a clear amplitude trend. Additionally, the excitation mode does not significantly affect the amplitude of the characteristic frequency at different load points. The data presented in Figure 17 further support the findings of Figure 10 and Table 3, where the defect width is not identified as a significant parameter regarding the characteristic frequency amplitudes. The data corresponding to the line-started 0.5 mm defect demonstrate an increased noise floor, causing the amplitudes to be at a slightly higher level. Note that the absolute amplitude values may show slight inaccuracies due to several phenomena such as parasitic load oscillations or bearing slippage, even if energy leakage is minimized by utilizing signal windowing.

Figures 18 and 19 show different start-up transient evolutions for different imposed resistant torques. The vibration signals were acquired at 12 o'clock and were plot after the initial shock appearance to better highlight variations across faulty components. Figure 18 shows the results of the study for the line-started actuation mode at 50% of the rated line-to-line voltage. This excitation mode is selected to avoid the fast slip swing observed at 100% of the rated line-to-line voltage. The analysis of the line-started results evidences the presence of bearing faulty components across different resistant torque levels and speeds. Both multiples of f_{BFO} and f_{BFI} are dominant for all load levels and defect widths during the transient start-up and once steady-state conditions are reached. The only effect of the load level is a minimal variation in the start-up duration, which is nearly imperceptible when analyzing the time–frequency maps.

Figure 19 shows the VFD-fed load analysis during a 20 s ramp start-up. This ramp is selected due to the increased transient signal section when compared with the 5 s start. The figure shows the dominance of the faulty characteristic frequencies across the studied load points. Note that even the amplitude levels are within the same ranges across all of the different loads and defect cases. The analysis presented in both Figures 18 and 19 indicates the load independence of the characteristic fault signature for all defect widths, load levels, and start-up modes and durations.

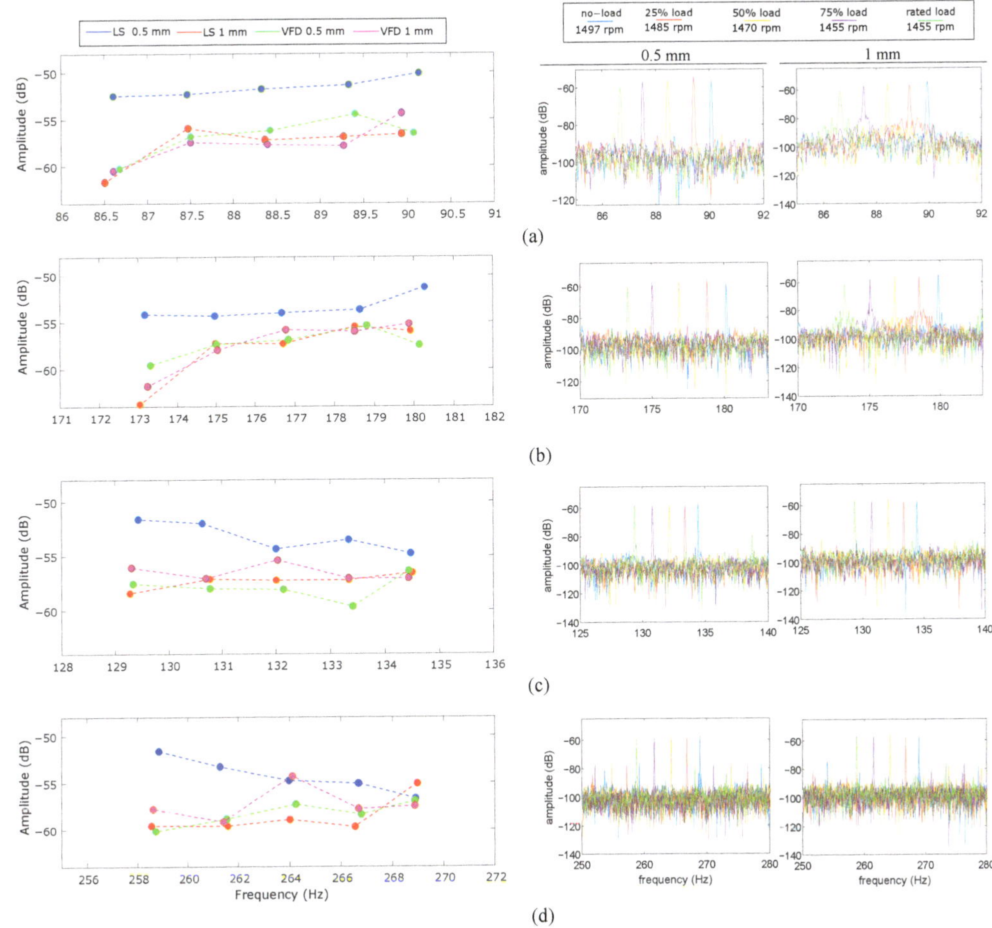

Figure 17. Load dependency steady-state analysis. (**a**) f_{BFO}, (**b**) $2f_{BFO}$, (**c**) f_{BFI}, (**d**) $2f_{BFI}$.

4.4. Comparison with HUST Dataset

The present subsection compares the acquired transient signature with a recently published open-source dataset. This is performed to further elucidate the effects of the mechanical transfer path and to allow for a deeper analysis of the results. The HUST dataset [53] includes vibration data acquired on a dedicated bearing module, thus avoiding vibrations of electromagnetic and mechanical origin from the machine itself. The dataset contains data from different faulty bearings at different loads imposed by a controlled powder brake. The bearings are damaged using electric discharge machining with a defect with of 0.2 mm. The same accelerometer (i.e., PCB325C33) was utilized to acquire the vibration data, which contains VFD-fed start-ups of a 5 s duration. Figure 20 shows the HUST dataset experimental set-up.

The qualitative comparison between datasets was performed under the closest possible conditions in terms of bearing dimensions and start-up duration. Thus, the inner and outer race defect data corresponding to the HUST bearing labelled as 6205 are used, since they possess the same number of rolling elements and similar inner, outer, and ball diameters to the bearing studied in the custom dataset. Moreover, the maximum load point is selected (i.e., 800 W load) in the case of the HUST dataset, providing a shaft speed of approximately 1370 rpm. The custom data are shown at the rated induction motor load point featuring a

1440 rpm shaft speed. Figure 21 shows the different time–frequency maps for the selected signal comparison.

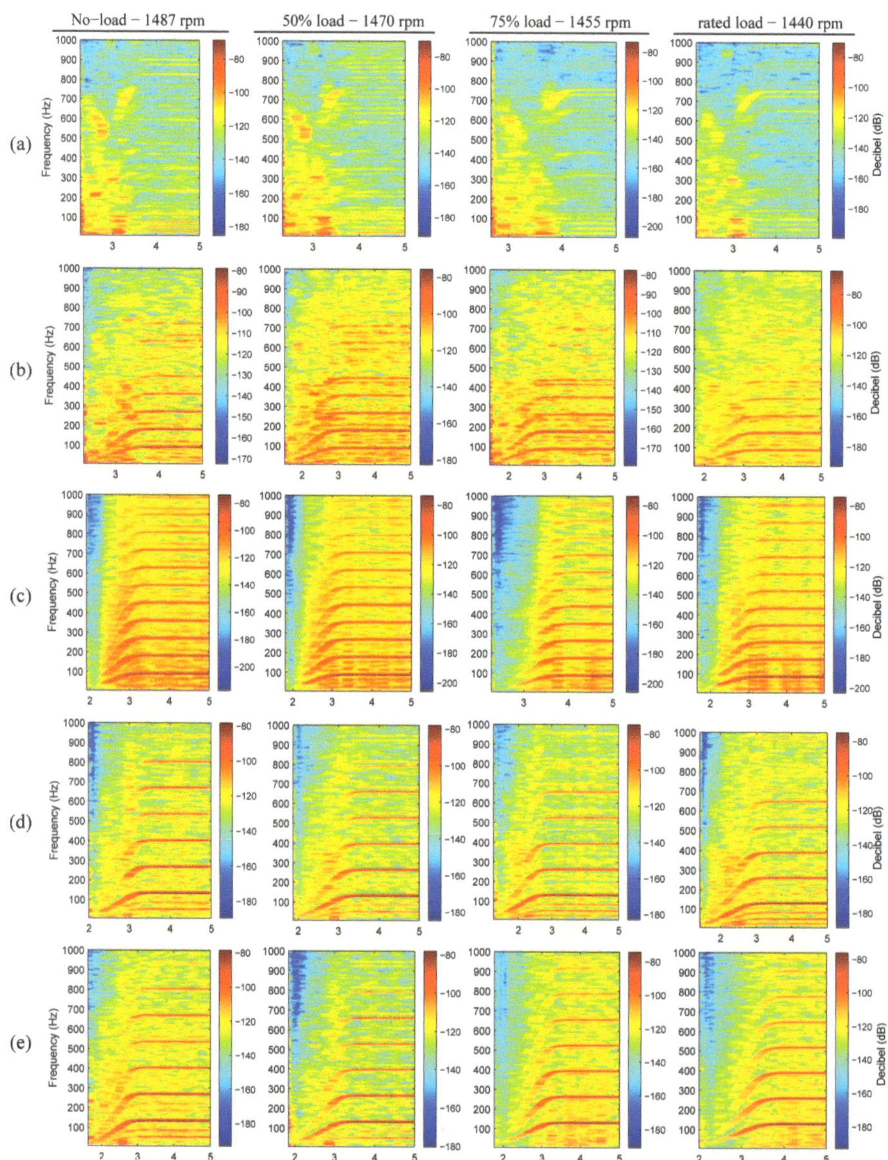

Figure 18. Load variation analysis during the line-started excitation mode at 50% rated line-to-line voltage. Vibration signals acquired at 12 o'clock. (**a**) Healthy bearing, (**b**) outer race 0.5 mm defect, (**c**) outer race 1 mm defect, (**d**) inner race 0.5 mm defect, (**e**) inner race 1 mm defect.

Figure 21 clearly shows the expected characteristic defect signatures for all the types of studied bearing defects during both transient start-up and steady-state sections. An immediate observation is the presence of increased rotating frequency modulations in Figure 21a when compared with the signals acquired in the motor housing. This is evidenced by the components $kf_{BPI} \pm f_r \quad \forall \quad k = 1, 2, 3, \ldots$, which are not observed in Figure 21b. The comparison between outer race defects proves the increased difficulty to discern characteristic faulty components when compared to the inner race case. Moreover, Figure 21c shows very small amplitudes in the beginning of the start-up evolution, which is in accordance with the signals shown in Figures 14 and 18. Overall, each set-up exhibits distinct characteristics; however, the studied defects are clearly identifiable during both the transient start-up and steady-state phases. Nevertheless, the comparison demonstrates that the mechanical transfer path plays a key role in elucidating the characteristic signature and its rotating frequency modulations. The significance of the mechanical transfer path limits the generalizability across different induction machine specimens with varying structural characteristics (i.e., mass, stiffness, and damping), and power ratings.

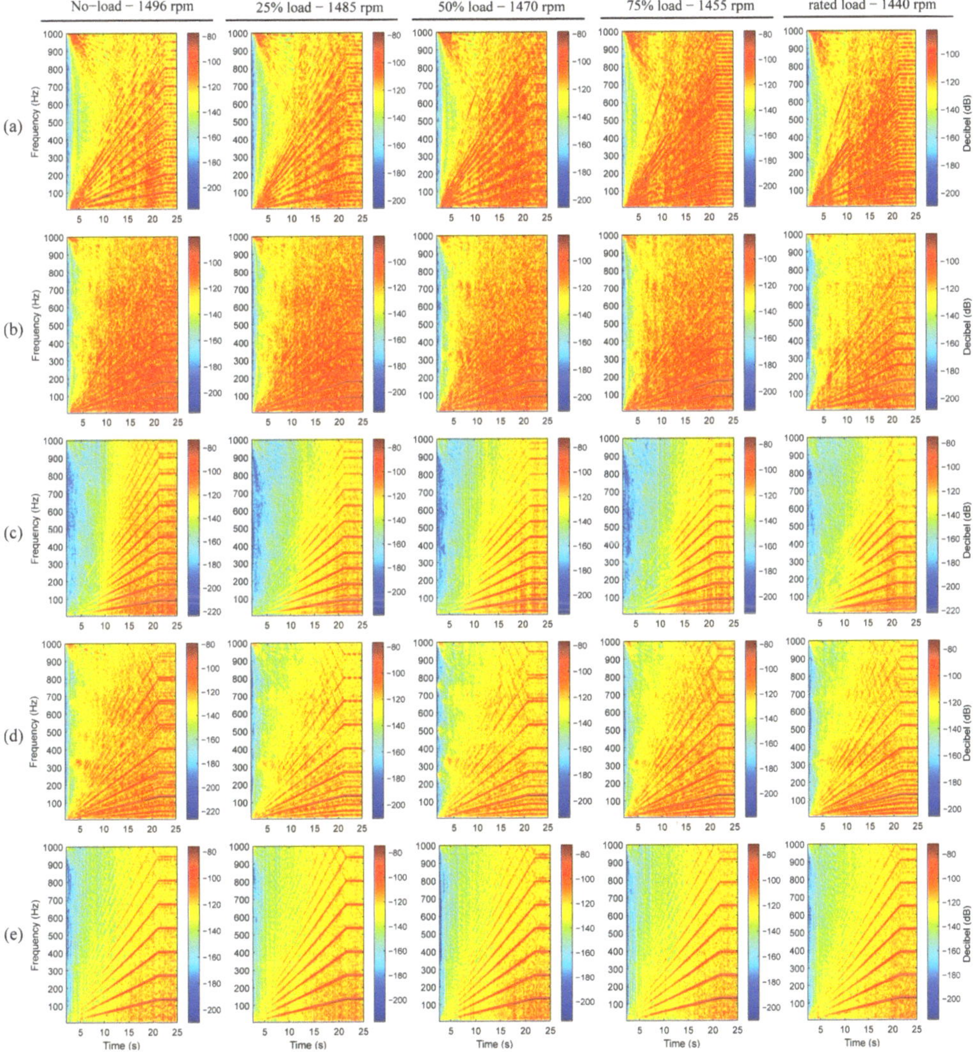

Figure 19. Load variation analysis during the VFD-fed excitation mode with 20 s ramp duration. (**a**) Healthy, (**b**) outer race 0.5 mm defect, vibration signals acquired at 12 o'clock, (**c**) outer race 1 mm defect, (**d**) inner race 0.5 mm defect, (**e**) inner race 1 mm defect.

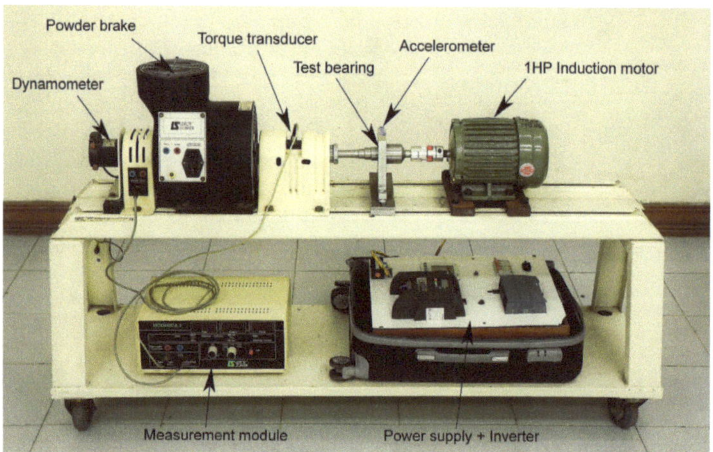

Figure 20. HUST dataset experimental test bench description [53].

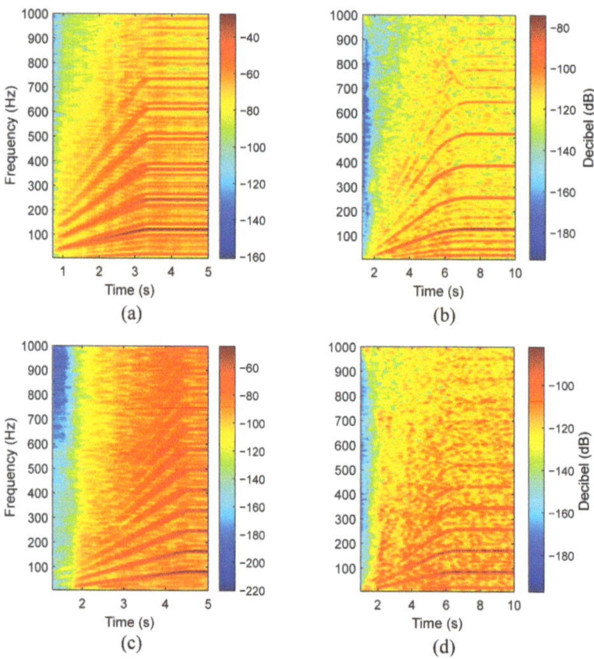

Figure 21. VFD-fed start-ups for inner and outer race defects, (**a**) HUST dataset inner race defect, (**b**) custom dataset inner race defect, (**c**) HUST dataset outer race defect, (**d**) custom dataset outer race defect.

5. Discussion

The analysis of low-frequency bearing defect characteristic frequencies performed in this work supports the detection feasibility of incipient localized bearing defects across different operating regimes and actuation modes in induction motors. These low-frequency components are identified using a simple and linear time–frequency transformation of the signal envelope. This indicates that the utilization of complex signal processing tools may not represent an efficient practice when a low-frequency characteristic fault signature is targeted. The utilized time–frequency transformation, which contains physical information about the fault, may be of interest for physics-informed data-driven approaches, which typically require the preprocessing of large volumes of data. The vibration envelope spectrum analysis of low-frequency characteristic fault components cannot properly discriminate between defect widths. Different defect widths in the range of $D_r < 1$ provide similar amplitudes across different load points and excitation modes. This is extended to time–frequency analysis of variable speed vibration signals. Finally, the observability of the characteristic frequency across different machine regimes aligns with signals from datasets obtained under slightly different mechanical conditions, as discussed in Section 4.4.

This study presents several limitations, such as the artificial implementation of defects, which may slightly differ from incipient naturally induced cracks. The presence and amplitude of outer race shocks heavily depends on the defect circumferential position, which is kept at the loaded area (i.e., 6 o'clock) for all studied faults in the present analysis. Outer race defects are only studied when they are located in the loaded zone. Moreover, the induction machine specimen is disassembled to change the bearing under test. This may slightly affect the mechanical features of the test bench, even if attention is paid to keep the assembly methodology as standard as possible. Finally, spurious mechanical shocks of a different nature may mask the bearing defect if these are included in the analyzed signal sections.

6. Conclusions

The present paper introduced an experimental study regarding localized bearing fault detection across different operating regimes in induction motors. First, this work introduced the relevant theoretical background to properly interpret the analysis and to describe the utilized time–frequency representation tool. Next, the experimental procedure was described in detail, including the induction motor test bench, the vibration signal acquisition system, the bearing fault implementation, and the presence of spurious mechanical shocks that may mask the characteristic fault signature. The analysis of the experimental results is presented in several steps. First, a classic steady-state analysis at the rated slip was conducted. Second, a broad start-up transient exploration at the rated slip, including different excitation modes, bearing defect widths, and acquisition circumferential positions, was performed. Third, the effects of load and thus light changes in shaft speed and imposed constant resistant torque were studied. To conclude the analysis, the acquired data were compared with the corresponding data from an existing open-source dataset. The utilized time–frequency envelope spectrum was used to identify the characteristic bearing fault signature independently of the start-up mode, defect width, and load point. This allows us to distinguish between defect locations during start-up in an efficient and straightforward manner across many operating scenarios. The mechanical transfer path of the induction machine influences the bearing defect signature transmission, as demonstrated by the comparison of signals acquired near the bearing with those measured at the motor housing.

Future research steps include the study under different lubrication states and outer race defects locations in the circumferential position. The applicability of the detection technique should be validated for machines with different masses, power ratings, and features in order to further generalize it. This may include a detailed structural analysis under different bearing defects and closely monitoring the test bench mechanical features. Moreover, the relation between defect width and rotating frequency modulating harmonics in the envelope spectrum should be explored in detail for both inner and outer race defects.

Author Contributions: Conceptualization, J.E.R.-S. and J.A.A.-D.; methodology, J.E.R.-S., J.A.A.-D. and C.M.; software, J.E.R.-S.; validation, J.E.R.-S.; formal analysis, J.E.R.-S. and J.A.A.-D.; investigation, J.E.R.-S.; resources, J.A.A.-D. and C.M.; data curation, J.E.R.-S.; writing—original draft preparation, J.E.R.-S.; writing—review and editing, J.A.A.-D. and C.M.; visualization, J.E.R.-S.; supervision, J.A.A.-D. and C.M.; project administration, J.A.A.-D. and C.M.; funding acquisition, J.A.A.-D. and C.M. All authors have read and agreed to the published version of the manuscript.

Funding: This research was funded in part by the European Commission (HORIZON program) within the context of the DITARTIS Project ("Network of Excellence in Digital Technologies and AI Solutions for Electromechanical and Power Systems Applications") under the call HORIZON-WIDERA-2021-ACCESS-03 (Grant Number 101079242), and in part by the Spanish "Ministerio de Ciencia e Innovación", Agencia Estatal de Investigación and FEDER program in the framework of the "Proyectos de Generación de Conocimiento 2021" of the "Programa Estatal para Impulsar la Investigación Científico-Técnica y su Transferencia", belonging to the "Plan Estatal de Investigación Científica, Técnica y de Innovación 2021–2023" (ref: PID2021-122343OB-I00).

Informed Consent Statement: Not applicable.

Data Availability Statement: The HUST dataset analyzed in this study is openly available from Mendeley Data at 10.1186/s13104-023-06400-4 [53].

Conflicts of Interest: The authors declare no conflicts of interest.

Abbreviations

The following abbreviations are used in this manuscript:

STFT	Short-Time Fourier Transform
CWRU	Case Western Reserve University
DWT	Discrete Wavelet Transform
EMD	Empirical Mode Decomposition
CWT	Continuous Wavelet Transform
WVD	Wigner–Ville Distribution
MUSIC	Multiple Signal Classification
DE	Drive End
FFT	Fast Fourier Transform
HT	Hilbert Transform
VFD	Variable Frequency Drive

References

1. Al-Badri, M.; Pillay, P. Modified Efficiency Estimation Tool for Three-Phase Induction Motors. *IEEE Trans. Energy Convers.* **2022**, *38*, 771–779. [CrossRef]
2. Orlowska-Kowalska, T.; Wolkiewicz, M.; Pietrzak, P.; Skowron, M.; Ewert, P.; Tarchala, G.; Krzysztofiak, M.; Kowalski, C.T. Fault diagnosis and fault-tolerant control of PMSM drives–state of the art and future challenges. *IEEE Access* **2022**, *10*, 59979–60024. [CrossRef]
3. Khan, M.A.; Asad, B.; Kudelina, K.; Vaimann, T.; Kallaste, A. The bearing faults detection methods for electrical machines—the state of the art. *Energies* **2022**, *16*, 296. [CrossRef]
4. Loparo, K. Bearing Vibration Data Set. 2003. Available online: https://engineering.case.edu/bearingdatacenter/download-data-file (accessed on 21 October 2024).
5. Lessmeier, C.; Kimotho, J.K.; Zimmer, D.; Sextro, W. Condition monitoring of bearing damage in electromechanical drive systems by using motor current signals of electric motors: A benchmark data set for data-driven classification. In Proceedings of the PHM Society European Conference, Bilbao, Spain, 5–8 July 2016; Volume 3.
6. Zhang, X.; Zhao, B.; Lin, Y. Machine learning based bearing fault diagnosis using the case western reserve university data: A review. *IEEE Access* **2021**, *9*, 155598–155608. [CrossRef]
7. Mian, Z.; Deng, X.; Dong, X.; Tian, Y.; Cao, T.; Chen, K.; Al Jaber, T. A literature review of fault diagnosis based on ensemble learning. *Eng. Appl. Artif. Intell.* **2024**, *127*, 107357. [CrossRef]
8. Kuemmlee, H.; Gross, T.; Kolerus, J. Machine vibrations and diagnostics the world of ISO. In Proceedings of the Industry Applications Society 60th Annual Petroleum and Chemical Industry Conference, Chicago, IL, USA, 23–25 September 2013; IEEE: Piscataway, NJ, USA, 2013; pp. 1–13.
9. Tsypkin, M. The origin of the electromagnetic vibration of induction motors operating in modern industry: Practical experience—Analysis and diagnostics. *IEEE Trans. Ind. Appl.* **2016**, *53*, 1669–1676. [CrossRef]
10. McCloskey, A.; Arrasate, X.; Hernández, X.; Gómez, I.; Almandoz, G. Analytical calculation of vibrations of electromagnetic origin in electrical machines. *Mech. Syst. Signal Process.* **2018**, *98*, 557–569. [CrossRef]
11. Blodt, M.; Granjon, P.; Raison, B.; Rostaing, G. Models for bearing damage detection in induction motors using stator current monitoring. *IEEE Trans. Ind. Electron.* **2008**, *55*, 1813–1822. [CrossRef]
12. Singh, S.; Kumar, N. Detection of bearing faults in mechanical systems using stator current monitoring. *IEEE Trans. Ind. Inform.* **2016**, *13*, 1341–1349. [CrossRef]

13. Duque-Perez, O.; Del Pozo-Gallego, C.; Morinigo-Sotelo, D.; Fontes Godoy, W. Condition monitoring of bearing faults using the stator current and shrinkage methods. *Energies* **2019**, *12*, 3392. [CrossRef]
14. Frosini, L.; Harlişca, C.; Szabó, L. Induction machine bearing fault detection by means of statistical processing of the stray flux measurement. *IEEE Trans. Ind. Electron.* **2014**, *62*, 1846–1854. [CrossRef]
15. Zamudio-Ramirez, I.; Osornio-Rios, R.A.; Antonino-Daviu, J.A.; Cureño-Osornio, J.; Saucedo-Dorantes, J.J. Gradual wear diagnosis of outer-race rolling bearing faults through artificial intelligence methods and stray flux signals. *Electronics* **2021**, *10*, 1486. [CrossRef]
16. Immovilli, F.; Bellini, A.; Rubini, R.; Tassoni, C. Diagnosis of bearing faults in induction machines by vibration or current signals: A critical comparison. *IEEE Trans. Ind. Appl.* **2010**, *46*, 1350–1359. [CrossRef]
17. Rai, A.; Upadhyay, S.H. A review on signal processing techniques utilized in the fault diagnosis of rolling element bearings. *Tribol. Int.* **2016**, *96*, 289–306. [CrossRef]
18. Randall, R.B.; Antoni, J. Rolling element bearing diagnostics—A tutorial. *Mech. Syst. Signal Process.* **2011**, *25*, 485–520. [CrossRef]
19. McFadden, P.; Smith, J. Model for the vibration produced by a single point defect in a rolling element bearing. *J. Sound Vib.* **1984**, *96*, 69–82. [CrossRef]
20. Rubini, R.; Meneghetti, U. Application of the envelope and wavelet transform analyses for the diagnosis of incipient faults in ball bearings. *Mech. Syst. Signal Process.* **2001**, *15*, 287–302. [CrossRef]
21. Yang, Y.; Yu, D.; Cheng, J. A fault diagnosis approach for roller bearing based on IMF envelope spectrum and SVM. *Measurement* **2007**, *40*, 943–950. [CrossRef]
22. Randall, R.B.; Antoni, J.; Chobsaard, S. A comparison of cyclostationary and envelope analysis in the diagnostics of rolling element bearings. In Proceedings of the 2000 IEEE International Conference on Acoustics, Speech, and Signal Processing, Istanbul, Turkey, 5–9 June 2000; Proceedings (Cat. No. 00CH37100); IEEE: Piscataway, NJ, USA, 2000; Volume 6, pp. 3882–3885.
23. Li, L.; Qu, L. Cyclic statistics in rolling bearing diagnosis. *J. Sound Vib.* **2003**, *267*, 253–265. [CrossRef]
24. Liu, D.; Cui, L.; Wang, H. Rotating machinery fault diagnosis under time-varying speeds: A review. *IEEE Sens. J.* **2023**, *23*, 29969–29990. [CrossRef]
25. Smith, W.A.; Randall, R.B.; du Mée, X.d.C.; Peng, P. Use of cyclostationary properties to diagnose planet bearing faults in variable speed conditions. In Proceedings of the 10th DST Group International Conference on Health and Usage Monitoring Systems, 17th Australian Aerospace Congress, Melbourne, Australia, 26–28 February 2017; pp. 26–28.
26. Liu, D.; Cheng, W.; Wen, W. Generalized demodulation with tunable E-Factor for rolling bearing diagnosis under time-varying rotational speed. *J. Sound Vib.* **2018**, *430*, 59–74. [CrossRef]
27. Guo, Y.; Liu, T.W.; Na, J.; Fung, R.F. Envelope order tracking for fault detection in rolling element bearings. *J. Sound Vib.* **2012**, *331*, 5644–5654. [CrossRef]
28. Sierra-Alonso, E.F.; Caicedo-Acosta, J.; Orozco Gutiérrez, Á.Á.; Quintero, H.F.; Castellanos-Dominguez, G. Short-time/-angle spectral analysis for vibration monitoring of bearing failures under variable speed. *Appl. Sci.* **2021**, *11*, 3369. [CrossRef]
29. Lee, D.H.; Hong, C.; Jeong, W.B.; Ahn, S. Time–frequency envelope analysis for fault detection of rotating machinery signals with impulsive noise. *Appl. Sci.* **2021**, *11*, 5373. [CrossRef]
30. Paliwal, D.; Choudhury, A.; Tingarikar, G. Wavelet and scalar indicator based fault assessment approach for rolling element bearings. *Procedia Mater. Sci.* **2014**, *5*, 2347–2355. [CrossRef]
31. Liu, W.; Han, J.; Jiang, J. A novel ball bearing fault diagnosis approach based on auto term window method. *Measurement* **2013**, *46*, 4032–4037. [CrossRef]
32. Yu, G.; Wang, Z.; Zhao, P. Multisynchrosqueezing transform. *IEEE Trans. Ind. Electron.* **2018**, *66*, 5441–5455. [CrossRef]
33. Ding, C.; Huang, W.; Shen, C.; Jiang, X.; Wang, J.; Zhu, Z. Synchroextracting frequency synchronous chirplet transform for fault diagnosis of rotating machinery under varying speed conditions. *Struct. Health Monit.* **2024**, *23*, 1403–1422. [CrossRef]
34. Chen, Y.; Hu, L.; Hu, N.; Zeng, J. A Synchrosqueezed Transform Method Based on Fast Kurtogram and Demodulation and Piecewise Aggregate Approximation for Bearing Fault Diagnosis. *Sensors* **2024**, *24*, 2502. [CrossRef]
35. Agrawal, S.; Mohanty, S.R.; Agarwal, V. Bearing fault detection using Hilbert and high frequency resolution techniques. *IETE J. Res.* **2015**, *61*, 99–108. [CrossRef]
36. Ruiz-Sarrio, J.E.; Antonino-Daviu, J.A.; Zamudio-Ramirez, I.; Osornio-Rios, R.A. Incipient Bearing Fault Signature Identification via Vibration Envelope Analysis and MUSIC. In Proceedings of the 2024 IEEE 33rd International Symposium on Industrial Electronics (ISIE), Ulsan, Republic of Korea, 18–21 June 2024; IEEE: Piscataway, NJ, USA, 2024; pp. 1–6.
37. Ruiz-Sarrio, J.E.; Antonino-Daviu, J.A.; Martis, C. Comprehensive diagnosis of localized rolling bearing faults during rotating machine start-up via vibration envelope analysis. *Electronics* **2024**, *13*, 375. [CrossRef]
38. Antonino-Daviu, J. Electrical monitoring under transient conditions: A new paradigm in electric motors predictive maintenance. *Appl. Sci.* **2020**, *10*, 6137. [CrossRef]
39. Lee, S.B.; Stone, G.C.; Antonino-Daviu, J.; Gyftakis, K.N.; Strangas, E.G.; Maussion, P.; Platero, C.A. Condition monitoring of industrial electric machines: State of the art and future challenges. *IEEE Ind. Electron. Mag.* **2020**, *14*, 158–167. [CrossRef]
40. Garcia-Calva, T.; Morinigo-Sotelo, D.; Fernandez-Cavero, V.; Romero-Troncoso, R. Early detection of faults in induction motors—A review. *Energies* **2022**, *15*, 7855. [CrossRef]
41. Yakhni, M.F.; Cauet, S.; Sakout, A.; Assoum, H.; Etien, E.; Rambault, L.; El-Gohary, M. Variable speed induction motors' fault detection based on transient motor current signatures analysis: A review. *Mech. Syst. Signal Process.* **2023**, *184*, 109737. [CrossRef]

42. Battulga, B.; Shaikh, M.F.; Goktas, T.; Arkan, M.; Lee, S.B. Vibration-Based Identification of Mechanical Defects in Induction Motor-driven Systems During the Starting Transient. In Proceedings of the 2024 International Conference on Electrical Machines (ICEM), Torino, Italy, 1–4 September 2024; IEEE: Piscataway, NJ, USA, 2024; pp. 1–7.

43. Ruiz-Sarrio, J.E.; Biot-Monterde, V.; Madariaga-Cifuentes, C.; Navarro-Navarro, A.; Antonino-Daviu, J.A. On the Utilization of Radial Vibration Transient Signals for Induction Machine Misalignment Diagnosis. In Proceedings of the 2024 International Conference on Electrical Machines (ICEM), Torino, Italy, 1–4 September 2024; IEEE: Piscataway, NJ, USA, 2024; pp. 1–6.

44. Konar, P.; Chattopadhyay, P. Bearing fault detection of induction motor using wavelet and Support Vector Machines (SVMs). *Appl. Soft Comput.* **2011**, *11*, 4203–4211. [CrossRef]

45. Haddad, R.Z.; Lopez, C.A.; Pons-Llinares, J.; Antonino-Daviu, J.; Strangas, E.G. Outer race bearing fault detection in induction machines using stator current signals. In Proceedings of the 2015 IEEE 13th International Conference on Industrial Informatics (INDIN), Cambridge, UK, 22–24 July 2015; IEEE: Piscataway, NJ, USA, 2015; pp. 801–808.

46. Boyanton, H.E.; Hodges, G. Bearing fluting [motors]. *IEEE Ind. Appl. Mag.* **2002**, *8*, 53–57. [CrossRef]

47. Singh, S.; Howard, C.Q.; Hansen, C.H. An extensive review of vibration modelling of rolling element bearings with localised and extended defects. *J. Sound Vib.* **2015**, *357*, 300–330. [CrossRef]

48. Harris, T.A.; Kotzalas, M.N. *Rolling Bearing Analysis-2 Volume Set*; CRC Press: Boca Raton, FL, USA, 2006.

49. Randall, R.B.; Antoni, J.; Chobsaard, S. The relationship between spectral correlation and envelope analysis in the diagnostics of bearing faults and other cyclostationary machine signals. *Mech. Syst. Signal Process.* **2001**, *15*, 945–962. [CrossRef]

50. Bastami, A.R.; Vahid, S. A comprehensive evaluation of the effect of defect size in rolling element bearings on the statistical features of the vibration signal. *Mech. Syst. Signal Process.* **2021**, *151*, 107334. [CrossRef]

51. Lindell, H.; Grétarsson, S.L.; Machens, M. High Frequency Shock Vibrations and Implications of ISO 5349: Measurement of Vibration, Simulating Pressure Propagation, Risk Assessment and Preventive Measures. 2017. Available online: https://www.diva-portal.org/smash/record.jsf?pid=diva2%3A1233230&dswid=-2115 (accessed on 21 October 2024).

52. Fernandez-Cavero, V.; Morinigo-Sotelo, D.; Duque-Perez, O.; Pons-Llinares, J. A comparison of techniques for fault detection in inverter-fed induction motors in transient regime. *IEEE Access* **2017**, *5*, 8048–8063. [CrossRef]

53. Thuan, N.D.; Hong, H.S. HUST bearing: A practical dataset for ball bearing fault diagnosis. *BMC Res. Notes* **2023**, *16*, 138. [CrossRef] [PubMed]

Article

Evaluation of Hand-Crafted Feature Extraction for Fault Diagnosis in Rotating Machinery: A Survey

René-Vinicio Sánchez [1,*], Jean Carlo Macancela [1], Luis-Renato Ortega [1,*], Diego Cabrera [2], Fausto Pedro García Márquez [3] and Mariela Cerrada [1]

1   GIDTEC, Universidad Politécnica Salesiana, Cuenca 010105, Ecuador; jmacancelap@est.ups.edu.ec (J.C.M.); mcerrada@ups.edu.ec (M.C.)
2   School of Mechanical Engineering, Dongguan University of Technology, Dongguan 523000, China; diego@dgut.edu.cn
3   Ingenium Research Group, Universidad Castilla-La Mancha, 13071 Ciudad Real, Spain; faustopedro.garcia@uclm.es
*   Correspondence: rsanchezl@ups.edu.ec (R.-V.S.); lortegal@est.ups.edu.ec (L.-R.O.)

Abstract: This article presents a comprehensive collection of formulas and calculations for hand-crafted feature extraction of condition monitoring signals. The documented features include 123 for the time domain and 46 for the frequency domain. Furthermore, a machine learning-based methodology is presented to evaluate the performance of features in fault classification tasks using seven data sets of different rotating machines. The evaluation methodology involves using seven ranking methods to select the best ten hand-crafted features per method for each database, to be subsequently evaluated by three types of classifiers. This process is applied exhaustively by evaluation groups, combining our databases with an external benchmark. A summary table of the performance results of the classifiers is also presented, including the percentage of classification and the number of features required to achieve that value. Through graphic resources, it has been possible to show the prevalence of certain features over others, how they are associated with the database, and the order of importance assigned by the ranking methods. In the same way, finding which features have the highest appearance percentages for each database in all experiments has been possible. The results suggest that hand-crafted feature extraction is an effective technique with low computational cost and high interpretability for fault identification and diagnosis.

Keywords: condition monitoring indicator; fault diagnosis; frequency domain; gears and bearings; hand-crafted features survey; signal processing; time domain

Citation: Sánchez, R.-V.; Macancela, J.C.; Ortega, L.-R.; Cabrera, D.; García Márquez, F.P.; Cerrada, M. Evaluation of Hand-Crafted Feature Extraction for Fault Diagnosis in Rotating Machinery: A Survey. *Sensors* **2024**, *24*, 5400. https://doi.org/10.3390/s24165400

Academic Editor: Jiawei Xiang

Received: 3 July 2024
Revised: 8 August 2024
Accepted: 13 August 2024
Published: 21 August 2024

1. Introduction

Industrial machinery, such as gearboxes, transmission shafts, or reciprocating compressors, are essential rotating equipment widely used in different industries due to their ability to generate force and movement, making them the heart of any mechanical system [1,2]. A machine designed to perform some specific function is expected to do so throughout its useful life. However, a machine may fail due to circumstances often outside our control. We can highlight mechanical parts such as gears, bearings, shafts, belts, or valves as standard components that may be susceptible to failure [2–5]. Failure or cessation of machine operation can represent significant monetary losses and affect the safety of plant personnel; therefore, the machine must be maintained to prevent such failures [6]. Three primary maintenance schemes are followed worldwide: Reactive, Preventive, and more recently, Predictive [7]. Preventive maintenance has traditionally been the most common maintenance policy in industries. Utilizing this strategy, components are replaced once a specific use time has elapsed. Predictive maintenance prevents failures by constantly monitoring the state of the system and identifying abnormal conditions of machine parts [8]. Predictive maintenance can be divided into Reliability-Centered Maintenance (RCM) and

Condition-Based Maintenance (CBM) [9]. CBM is a maintenance approach that uses advanced technologies to continuously monitor and assess the condition of the machinery. In addition, CBM plans necessary maintenance based on the information provided [10]. Due to the need for constant monitoring, the CBM uses a data-based paradigm of three stages: data acquisition, data processing, and diagnosis and decision-making [11,12]. CBM relies on robust and reliable fault diagnosis capabilities [13]. The main tasks of fault diagnosis as a final stage of CBM are indicator selection, identification and determining the cause of a problem or machine fault, intervention planning, monitoring, and evaluation [14,15].

The selection of suitable indicators is crucial for all these tasks to ensure that relevant problems and failures in machinery or equipment are detected. These condition indicators are usually physical or chemical measurements of the equipment used to assess machine conditions and predict the probability of failure. Some typical condition variables are vibration, temperature, and electric current, among others, in the form of temporal series or signals [16–21]. Fault diagnosis focuses on analyzing those signals of the machinery state. Changes in measured signals can indicate a problem or failure of the machinery. Constant monitoring of the machine's condition will support diagnostic processes, associating component failures with processed information. Once the signals for condition monitoring have been selected, the next step to reach a fault diagnosis is processing and analyzing the collected data. Different techniques can be used to analyze condition monitoring signals, such as statistical analysis, time series analysis, and artificial intelligence-based analysis, among others [22–26]. Each technique has advantages and disadvantages, depending on the signal type and the analysis's objectives. Of all the ways to diagnose rotating machinery failures, Machine Learning (ML) methods have gained the most relevance and growth recently [26–30]. Regardless of the algorithm used, all machine learning methodologies utilize a stage of information processing and reduction called Feature Extraction [31–37]. Feature extraction is a necessary digital signal processing (DSP) process to extract relevant information from a signal [38,39]. The goal is to reduce the dimensionality of the data and represent the signal in a smaller set of relevant features [40,41]. This process is often necessary as the original data may be poor quality, not in a format suitable for the algorithm, or contain redundant or irrelevant information.

Feature extraction can include techniques such as normalization, denoising, statistical value derivation, and one-hot coding, among others [42,43]. In fault diagnosis, feature extraction can reflect a change in machine components using DSP techniques and data analysis, such as vibration spectrum analysis, wavelet analysis, waveform analysis, and more [44,45]. These techniques transform condition-monitoring signals into a small set of features that a machine-learning algorithm can use to detect patterns and trends in signal values. There are mainly two ways of performing feature extraction: manual extraction of hand-crafted features and automatic extraction [46,47]. The use of hand-crafted features and automated feature extraction methods depends on different factors, such as the amount and complexity of the data, available resources, and the skills and time of the team. Despite advances in automated feature extraction methods, hand-crafted features remain valuable in fault diagnosis research for several reasons. Firstly, they offer high interpretability, allowing direct physical insights into fault mechanisms. Secondly, they provide flexibility in application across different types of rotating machinery. Thirdly, their computational efficiency makes them suitable for real-time monitoring applications. Lastly, hand-crafted features can capture domain-specific knowledge that might be overlooked by automated methods, potentially leading to more robust and generalizable fault diagnosis models. Automatic feature extraction is suitable when the data are extensive and complex, and there is no broad understanding of the phenomenon being investigated. However, the disadvantages of automatic feature extraction include complexity, difficult interpretability, and the need for an adequate volume of data [48–50]. On the other hand, hand-crafted features are more suitable when the data are small and the team has a broad understanding of the phenomenon being investigated. This methodology is more controllable and maintains simplicity in its implementation and execution, allowing a clear and physical interpretation in the context of the problem and a greater understanding of the features selected [51–54]. The final stage of the analysis and search for patterns for fault diagnosis is usually the classification task [55]. Fault classification is used to identify and categorize failures in a system. The goal is to classify an observation, typically a condition monitoring signal, into different fault categories, each representing a possible cause of failure. The features extracted from the signals are the inputs to the classification models, allowing machine learning algorithms to make a more precise separation between different failure categories [56–58]. The separability of the data for good class discrimination to occur by the various models and the interpretation of the classification results is directly related to the quality of the features extracted from the condition monitoring signals.

DSP-based hand-crafted features are created or designed to reveal and quantify specific information or behavior in the signals. Hand-crafted features can be extracted directly or indirectly by a human without the help of algorithms or automated tools. They may include methods such as manual calculation of statistical measures or definitions of specific knowledge about the phenomenon [42,54,59,60], in this case, the machine under study. Hand-crafted features can be more flexible and customized than automated methods. However, it can also be more labor-intensive and time-consuming in their search, design, implementation, function tests, and validations, especially in problems with many variables, for example, fault diagnosis of rotating machinery. The literature on these topics can become remarkably extensive and confusing since most focus on a specific application case and field of study. In Table 1, we show central reviews closer to manual feature extraction for condition monitoring on rotating machines. The table illustrates which main topics have been studied and how many Time features (TF), Frequency features (FF), Time-Frequency Features (T-FF), or Planetary Gearbox Features (PGF) have been mentioned or used for fault diagnosis purposes. Thus, highlighting a notable absence of a set of features.

Table 1. Main review publication closer to hand-crafted feature extraction in rotating machinery.

Ref.	First Author	Year	Main Topic	Feature Extraction
[22]	H. Yang	2003	Rotating Machinery	TF: 7
[44]	P. Večeř	2005	Gearbox	TF: 6
[61]	W. Yan	2008	Bearings	TF: 5
[62]	K. Tom	2010	Bearings and Gears	TF: 8
[63]	A. S. Sait	2011	Gearbox	TF: 7
[64]	Z. Y. Han	2013	Gearbox	TF: 11
[65]	X. Zhao	2013	Planet gear teeth fault	TF: 18, FF: 30, PGF: 15
[10]	K. L. Tsui	2014	Data-driven approaches in Prognostics and Health Management	TF: 10
[66]	V. Sharma	2016	Gear condition indicators	TF: 14, FF: 7
[67]	W. Caesarendra	2017	Low-Speed Slew Bearing	TF: 9, FF: 6
[68]	S. Riaz	2017	Vibration Feature Extraction	TF: 7
[69]	A. Ogundare	2017	Helicopter Gearbox	TF: 3
[70]	D. Goyal	2017	Fixed Axis Gearbox	TF: 13, FF: 9, T-FF: 1
[71]	T. Wang	2019	Wind turbine planetary gearbox	TF: 14, FF: 4
[72]	A. Stetco	2019	Wind turbine	TF: 13
[73]	X. Zhang	2021	Bearings, Case Western Reserve University Data	TF: 5, FF: 5

Table 1. *Cont.*

Ref.	First Author	Year	Main Topic	Feature Extraction
[74]	M. A. Khan	2022	Bearings of Electrical Machines	TF: 7
[75]	S. Zhang	2022	Vibration signal processing in gears	TF: 5
[15]	S. Gawde	2023	Industrial Rotating Machines	TF: 13, FF: 6
[76]	R. Pandit	2023	SCADA data for wind turbines	TF: 3
[77]	M. Romanssini	2023	Vibration Monitoring of Rotating Machinery	TF:4
[78]	X. Xu	2024	Wind turbine gearbox	TF: 10, FF: 11
This work	**This work**	**2024**	**Rotating machinery**	**TF: 123, FF: 46**

For this reason, in the current work, we present an exhaustive compilation of hand-crafted features based on mathematical and statistical calculations. These features can be computed on condition monitoring signals for both time and frequency domains and used together as a classical feature extraction process. Each feature's theoretical foundation is duly documented and presented as a mathematical formula.

This work aims to address several key research objectives: (1) To compile and standardize a comprehensive set of hand-crafted features from diverse fields for rotating machinery fault diagnosis; (2) To systematically evaluate the performance of these features across multiple databases and classification models; (3) To identify the most relevant and generalizable features for different types of rotating machines; (4) To assess the effectiveness of time-domain, frequency-domain, and fusion-based feature sets. By addressing these objectives, we seek to provide a thorough understanding of the role and effectiveness of hand-crafted features in modern fault diagnosis applications, supporting efforts to bridge the gap between traditional signal processing techniques and advanced machine learning approaches.

To address these research objectives, our study makes the following key contributions:

1. A compilation of 169 hand-crafted features for condition monitoring signals, including 123 features for the time domain and 46 features for the frequency domain, based on a literature review of two decades. The features come from various fields and have mathematical/statistical foundations. This includes a unification of the nomenclature of the formulas and a categorization based on the aspect they are intended to measure.
2. A rigorous evaluation of vibration signal features across seven feature ranking methods, three classification models, and seven datasets of vibration signals from gearboxes and bearings. One dataset is a public benchmark, while the others belong to our institution. Furthermore, features are evaluated under the same conditions.
3. Analysis of top selected features by ranking methods across multiple datasets of different rotating machinery types. This demonstrates the effectiveness of classical feature extraction and provides insight into the most useful features for fault diagnosis in various mechanical systems.

The general restriction that the hand-crafted feature must maintain in our collection is that it must be self-contained, and its mathematical or statistical foundation must allow the calculation to be performed directly on the signal in any of the domains. There are many more features, some so sophisticated that for their calculation, they require the implementation of their complete algorithms, which, for the reasons previously stated, are not the objective of this compilation. It also does not consider types of signal processing that return the same signal but are modified, such as filters or TSA (Time Synchronous

Averaging) [79]. Only simple hand-crafted features that seek to measure or quantify some specific phenomenon within the signal were collected, that is, that return a numerical value after the feature extraction process, converting the feature into a variable that can be analyzed. Finally, the results show high classification percentages for the various databases, showing that results comparable to deep learning methodologies can be obtained with adequate classical feature extraction. The evaluation shows that applying these hand-crafted features is helpful for any signal because they are calculated directly on them, and the vast majority do not have any configurable parameters, making the processing fast, optimizable, and with low computational cost.

The remainder of this paper is organized as follows: Section 1 provides a comprehensive compilation of hand-crafted features for classical feature extraction, including 123 time domain and 46 frequency domain features, along with their mathematical formulations and references. Section 2 describes the experimental data used for evaluating the performance of these features, detailing seven databases of rotating machinery faults. Section 3 outlines the proposed methodology for feature evaluation, including the feature extraction process, dimension reduction techniques, and classification techniques. Section 4 presents the results of the extensive experimentation, discussing the effectiveness of different feature subsets, ranking methods, and classifiers across various fault types and databases. This section also analyzes the most consistently useful features and their potential physical significance. Finally, Section 5 concludes the paper by summarizing the key findings, discussing the implications for fault diagnosis in rotating machinery, and suggesting directions for future research in this field.

2. Classical Feature Extraction

Condition monitoring is a systematic process that seeks to assess the current state of a system or component. In the case of rotating machinery, this may include vibration monitoring, which is the measurement of vibrations generated by the operation of the machine and the most widely used condition indicator for the diagnosis of machine faults [80–82]. Vibration is an oscillatory movement of an object. A vibration signal is a graphical or numerical representation of the vibration of an object recorded over time. The vibration signal can be measured with vibration sensors, such as accelerometers, and its analysis is performed using vibration analysis techniques, such as the Fourier transform or time domain analysis [83]. Vibration is an indirect way to measure machine health, as various factors such as wear, lubrication problems, structural problems, and other technical difficulties [84–86]. Consequently, these vibrations can be analyzed to detect possible machinery failures, known as vibration-based fault diagnosis. In fault diagnosis applications, feature values can be compared with predefined thresholds to determine normal or fault conditions, so selecting the appropriate feature is critical. In an ideal case, a significant feature is expected to distinguish normal conditions from fault conditions, establish a trend analysis, and avoid the influence of other equipment operating parameters [62,87]. However, the features are not necessarily informative for all cases or types of faults. For this reason, it is necessary to combine different types of features, so that information with the most remarkable possible diversity is obtained and can be used to reflect the machine's condition, making it useful for fault diagnosis.

2.1. Signal Processing-Based Hand-Crafted Features

Feature extraction is a critical process in vibration analysis and fault diagnosis. It consists of identifying and selecting the signals relevant to fault diagnosis, which can be used to distinguish between different types of faults. These features can be extracted manually or automatically and may include measurements such as frequency, amplitude, and shape of the vibration signal. A direct way to achieve this extraction is by using digital signal processing (DSP) since one of its main tasks is extracting useful information and manipulating and transforming signals. The goal of using these DSP techniques is to find a new form of simple, effective, and reduced representation of the original signals. In the

DSP, various transforms are used to analyze, manipulate, and represent signals differently. Some of the most common transforms include Laplace, Kalman, Wavelet, Hilbert, or Z transform. These are just some common transforms and techniques used in DSP. The choice of the appropriate transform depends on the signal's characteristics and the problem's specific needs to be solved because each method will reveal different information from the signal. In the present compilation, the search has been limited to using techniques that use simple Fourier transforms, such as the FFT and the PSD. This choice is because simple processes were sought and the possibility of reusing the transformation results in further processing. For these reasons, new representations of signals are sought, analyzed, and processed in the time or frequency domains. The time domain representation is the natural way a signal is presented and shows how the magnitude of a variable changes over time, whereas the frequency domain representation shows the energy distribution of a signal at different frequency components. Regardless of the domain used to represent and analyze a signal, a traditional way to extract information from it is through mathematical or statistical calculations. These calculations can be applied directly to the signal, using its data points as the statistic sample or input values to a function. Therefore, the computed values are then the features extracted from the signals, and a suitable set of features could efficiently represent a signal. This process is also a way to reduce the dimension of the data because now we use the feature set instead of the whole signal.

The information extracted from the signals will depend on the nature of the calculation or what is to be measured, quantified, or sought to be reflected with the features. In this way, we have some formulas (features) that measure the energy contained in the signal. In contrast, others seek to quantify the times the signal exceeds a certain threshold and many more types of information search. Thousands of processes, calculations, methodologies, and algorithms could be used to process a signal and extract some specific information from it. For instance, ref. [88] uses two approaches to align features in the time-frequency domain to classify faults. In addition, they use supervised and unsupervised techniques to improve the method. Besides, ref. [89] constructed a Wavelet auto-encoder, and then a deep approach to this model was performed. In [90], they apply a time-frequency transformation to express the signals. Then, a multi-scale TransFusion model is used to classify the features of time-frequency signals. Most groups of methods or techniques to extract features are applied in a specific way in different fields of application; the same ones seek all kinds of information and not one in particular. Many even depend on the specific application case where they will be used. For this reason, it is challenging to accurately categorize all of them based on the type of information they seek. To find the most significant number of hand-crafted features that can be calculated directly from the signal, an exhaustive compilation of formulas and calculations used as classical feature extraction methods on signals and applied in machine learning tasks has been carried out. These hand-crafted features have been chosen over the automatic ones mainly because, as an approximation to the DSP, they carry a direct physical significance with the phenomenon measured from the signals. This fact notably helps increase the ability to interpret the results within the context of the problem, in this case the diagnosis of failures in rotating machinery. We can also mention other significant advantages such as its simplicity, the control of the choice it offers the user, and its low computational cost in the execution. Other filtering criteria to accept features in this collection are that they are complete and self-contained, that their process can be represented in simple mathematical formulas, and that their calculation is conducted directly on the signal, whether in the time or frequency domain. Furthermore, we chose features that directly return real scalar numeric values after their computation and not processing styles that return the same but modified signal, such as filters, TSA, or noise cancellation techniques that are very common in DSP.

Features of various fields where signal processing is performed have been collected. Biomedical signal processing, such as Electrocardiography (ECG), Electromyography (EMG), or Electroencephalography (EEG), is a great exponent of fields where feature extraction is necessary. Other fields, such as audio signal analysis, roughness profiles, and entropy measurement, also require signal processing and are different from the field of fault diagnosis. Our compilation has shown that hand-crafted features follow certain trends in quantifying specific information they pursue according to the signal representation domain they use. In Figure 1, we present our proposed summary of the main information trends for which hand-crafted features try to quantify signals in the time and frequency domains. In this way, we can see that the information searched by the feature is closely linked to the domain to which it is applied. Signal analysis in different domains can help identify patterns and trends that are not evident in the original domain, which can be helpful in fault diagnosis, vibration analysis, anomaly detection, and other similar applications. The main trends in the time domain allude to the detection of patterns related to the waveform, while those in the frequency domain tend to search for changes in the frequency components of the spectra.

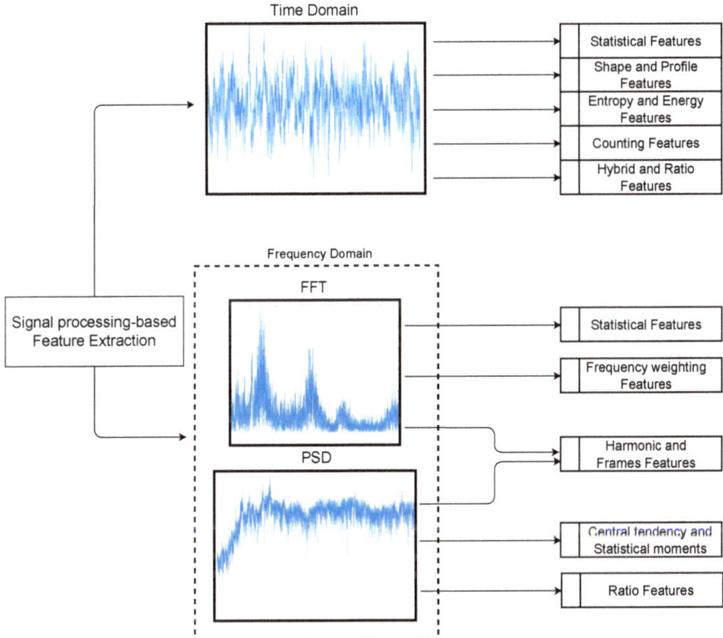

Figure 1. Main trends in information search for hand-crafted signal processing-based features in the time and frequency domain.

Some features search for or quantify information, as established in Figure 1, but do so in frames, bands, or segments of the signal in any domain. This is conducted to focus the search for information based on priori knowledge that a certain phenomenon should manifest in a specific signal locality, reducing the number of signal data points that would be used to calculate the feature. This would mainly reduce the "noise" of statistical features and make the calculation more relevant or discriminating. This happens more frequently with features from fields such as music or roughness since they seek to verify whether any variation manifests in specific parts of waveform or frequency spectrum segments. This hand-crafted feature calculation is mainly applied in the time domain when the signals are not stationary or their pattern varies over time, and in the frequency domain when it is known in advance that a band or frequency component should be altered for some reason. Because the literature reflects this type of emphasis in certain fields where feature extraction is applied, in the collection shown further, some features with the same formulation are counted as different, often depending on when applied to the complete signal or by bands or frames. As a substantial part of the contribution of this paper, a unification of the mathematical nomenclature has been carried out so that the calculation of the features and the information they quantify may be more understandable.

The unification of mathematical nomenclature presented in this paper is a significant contribution to the field. It addresses the inconsistencies in terminology across different domains and applications, providing a standardized framework for feature calculation and interpretation. This unified approach improves the clarity and comparability of feature extraction methods, facilitating better understanding and more effective use of these techniques in various disciplines.

For this reason, Table 2 shows a glossary of terminology. On many occasions, different authors use one version or another of the feature. However, they have different effects when classifying faults. This fact can be verified in the following sections, where the features are evaluated. The following subsections will detail the tables of the features collected for each domain.

Table 2. Feature nomenclature.

Symbol	Description
y	is a signal in the time domain
y_i	is the i-th element of y
N	is the total number of samples of y
p	is a threshold value
η	is a scale factor
m_t	is an integer value of the temporal moment order
Rp_i	is the peak average of the signal
Rv_i	is the valley average of the signal
$R3z_iT$	is the peak-to-peak value of the third ridge and valley
ROT	are the regions over a threshold
SBP	is the spacing between peaks
N_{SBP}	is the total number of spaces between peaks
N_{HSC}	is the total number of regions above threshold
$g(x)$	is a custom function
$PPCM$	is the spaces between profile peaks crossing the midline
lag	is a period between one event and another.
SAM	is a vector containing the lengths of consecutive signal samples above the mean.
SBM	is a vector containing the lengths of consecutive signal samples below the mean.
lb	is a lower bound
ub	is an upper-bound
r	is a multiplier of the standard deviation
l	is an integer number corresponding to the index of a time frame
N_{hop}	is the number of samples per analysis section
L	is L is the total number of time frames
y_l	is the l-th time frame from the y signal
STE_i	is the i-th of energy in a short period
$avSTE$	is the average STE over a 1 s window
N_w	is the number of points of a sample frame of the original signal y_i
f_0	is the minimum fundamental frequency to be analyzed

Table 2. *Cont.*

Symbol	Description
Y	is the frequency amplitude spectrum of y
Y_k	is the k-th measurement of the frequency amplitude spectrum (Y)
K	is the number of lines in the frequency spectrum
P	is the power spectrum frequency of y
P_k	is the k-th measurement of the power spectrum frequency (P)
K_P	is the total number of spectrum lines in the power spectrum
f_k	is the frequency value of the k-th spectrum line
ULC	Upper-cutoff frequency of the low-frequency band.
LLC	Lower-cutoff frequency of the low-frequency band.
UHC	Upper-cutoff frequency of the high-frequency band.
LHC	Lower-cutoff frequency of the high-frequency band.
$f_{h,l}$	is the frequency of harmonic peak h-th in frame l-th.
N_H	is the number of harmonics that is considered.
$A_{h,l}$	is the amplitude of harmonic h-th in frame l-th.
$S_l(k)$	is the l-th frame of a frequency spectrum
N_{FT}	is the number of points in the current frame of the spectrum
δ	is a small parameter to avoid calculation overflow.
b	is the band number
loK_b	is the integer frequency bin corresponding to the lower Edge of the band $loFb$.
hiK_b	is the integer frequency bin corresponding to the higher Edge of the band $hiFb$.
$*$	is the multiplication symbol

2.2. Time Domain Features Recopilation

Computational analysis in the time domain is less expensive in terms of processing, and the only preprocessing required is signal conditioning. On several occasions, a visual inspection of various signal parts will be enough to detect abnormal behavior. However, there are non-stationary, chaotic, and noisy signals, such as vibration signals, where the detection of patterns visually is practically impossible. In those cases, it is essential to use features that help us to describe analytically the behavior of the signal. Time domain signal analysis is a natural technique for processing and analyzing the waveform evolution of signals over time, mainly because most sensors deliver measurements or waveforms over time.

Any technique used for feature extraction in the time domain will aim to analyze and characterize the waveform of the signal in terms of its temporal behavior. The choice of features to extract depends on the specific application and the type of signal to be analyzed. Examples of time-domain DSP techniques from which one or more features can be extracted and quantized include peak and valley detection, filtering, period inspection, amplitude and phase, distortion, denoising, and segmentation, such as identifying specific sections of the signal. Combining these techniques in a time-domain signal analysis process allows valuable information to be extracted and used for fault diagnosis, condition monitoring, and pattern detection in vibration signals. A complete collection of 123 handcrafted features in the time domain (T1 to T123) is presented in Table 3 to perform the classical feature extraction. The nomenclature of the formulas shown can be seen in Table 2, while references to the characteristics can be found in Table 4. The general trends of the information extracted by this type of feature are statistical values, waveform patterns, signal integration, entropy, event count, ratios, and hybrid values from the combination of several calculations.

Table 3. Summary of the Time Feature Set.

Feature Name	Formula
Mean	$T1 = \frac{1}{N} \sum_{i=1}^{N} y_i$
Variance	$T2 = \frac{1}{N} \sum_{i=1}^{N} (y_i - T1)^2$
Standard deviation (STD)	$T3 = \sqrt{\frac{1}{N} \sum_{i=1}^{N} (y_i - T1)^2}$
Root mean square (RMS)	$T4 = \sqrt{\frac{1}{N} \sum_{i=1}^{N} (y_i)^2}$
Max value	$T5 = \max(y)$
Kurtosis	$T6 = \frac{N \sum_{i=1}^{N} (y_i - T1)^4}{\left[\sum_{i=1}^{N} (y_i - T1)^2 \right]^2}$
Skewness	$T7 = \frac{N \sum_{i=1}^{N} (y_i - T1)^3}{T3^3}$
Energy operator (EO)	$T8 = \frac{N^2 \sum_{i=1}^{N} \left((y_{i+1})^2 - (y_i)^2 - \text{mean}\left((y_{i+1})^2 - (y_i)^2 \right) \right)^4}{\left[\sum_{i=1}^{N} \left((y_{i+1})^2 - (y_i)^2 - \text{mean}((y_{i+1})^2 - (y_i)^2) \right)^2 \right]^2}$
Mean of absolute values (Mean abs.)	$T9 = \frac{1}{N} \sum_{i=1}^{N} |y_i|$
Square root amplitude value (SRAV)	$T10 = \left(\frac{\sum_{i=1}^{N} \sqrt{|y_i|}}{N} \right)^2$
Shape factor (SF)	$T11 = \frac{T4}{T9}$
Impulse factor (IF)	$T12 = \frac{T5}{T9}$
Crest factor	$T13 = \frac{T5}{T4}$
Clearance factor	$T14 = \frac{T5}{\frac{1}{N} \sum_{i=1}^{N} (y_i)^2}$
CPT1	$T15 = \frac{\sum_{i=1}^{N} \log(|y_i| + 1)}{N \log(T3 + 1)}$
CPT2	$T16 = \frac{\sum_{i=1}^{N} \exp(y_i)}{N * \exp(T3)}$
CPT3	$T17 = \frac{\sum_{i=1}^{N} \sqrt{|y_i|}}{N * T2}$
Mean Square Error (MSE)	$T18 = \frac{1}{N} \sum_{i=1}^{N} (y_i - T1)^2$
Log-log ratio (LLR)	$T19 = \frac{1}{\log(T3)} \sum_{i=1}^{N} \log(|y_i| + 1)$
Standard Deviation Impulse Factor (SDIF)	$T20 = \frac{T3}{T9}$
5th statistical Moment (FIFTHM)	$T21 = \sum_{i=1}^{N} (y_i - T1)^5$
6th statistical Moment (SIXTHM)	$T22 = \sum_{i=1}^{N} (y_i - T1)^6$
5th norm. moment (NM)	$T23 = \frac{\frac{1}{N} \sum_{i=1}^{N} (y_i - T1)^5}{\sqrt{\left(\frac{1}{N} \sum_{i=1}^{N} (y_i - T1)^2 \right)^5}}$
Kth central moment (KTHCM)	$T24 = \text{mean}\left[(y_i - T1)^k \right]$ k is set to 3
Pulse index (PI)	$T25 = \frac{T5}{T1}$
Margin index (MI)	$T26 = \frac{T5}{\left(\frac{1}{N} \sum_{i=1}^{N} \sqrt{|y_i|} \right)^2}$
Mean Deviation Ratio (MDR)	$T27 = \frac{T1}{T3}$
Difference absolute variance value (DVARV)	$T28 = \frac{1}{N-2} \sum_{i=1}^{N-1} (y_{i+1} - y_i)^2$
Min value	$T29 = \min(y_i)$
Peak Value	$T30 = \frac{1}{2} [T5 - T29]$
Peak to peak	$T31 = T5 - T29$

Table 3. *Cont.*

Feature Name	Formula
Hist. lower bound (Hist.LB)	$T32 = T29 - \frac{1}{2}\frac{T5-T29}{N-1}$
Hist. upper bound (Hist.UB)	$T33 = T5 + \frac{1}{2}\frac{T5-T29}{N-1}$
Latitude factor (LF)	$T34 = \dfrac{\max(|y_i|)}{\left(\frac{1}{N}\sum_{i=1}^{N}\sqrt{|y_i|}\right)^2}$
Norm. N. Neg. Likelihood (NNNL)	$T35 = \ln\left(\frac{T3}{T4}\right)$
Waveform indicators (WI)	$T36 = \frac{T4}{T1}$
Shannon entropy	$T37 = -\sum_{i=1}^{N} y_i^2 * \log(y_i^2)$
Log energy entropy (LEE)	$T38 = \sum_{i=1}^{N} \log(y_i^2)$ where $\log(0) = 0$
Threshold entropy	$T39 = \begin{cases} 1, & if\ |y_i| > p, and \\ 0, & elsewhere \end{cases}$ p is set to 0.2
Sure entropy	$T40 = N - \#\{i\ \text{such that}$ $|y_i| \le p\} + \ldots\ldots + \sum_i \min(y_i^2, p^2)$, p is set to 0.2
Norm entropy	$T41 = \sum_{i=1}^{N} |y_i|^p$, p is set to 0.2
Slope sign change (SSC)	$T42 = \sum_{i=2}^{N} g[(y_i - y_{i-1}) * (y_i - y_{i+1})]$ $g(y) = \begin{cases} 1, if & y \ge p \\ 0, if & otherwise \end{cases}$
Zero crossing (ZC)	$T43 = \sum_{i=1}^{N} \text{step}[\text{sign}(-y_i * y_{i+1})]$ $\text{sign} = \begin{cases} 1, if & y > 0 \\ 0, if & y = 0 \\ -1, if & y < 0 \end{cases}$ $\text{step} = \begin{cases} 1, if & y > 0 \\ \frac{1}{2}, if & y = 0 \\ 0, if & y < 0 \end{cases}$
Wilson amplitude	$T44 - \sum_{i=1}^{N} g(|y_i - y_{i+1}| - p)$ $g(y) = \begin{cases} 1, if & y \ge 0 \\ 0, if & y < 0 \end{cases}$ p is set to 0.2
Myopulse percentage rate (MYOP)	$T45 = \frac{1}{N}\sum_{i=1}^{N}[g(y_i)]$; $g(y) = \begin{cases} 1, & if & y \ge p \\ 0, & if\ otherwise \end{cases}$ p is set to 0.2
Wavelength	$T46 = \sum_{i=1}^{N} |y_{i+1} - y_i|$
Log detector	$T47 = \exp(\frac{1}{N})\sum_{i=1}^{N} \log|y_i|$
Mean of amplitude (MA)	$T48 = \sum_{i=1}^{N} (y_{i-1} - y_i)$
Energy	$T49 = \sum_{i=1}^{N} |y_i|^2$
Integrated signal	$T50 = \sum_{i=1}^{N} |y_i|$
Modified mean absolute value 1	$T51 = \frac{1}{N}\sum_{i=1}^{N} W_i|y_i|$ $W_i = \begin{cases} 1, if & 0.25N \le i \le 0.75N \\ 0.5, & if & otherwise \end{cases}$
Modified mean absolute value 2	$T52 = \frac{1}{N}\sum_{i=1}^{N} W_i|y_i|$ $W_i = \begin{cases} 1, & if & 0.25N \le i \le 0.75N \\ \frac{4i}{N}, & if & i < 0.25N \\ \frac{4(i-N)}{N}, & if & i > 0.75N \end{cases}$

Table 3. *Cont.*

Feature Name	Formula
Mean absolute value slope (MAVSLP)	$T53 = T9_{i+1} - T9_i$
Delta RMS (DRMS)	$T54 = T4_{i+1} - T4_i$
Root sum of squares (RSSQ)	$T55 = \sqrt{\sum_{i=1}^{N} |y_i|^2}$
Weighted SSR absolute (WSSRA)	$T56 = \frac{1}{N}\left(\sum_{i=1}^{N}\sqrt{|y_i|}\right)^2$
Log RMS	$T57 = \log(T4)$
Conduction velocity of Signal (CVS)	$T58 = \frac{1}{N-1}\sum_{i=1}^{N} y_i^2$
Average amplitude change (AAC)	$T59 = \frac{1}{N}\sum_{i=1}^{N-1} |y_{i+1} - y_i|$
Weibull negative log-likelihood (WNLL)	$T60 = -\sum_{i=1}^{N}\log[(T11*\eta)^{-T1}*\ldots$ $\ldots*|y_i|^{T1-1}\exp\left(\frac{|y_i|}{\eta}\right)^{T1}]$
V-ORDER 3	$T61 = \sqrt[3]{\frac{1}{N}\sum_{i=1}^{N} y_i^3}$
Maximum Fractal Length (MFL)	$T62 = \log_{10}\sqrt{\sum_{i=1}^{N-1}(y_i - y_{i+1})^2}$
Difference Absolute STD (DASDV)	$T63 = \sqrt{\frac{1}{N-1}\sum_{i=1}^{N-1}(y_{i+1} - y_i)^2}$
Higher order Temp. Moments (TM)	$T64 = \left|\frac{1}{N}\sum_{i=1}^{N} y_i^{m_t}\right|,$ m_t is set to 3 as default.
Autocorrelation function (ACF)	$T65 = \frac{1}{N-1}\sum_{i=1}^{N} y_i * y_{i-1}$
Amplitude density function (ADF)	$T66 = \sqrt[2]{2\pi * T3} * \exp\left(\frac{T31}{2*T3^2}\right)$
High spot count (NROT)	$T67 = \frac{1}{N}\sum(\text{ROT}),$ the threshold can be set to 70% of the maximum value
Mean slope of the profile (SOP)	$T68 = \frac{1}{N-1}\sum_{i=1}^{N} |y_{i+1} - y_i|$
Average wavelength (meanWavelenght)	$T69 = \frac{2\pi * T1}{T68}$
Mean spacing of adjacent peaks (MSBP)	$T70 = \frac{1}{N_{SBP}}\sum SBP$
Peaks mean values (NRZDIN)	$T71 = \frac{1}{2N}\left(\sum_{i=1}^{N} Rp_i - Rv_i\right)$
Mean height of peaks (Rpm)	$T72 = \frac{1}{N}\sum_{i=1}^{N} Rp_i$
Mean depth of valleys (Rvm)	$T73 = \frac{1}{N}\sum_{i=1}^{N} Rv_i$
Third point rugosity mean (RHZ)	$T74 = \frac{\sum_{i=1}^{N} R3z_i}{N}$
Number of peaks in profile (NPP)	$T75 = \text{sum of peaks}$
Mean spacing in the mean line (Sm)	$T76 = \text{mean}(PPCM)$
Peak count (PC)	$T77 = \frac{1}{T76}$
Profile solidity factor (PSF)	$T78 = \frac{T29}{T5}$
Relative length of the profile (RLP)	$T79 = \frac{1}{N}\sum_{i=1}^{N}\sqrt{(y_{i+1} - y_i)^2 + 1}$
Mean peak radius of curvature (Rp)	$T80 = \frac{1}{N-2}\sum_{i=1}^{N-2}\frac{2y_i - y_{i-1} - y_{i+1}}{N^2}$
Stepness factor of the profile (SFP)	$T81 = \frac{T1}{T76}$
RMS slope of the profile (SlopeRMS)	$T82 = \sqrt[2]{\left(\frac{1}{N-1}\right)\sum_{i=1}^{N-1}(y_i - y_{i-1} - \Theta m)^2},$ where $\Theta m = \frac{1}{N-1}\sum_{i=1}^{N-1} y_i - y_{i-1}$

Table 3. *Cont.*

Feature Name	Formula
Mean spacing at mean line (SMR)	$T83 = \frac{1}{N-1} \sum_{i=1}^{N-1} \tan^{-1}(y_{i+1} - y_i)$
Mean of inflection points	$T84 = \frac{1}{N} \sum (\text{total of inflection points})$
Height of irregularities (NRZISO)	$T85 = \frac{1}{N}(\sum_{i=1}^{N} Rp_i - Rv_i)$
Waviness factor of the profile (WF)	$T86 = \frac{1}{T1} \sum_{i=1}^{N-1} \sqrt[2]{(y_{i+1} - y_i)^2 + 1}$
Estimation of the Autocorrelation (AGGA)	$T87 = \frac{1}{(N-lag)(T2)^2} \sum_{i=1}^{N-lag} (y_i - T1)(y_{i+lag} - T1)$
C3	$T88 = \frac{1}{(N-2lag)} \sum_{i=1}^{N-2lag} ((y_{i+2lag})^2 (y_{i+lag})(y_i))$
Count above mean (CAM)	$T89 = \sum_{i=1}^{N} (y_i > T1)$
Count below mean (CBM)	$T90 = \sum_{i=1}^{N} (y_i < T1)$
First location of maximum (FLOM)	$T91 = \frac{T5}{N}$
First location of minimum (FLOMIN)	$T92 = \frac{T29}{N}$
Has duplicate (HD)	$T93 = \sum_{i=1}^{N} (Vect_{unique}(y_i) \sim= 0) \sim= N$
Has duplicate max (HDMAX)	$T94 = \begin{cases} 1 \, or \, True, & if \, (T5 = y_1 \ldots y_N), \text{"Twice"} \\ 0 \, or \, False, & else \end{cases}$
Has duplicate min (HDMIN)	$T95 = \begin{cases} 1 \, or \, True, & if \, (T29 = y_1 \ldots y_N), \text{"Twice"} \\ 0 \, or \, False, & else \end{cases}$
Large standard deviation (LSD)	$T96 = \begin{cases} 1 \, or \, True, & if \, (T3 > r(T5 - T29)) \\ 0 \, or \, False, & else \end{cases}$
Last location of maximum (LLOM)	$T97 = \frac{\text{index}(T5)}{N}$
Last location of minimum (LLOMIN)	$T98 = \frac{\text{index}(T29)}{N}$
Longest strike above mean (LSAM)	$T99 = \max(SAM)$
Longest strike below mean (LSBM)	$T100 = \max(SBM)$
Mean second derivate central (MSDC)	$T101 = \frac{1}{N} \sum_{i=1}^{N-1} \frac{1}{2}((y_{i+2}) - 2(y_{i+1}) + (y_i))$
Percentage of recurring data points all datapoints (PRDAD).	$T102 = \frac{\text{length(Repeated values in signal)}}{\text{length(signal)}}$
Percentage of recurring values to all values (PRVAV).	$T103 = \frac{\text{length(Repeated values in signal)}}{\text{length}(Vect_{unique}(signal))}$
Range count (RC)	$T104 = \sum((y_i \geq lb)(y_i < ub))$
Ratio beyond r sigma (RBRS)	$T105 = \frac{1}{N} \sum(|y_i - T1| > (r * T3))$
Sum of recurring data points (SORDP)	$T106 = \text{Sum of non-unique values}$
Sum of recurring values (SORV)	$T107 = \text{Sum of the non-unique values count}$
Factor B (ffbb)	$T108 = \frac{T6*T13}{T3}$
Talaf	$T109 = \log(T6 + T4)$
Thikat	$T110 = \log(T6^{T13} + T4^{T30})$
Siana	$T111 = \log(\frac{T13^{T6}}{T30^{T4}})$
Inthar	$T112 = \log(\frac{T30^{T6}}{T13^{T4}} * T12)$
Audio power (AP)	$T113_l = \frac{1}{N_{hop}} \sum_{i=1}^{N_{hop}-1} |y_{(i+lN_{hop})}|^2$
Temporary bell (Env)	$T114_l = \sqrt{T114_l}$
Zero Crossing Rate for each Nhop (ZCRn)	$T115 = \frac{1}{2N_{hop}} \sum_{i=1}^{N_{hop}} \text{sign}(y_{Nhop_{i-1}} - y_{Nhop_{i-2}}) * F_s$
Simple quadratic integral (SSI)	$T116 = \sum_{i=1}^{N} |y_i|^2$

Table 3. *Cont.*

Feature Name	Formula
Runtime log (LAT)	$T117 = \log_{10}(\text{index}(T5) - \text{index}(T29))$
Temporal centroid (TC)	$T118 = \dfrac{N_{hop}}{F_s} \dfrac{\sum_{l=1}^{L}(lT115_l)}{\sum_{l=1}^{L}(T115_l)}$
Harmonic Ratio (HR)	$\Gamma_m = \dfrac{\sum_{i=1}^{Nw}(y_i)*(y_{i-m})}{\sqrt{\sum_{i=1}^{Nw}(y_i)^2 * \sum_{i=1}^{Nw}(y_{i-m})^2}},$ $(1 \le m \le M\,;\, 0 \le l \le (L-1)),$ where $M = \dfrac{F_s}{(f_0)^2}$, is set to 2 lags $T119 = max(\Gamma_m)\,(1 \le m \le M)$
High Zero-Crossing Rate Ratio (HZCRR)	$T120 = \dfrac{1}{2L}\sum_{i=1}^{L}[\text{sign}(T116 + \ldots$ $\ldots - 1.5 * \text{mean}(T116)) + 1]$
Low energy ratio in the short term (LSTER)	$T121 = \dfrac{1}{2L}\sum_{i=1}^{L}[\text{sign}(0.5 * avSTE - STE_i) + 1]$
Health indicators (INDI)	$T122 = \dfrac{T4}{T3}$
Factor A (ffaa)	$T123 = \dfrac{T5}{T3*T2^2}$

2.3. Frequency Domain Features Recopilation

The representation of a signal in other domains allows us to analyze and understand its properties. For example, the transition from the time domain to the frequency domain allows us to analyze the signal in terms of the frequency and amplitude of its frequency components. This type of transformation can be performed using different techniques, such as the Fourier Transform, which converts a signal in time in its spectrum, representing the signal in the frequency domain. Analysis of signals in the frequency domain allows the identification of patterns and traits that are not evident in the time domain. This analysis is helpful for various applications, including the detection of abnormalities in the system that generates the signal, the identification of sources of interference and the identification of signal quality problems. In addition, it is also used in condition monitoring and fault diagnosis, as abnormal frequency patterns can indicate problems in mechanical, electromechanical, or electronic components of rotating machinery or systems. However, temporary information is lost in the transformation, which can be a disadvantage in some applications.

Based on the Fourier spectrum, other spectra analyses such as power spectral density (PSD) [91], bispectrum [92], trispectrum [93], and cepstrum [94], among others, have also been used in the diagnosis of rotating machinery faults. All the techniques mentioned above, excluding the frequency spectrum and PSD, are beyond the scope of this compilation study. These techniques involve more complex transformations that process or abstract the signal to various levels, some having multiple definitions or versions. The frequency spectrum represents the amplitude of the different frequencies that make up a signal. On the other hand, the spectral density or power spectrum is a graphical representation of the energy distribution in a signal as a function of frequency. When there is some alteration in a mechanical component, the frequency and power components also change; therefore, the position of the central spectrum peak will also change. This phenomenon can be reflected analytically using features that quantify those changes and reflect the condition of a machine. Table 5 presents 46 hand-crafted features in the frequency domain (F1 to F46). The nomenclature of the formulas shown can be seen in Table 2, while the references to the features can be found in Table 4. The general trends of the information extracted by this type of feature are values and statistical moments, the weighting of specific frequency components and ratios, and the search for values isolating harmonics or frames.

Table 4. References of collected features.

Features	Reference	Features	Reference	Features	Reference	Features	Reference
T1–T14	[44,63,95–111]	T29–T33	[97–101,103,112]	T65–T86	[113–117]	F1, F6, F7	[95,101,118,119]
T15–T17	[120]	T34–T36	[121–123]	T87–T107	[124]	F2–F4, F8–F12	[95]
T18–T20	[125,126]	T37–T41	[127]	T108	[128]	F5	[118,129]
T21–T22	[129–134]	T42–T45	[107,108,135–137]	T109	[138]	F13–F15	[129,137,139]
T23	[121]	T46–T53	[108,125,135]	T110–T113	[140]	F16–F25	[137,141]
T24–T25	[98]	T54	[44,101]	T114–T120	[129]	F26–F31, F40–F45, F47–F48	[129]
T26–T27	[122,123]	T55–T63	[126,136,137,142–144]	T121–T122	[129–134]	F32–F39	[67,138]
T28	[107]	T64	[107]	T123–T125	[67,138,145]	F46	[146]

Table 5. Summary of the Frequency Feature Set.

Feature Name	Formula
Mean Frequency	$F1 = \frac{\sum_{k=1}^{K} Y_k}{K}$
Variancef	$F2 = \frac{\sum_{k=1}^{K} (Y_k - F1)^2}{K-1}$
Skewnessf	$F3 = \frac{\sum_{k=1}^{K} (Y_k - F1)^3}{K(\sqrt{F2})^3}$
Kurtosisf	$F4 = \frac{\sum_{k=1}^{K} (Y_k - F1)^4}{K(F2)^4}$
Central Frequency	$F5 = \frac{\sum_{k=1}^{K} f_k Y_k}{\sum_{k=1}^{K} Y_k}$
STDF	$F6 = \sqrt{\frac{\sum_{k=1}^{K} (f_k - F5)^2 Y_k}{\sum_{k=1}^{K} Y_k}}$
RMSF	$F7 = \sqrt{\frac{\sum_{k=1}^{K} f_k^2 Y_k}{\sum_{k=1}^{K} Y_k}}$
CP1	$F8 = \frac{\sum_{k=1}^{K} (f_k - F5)^3 Y_k}{K(F6)^3}$
CP2	$F9 = \frac{F6}{F5}$
CP3	$F10 = \frac{\sum_{k=1}^{K} (f_k - F5)^{\frac{1}{2}} Y_k}{K\sqrt{F6}}$
CP4	$F11 = \frac{\sum_{k=1}^{K} (f_k - F5)^3 Y_k}{F6^2 K}$
CP5	$F12 = \sqrt{\frac{\sum_{k=1}^{K} f_k^4 Y_k}{\sum_{k=1}^{K} f_k^2 Y_k}}$
Spectral Centroid	$F13 = \frac{\sum_{k=1}^{K} k Y_k}{\sum_{k=1}^{K} Y_k}$
Spectral Spread	$F14 = \sqrt{\frac{\sum_{k=1}^{K} (k - F13)^2 Y_k}{\sum_{k=1}^{K} Y_k}}$
Spectral Entropy	$F15 = -\sum_{k=1}^{K-1} P_n(k)\log_2[P_n(k)]$ where: $P_n(k) = \frac{Y_k}{\sum_{k=1}^{K} Y_k}$
Total power	$F16 = \sum_{k=1}^{K_P} P_k$
Median Frequency	$F17 = \frac{1}{2}\sum_{k=1}^{K_P} P_k$
Peak frecuency (PKF)	$F18 = \max(P)$
First Spectral Moment	$F19 = \sum_{k=1}^{K_P} P_k f_k$
Second Spectral Moment	$F20 = \sum_{k=1}^{K_P} P_k f_k^2$
Third Spectral Moment	$F21 = \sum_{k=1}^{K_P} P_k f_k^3$
Fourth Spectral Moment	$F22 = \sum_{k=1}^{K_P} P_k f_k^4$
Variance of central frequency(VCF)	$F23 = \frac{F20}{F16} - \left(\frac{F19}{F16}\right)^2$
Frequency Deformation	$F24 = \frac{\sqrt{F20/F16}}{F19/F16}$

Table 5. *Cont.*

Feature Name	Formula
Frequency ratio (FR)	$F25 = \sum_{LLC=f_{min}}^{ULC=f_{max}/2} P_k \,/\, \sum_{LHC=\frac{f_{max}}{2}+1}^{UHC=f_{max}} P_k$
Harmonic Spectral Centroid (HSC)	$F26 = \frac{1}{L}\sum_{l=1}^{L} LHSC_l,$ where $LHSC_l = \dfrac{\sum_{h=1}^{N_H}(f_{h,l}A_{h,l})}{\sum_{h=1}^{N_H}(A_{h,l})}$
Harmonic Spectral Deviation (HSD)	$F27 = \frac{1}{L}\sum_{l=1}^{L} LHSD_l,$ where $LHSD_l = \dfrac{\sum_{h=1}^{N_H} \lvert \log_{10}(A_{h,l}) - \log_{10}(SE_{h,l})\rvert}{\sum_{h=1}^{N_H} \log_{10}(A_{h,l})},$ $SE_{h,l} = \begin{cases} \frac{1}{2}(A_{h,l} + A_{h+1,l}), & if\ h = 1 \\ \frac{1}{3}(A_{h-1,l} + A_{h,l} + A_{h+1,l}), & if\ 2 \le h \le N_H - 1 \\ \frac{1}{2}(A_{h-1,l} + A_{h,l}), & if\ h = N_H \end{cases}$
Harmonic Spectral Spread (HSS)	$F28 = \frac{1}{L}\sum_{l=1}^{L} LHSS_l,$ where $LHSS_l = \dfrac{1}{LHSC_l}\sqrt{\dfrac{\sum_{h=1}^{N_H}\left[(f_{h,l}-LHSC_l)^2 A_{h,l}^2\right]}{\sum_{h=1}^{N_H} A_{h,l}^2}}$
Harmonic Spectral Variation (HSV)	$F29 = \frac{1}{L}\sum_{l=1}^{L} LHSV_l,$ where $LHSV_l = 1 - \dfrac{\sum_{h=1}^{N_H}(A_{h,l-1}A_{h,l})}{\sqrt{\sum_{h=1}^{N_H} A_{h,l-1}^2}\sqrt{\sum_{h=1}^{N_H} A_{h,l}^2}}$
Spectral Flux (SF)	$F30 = \frac{1}{L*N_{FT}}\sum_{k=1}^{L}\sum_{k=1}^{N_{FT}}[\log(\lvert S_l(k)\rvert + \delta) - \log(\lvert S_{l-1}(k)\rvert + \delta)]^2$
Frequency Centre (FC)	$F31 = \dfrac{\sum_{k=2}^{K} x_{k'} x_k}{2\pi \sum_{k=1}^{K} x_k^2},$ where $x_{k'} = x(k+1) - x_k$
Mean square Frequency (MSF)	$F32 = \dfrac{\sum_{k=2}^{K}(x_{k'})^2}{4\pi^2 \sum_{k=1}^{K} x_k^2}$
Root Mean square Frequency (RMSF)	$F33 = \sqrt{F32}$
Grand mean (GM)	$F34 = \dfrac{\sum_{k=1}^{K} f_k Y_k}{\sum_{k=1}^{K} Y_k}$
Standard Deviation (STDA)	$F35 = \sqrt{\dfrac{\sum_{k=1}^{K}(f_k - F35)^2 Y_k}{K}}$
C Factor (ffcc)	$F36 = \sqrt{\dfrac{\sum_{k=1}^{K} f_k^2 Y_k}{\sum_{k=1}^{K} Y_k}}$
D Factor (ffdd)	$F37 = \sqrt{\dfrac{\sum_{k=1}^{K} f_k^4 Y_k}{\sum_{k=1}^{K} f_k^2 Y_k}}$
E Factor (ffee)	$F38 = \dfrac{\sum_{k=1}^{K} f_k^2 Y_k}{\sqrt{\sum_{k=1}^{K} Y_k \sum_{k=1}^{K} f_k^4 Y_k}}$
G Factor (ffgg)	$F39 = \dfrac{F36}{F35}$
Audio Spectrum Envelope (ASE)	$F40 = \sum_{k=1}^{N_{FT}} P_b(k),\ \text{for } 1 \le b \le B$
Audio Spectrum Flatness (ASF)	$F41 = \dfrac{^{hiK_b' - loK_b' + 1}\sqrt{\prod_{k'=loK_b'}^{hiK_b'} P_b(k)}}{\frac{1}{hiK_b' - loK_b' + 1}\sum_{k'=loK_b'}^{hiK_b'} P_b(k)},\ \text{for } 1 \le b \le B$
Audio Spectrum Spread (ASS)	$F42 = \sqrt{\dfrac{\sum_{k'=0}^{(N_{FT}/2)-K_{low}}\left[\log_2\left(\frac{f'(k')}{1000}\right) - F42\right]^2 P'(k')}{\sum_{k'=0}^{(N_{FT}/2)-K_{low}} P'(k')}}$
Power Spectral Centroid Segment (SC)	$F43 = \dfrac{\sum_{k=0}^{N_{FT}} f(k)P_s(k)}{\sum_{k=0}^{N_{FT}} P_s(k)},\ \text{where}$ P_s is the estimated power spectrum for the segment.
SNR	$F44 = \dfrac{F1}{F36}$

Table 5. *Cont.*

Feature Name	Formula
Spectral Rolloff Frequency (SRF)	$F45 = 0.85 \sum_{k=0}^{N_{FT}} \lvert Y_k \rvert$
Upper Limit of Harmonicity (ULH)	$F46 = \log_2\left(\frac{f_{ulh}}{1000}\right),$ where f_{ulh} is defined by: $$f_{ulh} = \begin{cases} 31.25, & for\ k' = 0 \\ f(k' + K_{low}), & for\ 1 < k' < \frac{N_{FT}}{2} - K_{low} \end{cases}$$

The hand-crafted features in our study are designed to capture various aspects of vibration signals that are indicative of machinery conditions. In Figure 2, we illustrate a vibration signal in the time domain in which we exemplify the computation of features. In particular, this figure shows in yellow the computation of the T74 feature in which the magnitude between the third ridge and valley is identified. In purple, we see PPCM, which is the number of spaces between profile peaks crossing the midline to compute the Mean spacing in the mean line (Sm) or T76 feature. In navy blue, we identify the space between peaks SPB to compute the Mean spacing of adjacent peaks (MSBP) or T70 feature. In light blue, we show a region over a threshold (ROT), which is used to compute the High spot count (NROT) or T67 feature. We present other parameters as the number of peaks and valleys per l time frame, upper and lower thresholds, N that means the total number of times samples and L that means the total number of time frames as was established in Table 2. Naming the remaining parameters from features is vast, and it is not the purpose of this article.

These features can be broadly classified according to the signal characteristics they reflect:

- Amplitude-based features (e.g., peak value, crest factor): These reflect the magnitude of vibrations, which can indicate the severity of faults.
- Statistical features (e.g., mean, variance, skewness, kurtosis): These describe the distribution of vibration amplitudes, which can change with different fault types.
- Energy-based features (e.g., signal energy, entropy): These quantify the energy content and complexity of the signal, which often increase with fault severity.
- Time-series features (e.g., zero-crossing rate, autocorrelation): These reflect the temporal behavior of the signal, which can indicate periodicity or irregularities in vibrations.
- Frequency-based features (e.g., spectral centroid, frequency ratio): These capture the distribution of energy across different frequencies, helping identify characteristic fault frequencies.
- Shape-based features (e.g., impulse factor, margin factor): These describe the shape of the waveform, which can change with different fault types.

Each feature is sensitive to different aspects of the vibration signal, allowing for a multi-faceted analysis of machinery conditions. The combination of these features provides a comprehensive representation of the vibration signal, enabling effective fault diagnosis across various types of rotating machinery.

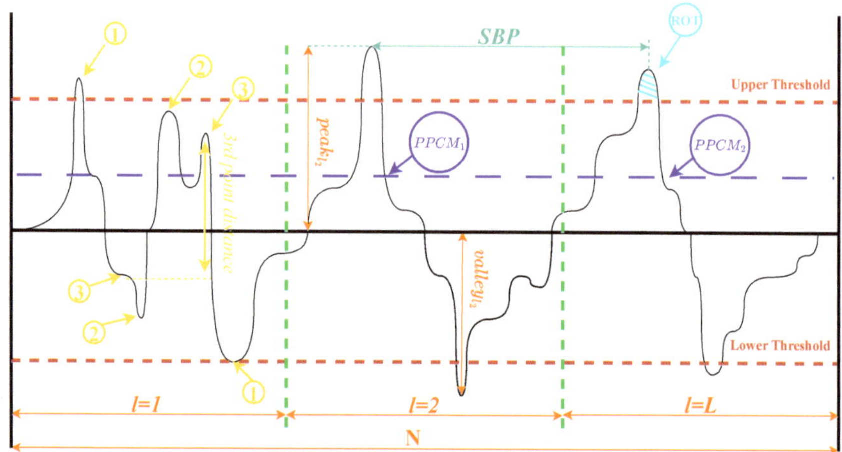

Figure 2. Illustration of some feature parameters on a vibration signal.

3. Experimentation Data for Hand-Crafted Features Performance Evaluation

The data employed to evaluate hand-crafted features and test their effectiveness in fault classification tasks add up to seven databases. The databases were obtained from different rotating machines, such as gearboxes and motors with shafts and bearings. Each database has its study elements with normal conditions and induced faults. The experimental configurations of the databases are diverse, have specific mechanical study elements, and reflect specific phenomena. An organizational scheme of all the rotating machines used, their derived databases, and study elements are shown in Figure 3. Some databases include multiple levels of fault severity in the elements under study. In contrast, others include various types of faults or fault combinations in their elements, known as multi-fault databases. Regardless of the specific phenomenon, machine, or element under investigation, these databases serve as valuable resources for conducting fault diagnosis studies. They are particularly useful because their data were collected using experimental setups designed to simulate various modes and effects of failures. As a result, these databases provide comprehensive, labeled information on different types of faults, making them ideal for research and analysis in the field of fault diagnosis.

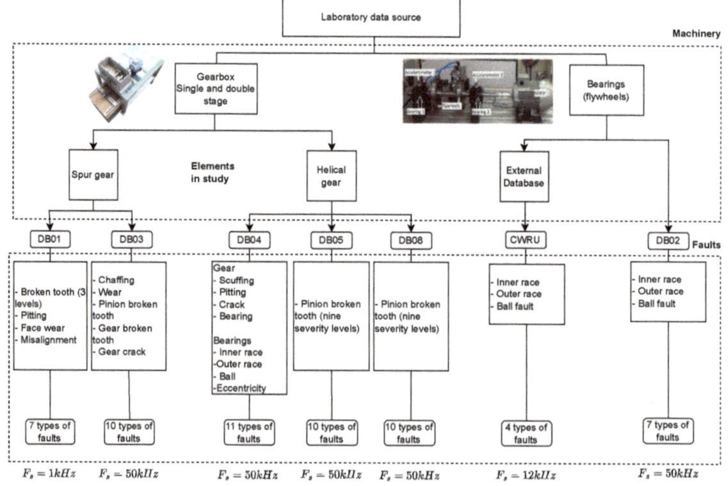

Figure 3. Schematic description of the databases used.

The internal mechanical components of rotating machinery are diverse. However, some of these components are more prone to failure than others. Therefore, the most common gearbox faults that could appear in gears are tooth breakage, cracking, pitting, wear, chafing, and scuffing. Another type of fault is misalignment. Meanwhile, the main bearing fault is scratch and occurs mainly in the inner race, outer race, and rolling element. Another type of fault is eccentricity. In reciprocating compressors, failures occur mainly in valves and bearings. Figure 4 shows a general compilation of the main elements subject to failure of the different rotating machines and examples of the appearance of the associated fault. Another consideration is that six of the seven databases were produced by the Industrial Technology Research and Development Group (GIDTEC) of the Salesian Polytechnic University of Cuenca-Ecuador. For an external comparison option, the remaining database belongs to the Bearing Data Center of Case Western Reserve University, which is a famous benchmark used to test new techniques and methodologies for fault diagnosis.

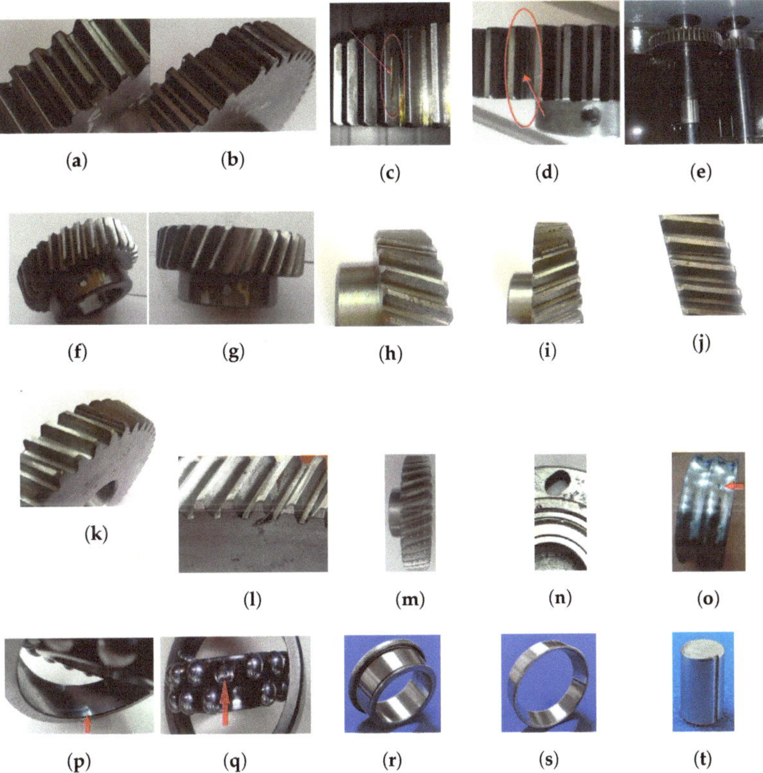

Figure 4. Compilation of the main mechanical components prone to failures under study in the different databases. (**a**) Broken tooth 25%. (**b**) Broken tooth 100%. (**c**) Scuffing. (**d**) Wear. (**e**) Misalignment. (**f**) Broken tooth 11.5%. (**g**) Broken tooth 100%. (**h**) Scuffing. (**i**) Scuffing 10 mm large. (**j**) Scuffing one stripe. (**k**) Scuffing two stripes. (**l**) Crack. (**m**) Pitting. (**n**) Eccentricity. (**o**) Inner race fault. (**p**) Outer race fault. (**q**) Rolling element fault. (**r**) Inner race fault. (**s**) Outer race fault. (**t**) Rolling element fault.

The most relevant information about each database utilized to test the methodology followed in this study is briefly described below:

- **DB01:** This experiment involves a gearbox connected to an electric motor on the input shaft and an electromagnetic brake on the output shaft. The gearbox is a single-stage reduction type. The gears used are spur gears. Seven fault conditions are introduced: three broken tooth severities, misalignment condition, Pitting, and Pitting with face wear. More information on this database can be found at [147]. This database consists of vibration signals sampled at 1 Ks/s with a duration of 2 s, seven classes, and 150 observations for each class.

- **DB03:** This experiment involves a gearbox with ten health conditions, including chaffing tooth, worn tooth, broken tooth at three levels (25%, 50%, and 100%), 25% and 100% gear crack, 25% pinion broken tooth and 25% gear crack, 50% gear chaffing and normal condition in spur gears (number of teeth $Z_1 = 53$ and $Z_2 = 80$ with modulus 2.25 and impact angle 20°). The experiments were carried out with three loads, three constant speeds and three variable speeds, in each of them, five samples were collected in a duration of 10 s [148]. This database consists of vibration signals sampled at 50 Ks/s with ten classes and 90 observations for each class.

- **DB04:** The experimental setup consists of a gearbox connected to an electric motor on the input shaft and an electromagnetic brake on the output shaft. The gearbox is a two-stage reduction gear. The gears are of helical type. Ten different fault conditions are introduced, which include issues such as Scuffing, Pitting, and Crack in gears. In the case of bearings, the faulty components include the inner race, outer race, and rolling element. In addition, an eccentricity fault is introduced in the bearing housing. This database consists of vibration signals sampled at 50 Ks/s with a duration of 10 s, eleven classes, and 45 observations for each class [149,150].

- **DB05:** This experiment consists of nine severity levels of a broken pinion tooth. The gearbox is connected to an induction motor and an electromagnetic brake. The gearbox is a single-stage and is of the reduction gear type. The gears used are helical. The experiments were carried out with three loads produced by the brake on the gearbox's output shaft. In each of them, five samples were collected for 10 s. Additionally, the experiment was performed at three constant speeds. This database consists of vibration signals sampled at 50 Ks/s with ten classes and 75 observations for each class [151].

- **DB08:** This is also a study of nine fault severity levels of the helical gearbox dataset. The gearbox is single-stage and is of reduction gear type. The experiments were carried out with three loads produced by the brake on the gearbox's output shaft. The main difference between the DB05 and the DB08 is that the latter is more recent and captures more condition signals and observations. The experimental setup is the same, with differences in the materials of the elements and the procedures in the box assembly [152]. This database consists of vibration signals sampled at 50 Ks/s with a duration of 10 s, ten classes, and 180 observations for each class.

- **CWRU:** This database is a famous benchmark for testing bearing fault diagnostic techniques. Their page has information on the bearings used, the diameters and depths of the failures, and the configuration of the equipment assembly for the experiments. The experiment has four levels of speed and load. It has a normal class and induces failures in the outer race, the inner race, and the rolling element. Another important fact is that the bearing vibration signals were captured at a 12 kHz sampling rate. For the experiment in this paper, the 12 k fan end accelerometer data were used [153]. This database consists of vibration signals with a duration of 0.1667 s, four classes, and 80 observations for each class.

- **DB02:** This dataset aims to study bearing behavior. It involves a $\varnothing$30 mm shaft with flywheels mounted. The shaft is connected to an induction motor and is seated on the bearings and their housings. The acquired signals belong to an accelerometer placed in a vertical position near the movement source. Details of the experimental setup are described in [151]. This database consists of vibration signals sampled at 50 Ks/s with a duration of 20 s, seven classes, and 45 observations for each class.

Accelerometers are strategically placed closest to the source of motion on the machine across all databases to capture data effectively. All signals from all experiments were stored as a group in a database, differentiating according to the failure labels they contained. The condition monitoring signals used for this experiment have been limited to vibration signals since vibration analysis has traditionally been the most widely used means of detecting anomalies in rotating machines. Especially when combined with spectral analysis, it provides a direct association of the characteristic frequency components that a machine should have based on the movement or rotation of the elements that compose it. Figure 5 shows a compilation of vibration signals taken from the normal condition of each database used, compared to their respective frequency spectrum. When analyzing the vibration signals over time, it can be verified that certain signals present a slightly more cyclical behavior, as could be the case for DB03 and DB05. On the other hand, the vibration signatures of the other databases seem much more chaotic. On the other side of the frequency spectrum, one can observe clear differences in their different frequency components. The DB02 and the CWRU clearly show characteristic signatures with frequency sidebands, while the other spectra show that most of their components are housed in the lower frequencies. Clear differences are distinguished between the different time and frequency graphs among all databases. These differences emphasize that each machine has many internal behaviors that must be characterized to correctly diagnose a fault.

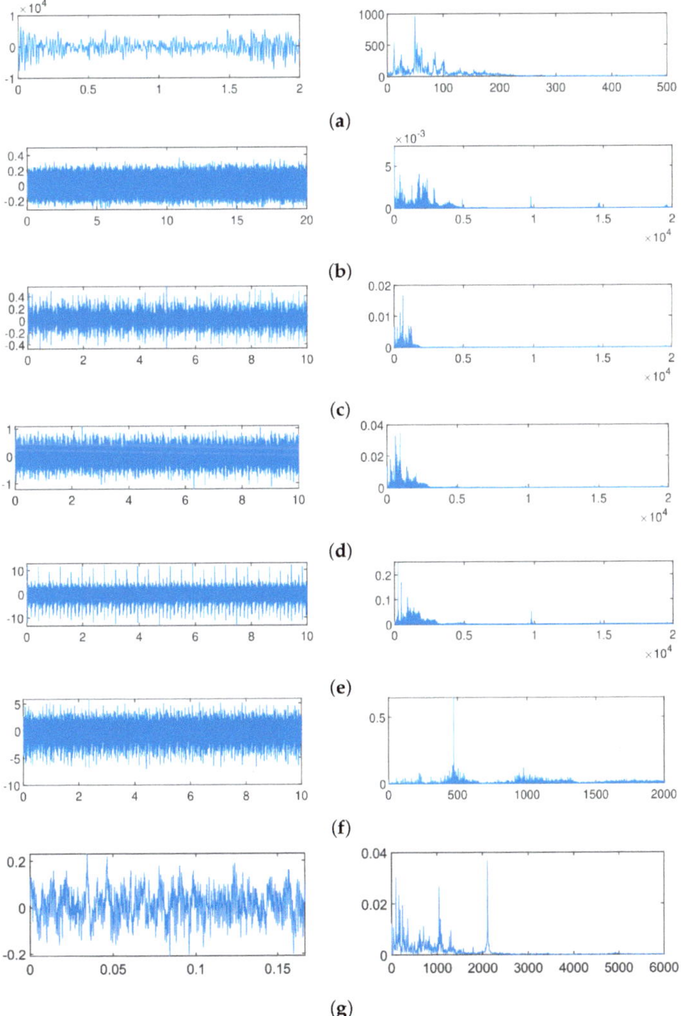

Figure 5. Example of a time domain signal alongside their respective frequency spectrum per each database. (**a**) DB01. (**b**) DB02. (**c**) DB03. (**d**) DB04. (**e**) DB05. (**f**) DB08. (**g**) CWRU.

4. Proposed Feature Evaluation Methodology

After the compilation of hand-crafted features for their application in condition monitoring signals, a systematic methodology is proposed to establish an adequate reference framework to carry out the exhaustive feature evaluation in fault classification tasks in rotating machinery. Figure 6 shows graphically the sequence of steps followed in the proposed methodology.

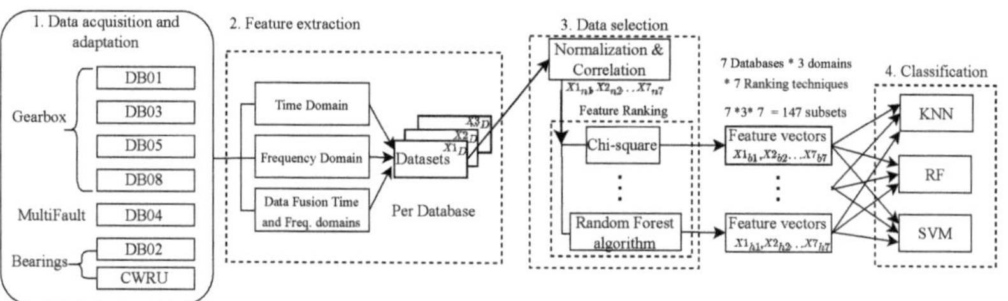

Figure 6. Diagram of the methodology employed

1. **Data acquisition and adaptation:** Different vibration signals were collected from seven databases and stored separately. The signals are organized in local directories, separating the normal condition and its different fault modes for each database. Seven original datasets with vibration signals were acquired from different machines with different fault configurations, number of observations, acquisition rates, and signal sizes.

2. **Feature extraction step:** In this phase, the calculations of the hand-crafted features were applied according to the mathematical formulations shown in Tables 3 and 5. The 123 time-domain and 46 frequency-domain hand-crafted features were implemented as functions of the MATLAB software R2023b, and an algorithm was created that is responsible for iterating them for each signal from all databases. To exhaustively evaluate the features, they were organized into three large evaluation groups: one with only time domain features, another with only frequency domain features, and the last group containing a data fusion scheme. The data fusion consists of concatenating time domain and frequency domain features into a single-dimensional feature vector for each signal. Thus, Group 1 evaluates 123 features, Group 2 evaluates 46 features, and Group 3 evaluates 169 features. Each group contains separately the calculated hand-crafted features for all databases. Subsequently, the features were organized in a matrix format for the later machine learning process, obtaining 21 data corpora, seven per evaluation group corresponding to one per original dataset.

3. **Feature selection:** Features were normalized and filtered by correlation. Subsequently, seven feature ranking methods are applied to each of the 21 data corpora to obtain the ten most important features of each one. The feature ranking methods used in this feature selection step were as follows: RelieF Algorithm [154] (RA), Chi-Square [155] (CS), Information Gain [156] (IG), Pearson Correlation [157] (PC), Fisher Score [158] (FS), Gain Ratio [159] (GR), and Random Forest [160] (RF). A total of 147 data subsets were generated, i.e., 49 subsets per evaluation group, based on the ten most relevant features. Part of the exhaustive evaluation of the hand-crafted features is also focused on knowing their relative importance within the datasets used to classify rotating machinery faults, measuring the individual contribution of each one, and observing the relative persistence of each feature. For this reason, several ranking methods were used to enrich this process. Each ranking method has different mathematical foundations, as do the calculated features, which would allow valuable conclusions to be drawn in case of finding patterns within the top 10 lists of selected features. We have used the ranking methods that we consider to be the most important and reported in the literature.These methods were selected based on their use in mechanical systems.

For researchers seeking a more comprehensive understanding of each approach, references to seminal works are provided.

4. **Classification step:** In this final phase, each of the 147 data subsets was evaluated with three different machine learning classification models: Support Vector Machines [161] (SVM), k-Nearest Neighbors [162] (KNN), and Random Forest [163] (RF). The results were compared in terms of the best performance achieved (accuracy percentage) and the number of features needed to reach that value. As in the previous step of feature selection, in this last step, we try to enrich the evaluation process using three classification models. As previously mentioned, each feature and ranking method has its mathematical foundation, and in the same way, it happens with the classifiers. In this way, we have a parametric classifier, one based on proximity and one based on trees. The diversity of classifiers and ranking methods contributes to a comprehensive feature evaluation. The aim is to compare the classification results and find trends with the values achieved, the feature groups used in the process, and the frequency of appearance of certain features in the top 10 list.

An additional fact to mention is that a standard and fixed configuration was used in the configuration of hyperparameters used by the classifiers and ranking methods. Therefore, SVM uses a linear kernel with a penalty of 10, KNN uses k = 3 with an s-euclidean distance metric, while RF uses a tree number of 40. All classifiers were trained following the five folds (k-folds) validation schema. No hyperparameter search, optimization step, or process is performed because the methodology seeks to evaluate the feature's ability to separate classes and not the power of the classification model. The developed methodology has been applied transversally to different types of rotating machines. With the successive application of these steps, this methodology is expected to allow a more in-depth understanding of fault patterns and a systematic feature evaluation to validate their use in fault diagnosis tasks. The summary of results is grouped according to the mechanical components under study from the different databases. In this way, Table 6 shows the result condensation of the databases that use gearboxes with spur gears, Table 7 shows the results for gearboxes with helical gears, and Table 8 shows the results for bearings. The maximum classification percentage (Cl. %) achieved and its number of features (#F.) are shown.

The total summarized process of the methodology followed for the exhaustive evaluation of hand-crafted features can be seen in Figure 6. In this way, the different domains in which the features were calculated could be evaluated in an isolated manner, as well as their potential and contribution as a whole. The training was also carried out exhaustively. First, the most important feature is taken from a subset of data containing the ten best features ranked by some method. Then, the three classifiers are trained to obtain classification performances. The next step is to take the first two most significant features and retrain all three classifiers. This process is repeated iteratively, adding the next most important feature according to the ranking method in each iteration until a total of 10 features are reached to carry out the training. This process is repeated for the 147 data subsets, constantly training the three classifiers. This process is conducted to find the best result (highest classification performance) by a ranking-classifier combination, the number of features required to achieve said performance, and the individual contribution of each feature. The summaries of these results with the best and worst results can be viewed in Tables 9 and 10.

Table 6. Summary of classification results of all experimentation for spur gear gearboxes.

Database	Ranking Method	Only Time Features						Only Frequency Features						Fusion Time and Frequency Features					
		RF		KNN		SVM		RF		KNN		SVM		RF		KNN		SVM	
		Cl. %	#F.	Cl. %	#F.	Cl. %	#F.	Cl. %	#F.	Cl. %	#F.	Cl. %	#F.	Cl. %	#F.	Cl. %	#F.	Cl. %	#F.
DB01	RA	86.61	10	86.24	10	38.91	10	68.19	10	67.43	10	56.38	10	82.89	10	82.69	10	49.33	10
	CS	84.7	9	81.93	10	42.06	9	68.95	10	62.19	10	55.52	9	84.09	10	80.55	7	46.66	10
	IG	82.31	6	82.98	10	36.43	10	68.57	10	68.95	10	59.81	10	81.88	10	79.75	9	53	10
	PC	71.89	8	70.94	6	35.76	9	50.1	9	38.67	4	33.43	7	72.06	10	71.19	10	45.12	9
	FS	69.5	7	57.27	7	23.89	8	69.52	10	62.57	7	30.57	10	64.03	8	49.94	8	16.31	10
	GR	82.22	8	70.75	10	28.49	9	64.48	10	60.38	10	54.1	8	82.55	9	74.4	10	39.84	8
	RF	87.09	8	79.73	8	36.15	9	69.33	10	69.71	10	59.52	9	82.09	9	80.95	10	56.01	10
DB03	RA	97.66	9	96.88	7	72.94	10	94.11	8	95.33	5	60.89	9	64.59	9	57.61	8	56.08	8
	CS	94.99	8	89.09	10	67.15	10	93.78	10	86	10	48.11	10	68.85	8	55.8	7	56.18	8
	IG	97.33	10	98.77	10	78.51	10	94.56	8	96.56	9	62	10	69.89	10	44.45	1	46.57	10
	PC	74.83	8	65.03	8	33.41	10	49.67	7	42	9	35.11	10	58.91	6	45.41	2	53.96	10
	FS	94.65	8	93.21	8	78.06	10	93.89	7	86.89	3	52.22	10	65.76	10	51.61	9	50.24	10
	GR	93.21	10	90.43	10	71.05	10	90.56	8	86.56	6	42.11	9	60.13	10	45.62	2	50.03	10
	RF	94.32	9	93.43	10	70.38	10	90.78	10	91.56	4	61.11	8	68.04	9	56.62	10	54.11	10

Table 7. Summary of classification results of all experimentation for helical gear gearboxes.

Database	Ranking Method	Only Time Features						Only Frequency Features						Fusion Time and Frequency Features					
		RF		KNN		SVM		RF		KNN		SVM		RF		KNN		SVM	
		Cl. %	#F.	Cl. %	#F.	Cl. %	#F.	Cl. %	#F.	Cl. %	#F.	Cl. %	#F.	Cl. %	#F.	Cl. %	#F.	Cl. %	#F.
DB04	RA	96.16	7	97.58	10	75.76	10	85.86	5	89.9	4	55.56	10	96.97	10	97.37	9	78.99	10
	CS	97.17	9	97.17	10	75.35	10	82.02	8	73.54	8	53.94	9	98.18	9	98.59	9	72.53	10
	IG	96.57	8	97.37	9	78.38	9	84.04	7	77.17	7	53.13	10	97.17	9	97.37	10	80	10
	PC	97.37	9	97.78	7	71.92	9	71.92	8	45.66	8	34.34	10	93.13	10	82.63	10	64.85	10
	FS	96.57	9	96.36	8	75.56	9	76.36	9	64.44	4	46.87	9	96.77	9	96.77	9	75.76	10
	GR	97.58	9	95.76	9	73.33	10	68.69	5	55.76	5	42.63	9	97.17	10	95.96	9	73.33	10
	RF	97.98	9	98.59	8	78.59	10	83.43	9	75.35	7	55.56	10	96.97	9	97.58	9	86.26	10

Table 7. *Cont.*

Database	Ranking Method	Only Time Features						Only Frequency Features						Fusion Time and Frequency Features					
		RF		KNN		SVM		RF		KNN		SVM		RF		KNN		SVM	
		Cl. %	#F.	Cl. %	#F.	Cl. %	#F.	Cl. %	#F.	Cl. %	#F.	Cl. %	#F.	Cl. %	#F.	Cl. %	#F.	Cl. %	#F.
DB05	RA	93.07	10	87.2	10	41.07	10	70	7	66	7	23.47	9	86.53	10	87.47	10	44	10
	CS	90	10	81.73	10	38.27	10	61.87	9	52.27	9	23.6	10	89.47	10	84.93	9	52.13	10
	IG	86.13	9	79.07	9	44.53	10	72	7	65.87	7	20.8	10	86.93	10	84.4	10	55.33	10
	PC	90.13	9	83.33	10	36.8	10	44.13	6	34.4	3	18.93	9	87.47	9	87.2	9	50.93	10
	FS	86.67	10	77.07	9	42.93	10	46.53	10	40.67	10	18.93	9	84.53	10	80.53	8	54.4	10
	GR	77.87	9	66.27	10	34.93	10	32.13	6	30.8	6	17.73	10	85.73	10	62.53	8	28.93	8
	RF	91.87	7	93.07	7	37.2	8	71.87	8	65.47	8	21.47	8	84.13	10	86	10	43.33	9
DB08	RA	98	9	98	10	60.76	10	90.33	9	90.56	5	39.67	10	98.44	10	98.78	10	64.22	10
	CS	92.78	10	94.27	10	36.52	10	91.44	7	91.78	6	36	8	93	10	93.72	10	36.83	10
	IG	97.67	10	98.61	10	54.81	10	90.56	8	90.72	5	39.06	10	98	10	98.39	9	58.33	10
	PC	96	10	95.89	10	45.41	10	57.72	10	35.39	10	20.61	10	94.39	10	91.5	10	38.56	10
	FS	98	10	98.22	10	55.2	10	83.56	8	76	7	36.28	9	97.72	9	98.61	10	57.11	10
	GR	97.33	9	97.83	10	49.64	10	90.11	10	84.56	6	37.56	10	97.17	9	97.67	10	50.06	10
	RF	98.39	9	99	10	61.98	10	90.67	9	90.61	5	39.56	10	98	10	99	10	67.67	10

Table 8. Summary of classification results of all experimentation for rotating machinery bearings.

Database	Ranking Method	Only Time Features						Only Frequency Features						Fusion Time and Frequency Features					
		RF		KNN		SVM		RF		KNN		SVM		RF		KNN		SVM	
		Cl. %	#F.	Cl. %	#F.	Cl. %	#F.	Cl. %	#F.	Cl. %	#F.	Cl. %	#F.	Cl. %	#F.	Cl. %	#F.	Cl. %	#F.
CWRU	RA	99.37	6	99.05	7	98.75	8	99.69	7	100	10	99.69	5	99.36	2	99.68	8	99.68	6
	CS	98.74	10	99.05	3	97.81	9	99.69	8	99.38	3	99.69	8	99.69	7	99.06	8	100	9
	IG	99.68	8	98.73	9	99.06	8	99.69	7	100	9	99.69	10	100	9	100	10	99.38	9
	PC	98.1	4	98.11	6	98.44	7	100	7	100	10	100	9	100	8	100	5	100	10
	FS	100	5	99.36	9	99.06	6	100	5	100	7	100	4	99.68	8	100	6	100	4
	GR	99.68	7	99.05	7	99.06	8	99.69	4	99.38	5	99.69	6	100	9	99.05	8	99.69	10
	RF	99.69	8	99.06	4	98.75	7	99.69	9	99.38	10	100	9	100	3	100	2	100	6

Table 8. *Cont.*

| Database | Ranking Method | Only Time Features | | | | | | Only Frequency Features | | | | | | Fusion Time and Frequency Features | | | | | |
| | | RF | | KNN | | SVM | | RF | | KNN | | SVM | | RF | | KNN | | SVM | |
		Cl. %	#F.	Cl. %	#F.	Cl. %	#F.	Cl. %	#F.	Cl. %	#F.	Cl. %	#F.	Cl. %	#F.	Cl. %	#F.	Cl. %	#F.
DB02	RA	89.48	9	87.24	10	66.77	10	84.76	6	80.95	4	40	9	92.66	10	91.37	9	67.74	10
	CS	87.54	9	89.12	7	60.38	9	84.76	5	85.4	5	41.27	10	90.43	10	88.19	7	58.78	10
	IG	89.8	8	87.23	10	51.11	10	85.08	5	83.17	5	40	5	92.03	10	92.35	9	60.39	10
	PC	76.06	8	64.25	8	54.63	8	82.22	4	79.05	4	37.14	10	78.29	10	65.49	10	58.44	9
	FS	75.4	7	67.43	7	52.67	10	83.81	6	83.49	4	38.41	9	77.34	9	70.6	8	52.69	7
	GR	79.89	10	73.78	10	49.17	9	82.22	4	83.17	4	40.63	9	82.14	10	71.53	10	47.93	9
	RF	92.05	9	90.71	9	67.73	10	84.76	5	86.03	5	40	7	86.91	9	91.04	10	65.17	10

Table 9. Summary of the best results settings for each database.

Mech. Comp.	Spur Gear Gearboxes						Helical Gear Gearboxes						Bearings			
Database	DB01		DB03		DB04		DB05		DB08		DB02		CWRU	
Eval. group	Only time		Only time		Only time		Only time		Fusion		Fusion		Fusion	
Max class %	87.09		98.77		98.59		93.07		99.00		92.66		100.00	
Acc. std	1.61		0.25		1.15		2.14		0.54		2.67		0.00	
Class Model	RF		KNN		KNN		RF		KNN		RF		KNN	
Ranking Method	RF		IG		RF		RA		RF		RA		RF	
1st feat.	NPP	24.18	mean	39.31	WF	45.25	hzcrr	14.67	FR	20.39	rc	28.78	stda	99.05
2nd feat.	meanInflectPoints	36.42	NRZISO	69.26	NPP	69.49	hr	31.07	WF	47.39	PKF	58.17	**SRF**	**100.00**
3rd feat.	KTHCM	65.68	prvav	78.39	hr	87.27	skewness	63.73	NPP	78.33	FR	73.78	mean	99.37
4th feat.	hr	72.85	prdad	86.41	ZCRn	93.54	ZC	76.40	WI	92.28	meanInflectPoints	85.63	SFP	98.73
5th feat.	RLP	81.83	hr	94.10	SMR	95.96	rc	85.33	talaf	95.56	SFP	88.50	WF	98.75

Table 9. *Cont.*

Mech. Comp.	Spur Gear Gearboxes				Helical Gear Gearboxes				Bearings			
6th feat.	eo	83.17	Env_3	95.10	prdad	98.38	NRZISO	89.20	SM4	96.56	NPP	89.79	HSS	99.05
7th feat.	c3	84.22	kurtosis	96.33	PC	96.77	PSF	90.13	PKF	97.67	mean	90.43	ASE	98.74
8th feat.	**DVARV**	**87.09**	Env_4	96.44	**kurtosis**	**98.59**	cbm	88.53	SlopeRMS	97.56	lsbm	90.44	Env_5	99.37
9th feat.	SMR	86.52	MSP	98.22	hzcrr	97.58	talaf	91.33	skewness	98.83	prdad	91.39	ASS	99.69
10th feat.	ffaa	85.56	**meanIn-flectPoints**	**98.77**	SFP	97.37	**NPP**	**93.07**	**RHZ**	**99.00**	lsam	**92.66**	prvav	99.68

Results highlighted in bold indicate the highest classification percentage achieved up to the tenth feature.

Table 10. Summary of the worst results settings for each database.

Mech. Comp.	Spur Gear Gearboxes				Helical Gear Gearboxes						Bearings		
Database	DB01		DB03		DB04		DB05		DB08		DB02		CWRU
Best Group	Only time		Only time		Only Freq.		Only Freq.		Only Freq.		Only Freq.		Only time
Max class %	23.89		33.41		34.34		17.73		20.61		37.14		97.81
Acc std	4.87		3.55		5.76		1.12		2.23		6.79		4.89
Class Model	SVM		SVM		SVM		SVM		SVM		SVM		SVM
Ranking Method	FS		PC		PC		GR		PC		PC		CS
1st feat.	kurtosis	14.72	hr	18.82	skewnessf	15.96	HSC_5	10.13	HSS_5	11.89	freqDeform	14.60	MDR	71.21
2nd feat.	skewness	15.11	ZCRn_3	25.05	HSV_1	15.15	HSV_4	10.40	HSC_4	10.67	SM4	18.10	meanWa-velength	92.74
3rd feat.	eo	16.73	lsbm	25.73	ASS_1	15.15	VCF	11.47	SC_4	10.89	PKF	26.98	prvav	91.75
4th feat.	LLR	17.89	ZCRn_2	26.39	HSD_1	16.57	HSS_4	11.20	SF_2	11.39	VCF	33.33	PI	90.82
5th feat.	SDIF	22.18	ZCRn_4	28.18	freqDeform	20.81	HSC_3	13.73	freqDeform	13.00	CP4	31.11	Env_2	97.18
6th feat.	FIFTHM	20.65	lsam	27.73	rmsfk	23.23	PKF	12.27	HSS_1	14.06	SF_3	32.38	SIXTHM	97.17
7th feat.	SIXTHM	20.37	ZCRn_1	29.62	PKF	25.66	SF_3	13.73	SF_3	17.28	SF_4	31.43	Env_4	97.17
8th feat.	**NM**	**23.89**	ZCRn_5	27.95	VCF	33.94	ffdd	11.73	HSV_2	16.94	SF_5	36.51	ffaa	97.17
9th feat.	MDR	19.69	lster	32.29	ASS_2	33.33	SF_5	14.93	SM4	19.72	SF_2	34.92	**inthar**	**97.81**
10th feat.	KTHCM	21.22	**lat**	**33.41**	**ASS_5**	**34.34**	**HSC_4**	**17.73**	**ffgg**	**20.61**	**HSS_1**	**37.14**	hr	97.81

Results highlighted in bold indicate the highest classification percentage achieved up to the tenth feature.

5. Results and Discussion

This section presents the results obtained from the experimentation to evaluate the performance of different hand-crafted features in fault classification tasks over rotating machinery. A total of 147 subsets of the top 10 most important features were obtained by applying seven ranking methods to seven original databases with three evaluation groups. Each of these data subsets was exhaustively evaluated by three different classification models. In the first place, the general results of the experiment are presented and organized according to the mechanical components associated with the different databases, namely spur gear gearboxes, helical gear gearboxes, and rotating machinery bearings. The tables show the percentage of classification and the number of features needed to achieve that value. Additionally, the results are condensed by database, ranking methods, and evaluation group, allowing a direct comparison and identifying the combinations that provide the best and the worst classification results. In this way, Table 6 summarizes the results obtained in the three evaluation groups for the DB01 and DB03 databases, the same ones associated with gearboxes with spur gears. Table 7 shows the overall results obtained for the databases associated with gearboxes with helical gears, specifically DB04, DB05, and DB08. For bearings, Table 8 displays the results corresponding to the DB02 database and the external benchmark CWRU. The classification results are shown in Tables 6–8. They show that ranking methods significantly impact the effectiveness of fault diagnosis. For example, suppose that a ranking method-classifier combination shows a high classification percentage and a low number of features. In that case, that method is very effective in identifying the most critical features. Furthermore, if a particular database shows a high classification percentage with a particular ranking method, this would indicate that the method is especially effective in identifying essential features for that specific database. In this way, the fault diagnosis can be performed with fewer data and greater precision.

Tables 6–8 also show the results by evaluation groups. That is, it shows the classification results and the number of features associated with Group 1, in which only the features of the time domain were used in its evaluation (123 in total). For Group 2, they used only the frequency domain characteristics (46 in total). Group 3 uses a data fusion scheme, combining time and frequency features (169 in total). It is essential to highlight that the results reflect the capacity of the classifiers to carry out fault diagnoses in rotating machinery and the effectiveness and usefulness of hand-crafted features calculated in the time and frequency domains. It should also be mentioned that if we read the results of Tables 6–8 by rows within each evaluation group, the data subset used is the same and will produce different performances with each classifier. The set of features used varies when reading the results between rows since a different ranking method will choose a different set of features. Even with this, one or several features may be repeated in different evaluation groups or data subsets that were selected by different ranking methods. A high percentage of classification and a low number of features indicate good feature selection. Therefore, combining ranking methods with hand-crafted features is a viable way to perform fault diagnosis in various rotating machines. The next part of the results obtained refers to the best and worst results obtained from all the exhaustive experimentation. In this way, Table 9 shows the best classification results obtained for each database with the respective set of features used, the deviation of the precision in the training, and the individual contribution of each feature within that configuration. Similarly, Table 10 shows the same information with the difference that presents the worst results obtained in all experiments. Both tables highlight in bold the maximum precision obtained by the best-performing classifier. It can be seen in both cases that, on some occasions, the classifier reaches its maximum precision with a certain number of features and that the act of adding more is counterproductive to the overall performance of the model. Table 9 shows that the evaluation groups that produced the best features were Group 1 of time alone and Group 3 of data fusion. It can also be seen that the ranking method with the best feature subsets is Random Forest, followed by the Relief Algorithm. The vast majority of the contribution to accuracy is achieved by using the four most important features.

For DB01, the best resulting combination is the RF method and the RF classifier, reaching 87.09% with eight exclusive features of the time domain. For DB03, the best result is achieved again with a set of time-exclusive features in the combination of the IG method and the KNN classifier, reaching 98.77% with 10 features. In DB04, groups 1 and 3 achieved equal precision of 98.59% using the KNN classifier. However, group 1 used the RF (Random Forest) ranking method, while group 3 employed the CS (Chi-square) ranking method. The only substantial difference between these results is that the time-exclusive feature group achieves this percentage using only eight features, whereas the fusion group achieves it using nine features. DB05 has a tie again, this time within the same group, with the exclusive time features being the winners. A maximum classification percentage of 93.07% was achieved. The first combination was the RA method and the RF classifier using 10 features and an acc_std of 2.14. In comparison, the second combination was the RF method and the KNN classifier using seven features and an acc_std of 3.42. The relatively high dispersion in precision values in the second combination could be attributed to the features used. Therefore, the table will include the list of the 10 features from the first winning combination. DB08 has a tie between the time-only and fusion groups. In both cases, an accuracy of 99% is achieved with 10 features with a combination of the RF method and the KNN classifier. In the first case, using the time-only group, an acc_std of 1.01 was achieved, while in the second case, employing the fusion group, an acc_std of 0.54 was obtained. A detail to highlight is that in the time group feature list, the features of Hr, NPP, and meanInflectPoints appear in all the sets of features for all previous databases. The last three databases are for helical gear gearboxes, and the five databases analyzed so far are for gearboxes. For DB02, the maximum classification percentage of 92.66% with the combination RA method and the RF classifier with an acc_std of 2.67 requires 10 features to reach this value. Finally, in the case of the external CWRU database, an amalgamation of almost perfect results is obtained. The maximum classification percentage achieved is 100% with the combination RF method and the KNN classifier using only two features of the fusion group, with an acc_std of 0. In this case, many combinations are obtained that return excellent results for the three groups of assessment. The fusion group has been chosen as the winner for the CWRU because it is the one that obtains a combination that requires fewer features to achieve the best results. It is worth mentioning that, while it is true that a maximum result was achieved in the fusion group, the features used to achieve this level of classification are features that are calculated in the frequency domain. The total list of features that appear in this subset is placed in Table 9. In this table, DB01 contains the first five features, which are associated with the counting, statistical, hybrid and ratio, and shape and profile features types. In DB03, the first five features are associated with the statistical, shape and profile, and hybrid and ratio features types. In DB04, the first five features are associated with the hybrid and ratio, counting, and shape and profile features types. In DB05, the first five features are associated with shape and profile, hybrid and ratio, statistical, and counting feature types. In DB8, the first five features are associated with hybrid and ratio, counting, and shape and profile feature types. In DB02, the first five features are associated with counting, central tendency and statistical moments, statistical, and hybrid and ratio feature types. In the CWRU database, the first five features are statistical, and hybrid and ratio features.

On the other hand, reviewing Table 10, it can be seen that, in general, the frequency-only group has the worst overall performance in precision with the classifiers, although acceptable values are reached for certain databases. It can also be seen that the classifier with the worst global results is the SVM. This could be because a linear kernel was used in its configuration, and not all classes within the databases have to be linearly separable. This could also indicate the complexity of the data available to diagnose failures. We must consider the results obtained in Table 9 with other essential results to obtain a complete view of the experiment. We have created distributions and ratios of the occurrence graphs of the evaluation groups that appear in Table 9 only, since they are the groups that appear in the best results, that is, time-only and fusion groups. The bar charts in Figures 7 and 8 show the frequency of occurrence of features in the 49 data subsets with the ten most important features per only time and fusion evaluation groups, respectively. The graphs show features such as hr, NPP, WF, skewness, meanInflectPoints, rc, mean, SC, FR, and SFP, among others, as the features with the most occurrences or repetitions in the two evaluation groups. In the case of the only time group, Figure 7 shows how the hr feature has the most occurrences, totaling 30 out of 49. This figure also shows two more features, such as NPP and WF, that follow hr as the most used features in this experimentation group. The data fusion group shares many features that appear in Group 1, but it also includes certain features that are calculated in the frequency domain and have notorious appearance levels. Figure 8 also shows features such as FR, SFP, and SC that appear more than 15 of the 49 times. As a curious example, we have the SC feature, a version of the spectral centroid computed over frames of the power spectrum by averaging the power-weighted frequencies within the frame of the spectrum. This feature combines the trends of calculations based on frequency weighting with those seeking information in sections of the spectrum, as discussed in section II. The hr feature is another example of a calculation over frames, and it belongs to the time domain. In contrast, features such as FR and NPP are calculated over the whole signal in their respective domains.

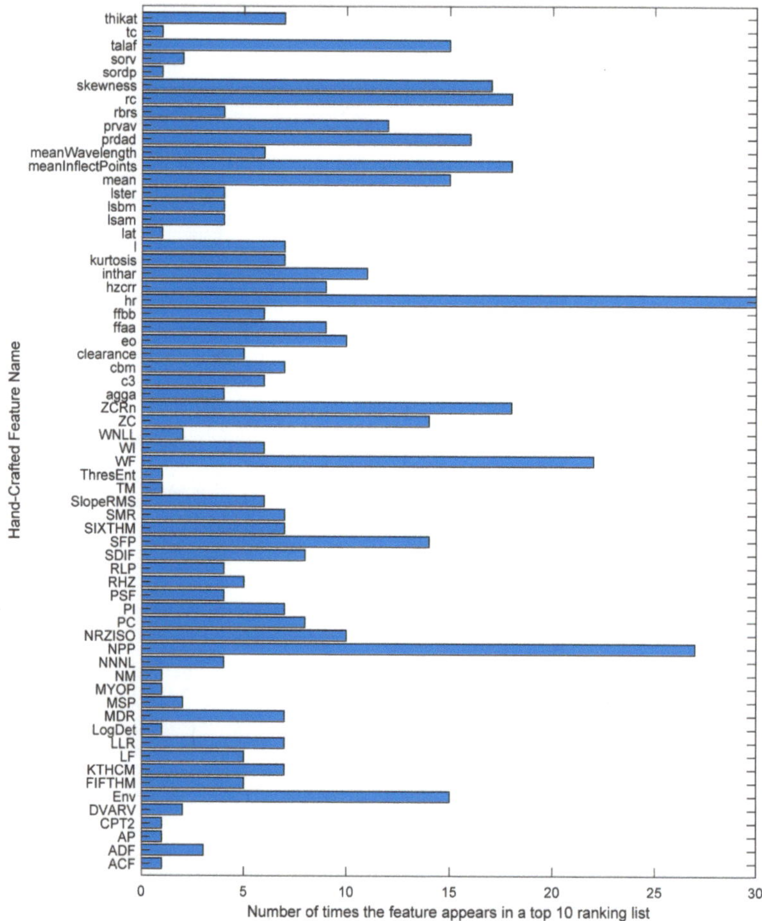

Figure 7. Hand-crafted feature occurrence distribution of the top ten ranking lists for all databases of the only time features group.

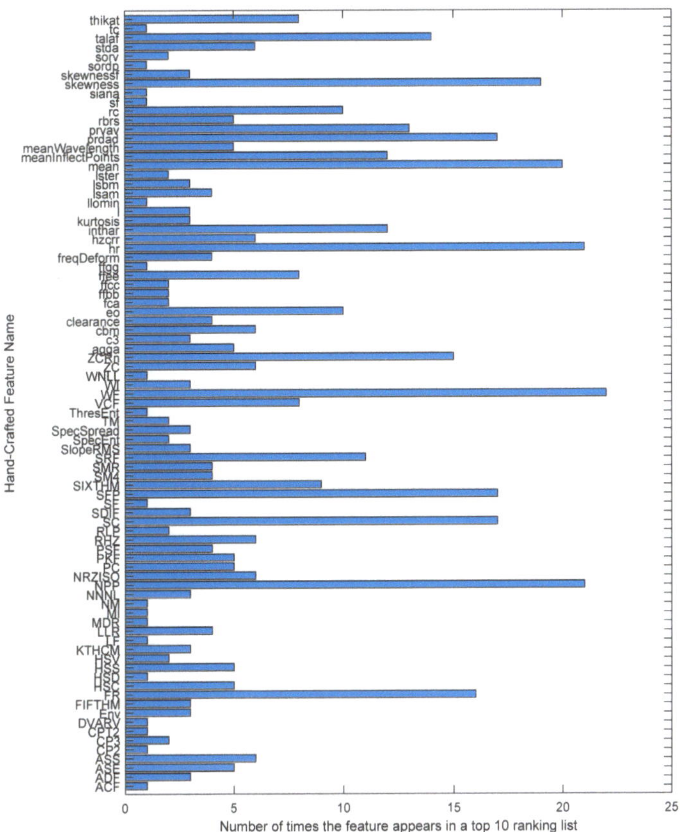

Figure 8. Hand-crafted feature occurrence distribution of the top ten ranking lists for all databases of the feature fusion group.

To accompany the feature occurrence distribution graphs, Figures 9 and 10 are shown, summarizing the features that appear the most times in the ranking lists for evaluation groups 1 and 3, respectively. These graphs show those features that appeared in more than three ranking lists. That is, more than three ranking methods put them somewhere in their top 10 most significant features. Virtually, the maximum value a feature could reach in these graphs is 7 since there is a maximum of seven ranking methods. However, for features that apply their calculation in frames or bands, there are cases where the same feature, calculated in different bands, appears several times in the same list of the top 10 most relevant features by some ranking method. This can be seen clearly with many examples in Table 10, and we have the Env and SRF features cases in Figures 9 and 10, respectively.

Analyzing Figures 7–10 together, we can see that some features appear more frequently than others. The count distribution shown in the bar graph in Figure 7 suggests that the features with the highest occurrences could be considered the most important in the evaluation performance experiment for Group 1. We can confirm this by looking at Figures 9 and 11 alongside Table 9. Taking DB03 as an example, Figure 11 shows that the hr feature appears in the first position of the rankings 29% of the time, and we already know from Figure 7 that this feature appears the most times in the time-only evaluation group. Inspecting Figure 9, we can see that the hr feature is chosen by six of the seven ranking methods used in the experiment. Finally, we can see in Table 9 that the best classification result for DB03 is achieved with the hr feature in the fifth place of importance. Additionally, even in Table 10 of the worst results, we can find the hr feature in the first position of importance in the configuration shown for DB03. Several other features follow this pattern of occurrences and are important in fault classification as hr. We have examples such as the WF for DB04 or the NPP for DB01 in the time-only group, while we see cases such as stda for the CWRU in the fusion group. There are also cases where a feature appears many times in a few ranking methods but is located in significant positions, as is the case of the FR feature for DB08 in the fusion group. In this example, the FR feature does not appear in Figure 10 but stands out in Figures 12 and 8, in addition to demonstrating its importance in the results of Table 9.

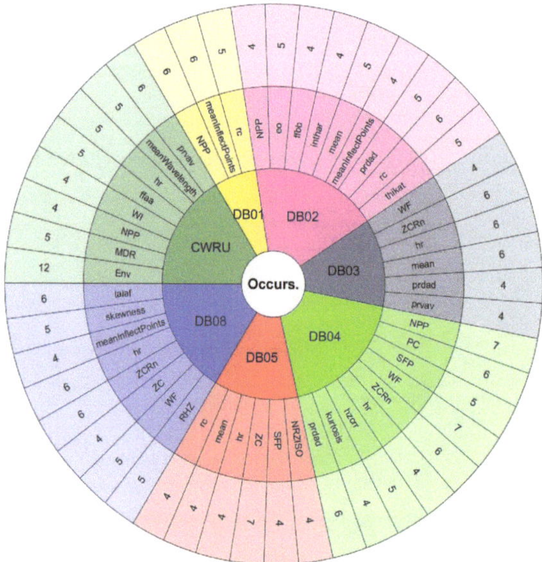

Figure 9. Occurrences in the top ten feature ranking lists of all databases for the group of only time features.

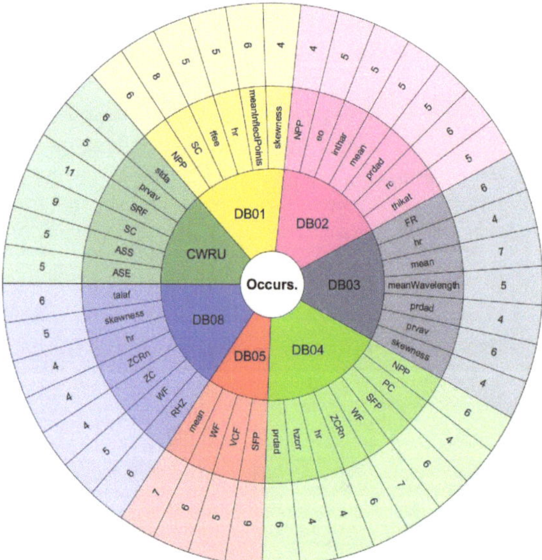

Figure 10. Occurrences in the top ten feature ranking lists of all databases for the group of fusion features.

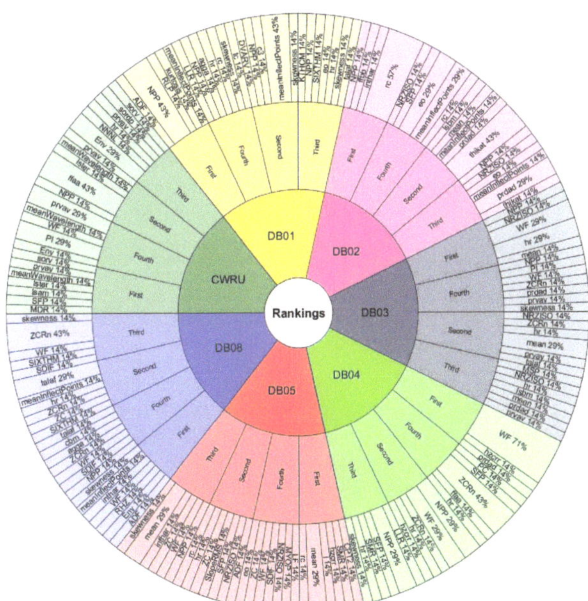

Figure 11. Sunburst plot of databases and feature apparition ratio in top four ranking importance in the only time features group. The second ring refers to the four most important positions in the rankings, and the third ring denotes the features placed there by the different ranking methods and the percentage of appearance in that position.

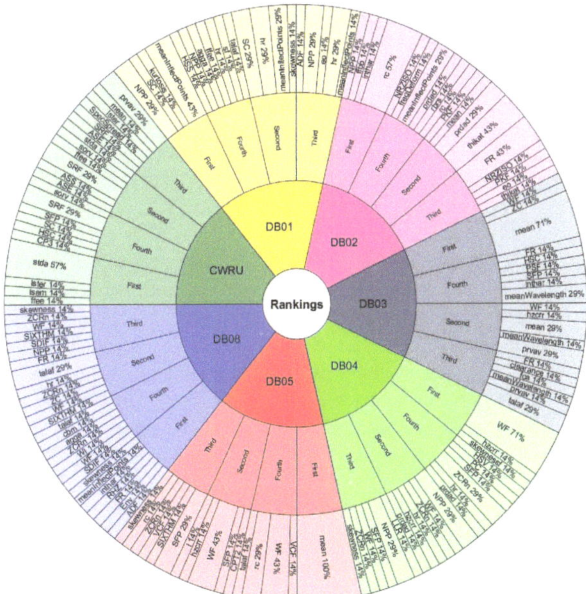

Figure 12. Sunburst plot of databases and feature apparition ratio in top four ranking importance in the fusion features group. The second ring refers to the four most important positions in the rankings, and the third ring denotes the features placed there by the different ranking methods and the percentage of appearance in that position.

The features with the most occurrences or selected by various ranking methods seem to show a great association with the study phenomenon. Therefore, they are the most useful in classification processes. Furthermore, the frequency of appearance of each feature in the top 10 lists may indicate the consistency of the ranking methods used to evaluate the performance of handcrafted features. If a feature frequently appears in the top 10 lists, it is evidence that it is considered essential by various ranking methods. On the other hand, if a feature appears infrequently on the top 10 lists, this suggests that it is considered less important or that the ranking methods used in the assessment need to be more consistent in their assessment of that feature.Figures 8 and 9 are valuable tools for understanding the relative importance of each feature in diagnosing rotating machinery faults. To better understand the relationship between the performance of the classifiers, the databases, and the features selected by the ranking methods, we must observe the results in Tables 9 and 10 with Figures 11–14. The sunburst graphs in Figures 11 and 12 show the top four critical features per database and their percentage of occurrences. Those graphs have been built using the results obtained from the first four places of the 49 lists of the top 10 most important features for each evaluation group and directly associating them with the databases to which they correspond. This type of visualization allows for a clear representation and an easy way to understand the data hierarchy, allowing us to see which features have a more significant influence within each of the databases and how they are related to each other. The fact that some features have high percentages of appearance in certain ranking positions within a database indicates that those features are critical for diagnosing failures in that specific rotating machine. This relation means that the repeatability in the feature ranking has allowed us to identify the most relevant features for each database and, therefore, focus efforts on understanding those specific features to detect faults efficiently and effectively.

Figures 13 and 14 present the other sunburst graphs that show the percentage of appearances of a feature in any position of the top four of importance, associated with the percentage of times that a ranking method places it in that particular position. The ring-shaped representation allows for a global vision of the performance of the features in the rankings and their ability to be identified as necessary by the different ranking methods used in the experimentation. It is essential to consider that the percentages of appearances of a particular feature in a position of the top 4 are global for all the experimentation; that is, the percentage of appearances marked by a feature is, in general, for all the databases used in the experiment for each evaluation group. If a feature has a high percentage of occurrences in some position in the top 4, it means that the ranking methods frequently identify it as an important feature in that position, making it valuable in a horizontal way; that is, its use is viable for many types of rotating machinery for fault diagnosis. Figure 13 shows interesting information, such as that the WF and NPP features have the highest occurrence percentages as the most important features with 16% and 10%, respectively, for the time-only evaluation group. On the other hand, Figure 14 shows the features of mean and WF as those with the highest percentage of appearance in the first place of importance in the fusion evaluation group, being chosen by almost all ranking methods at least once. If we analyze those graphs together with Figures 7 and 8, we can see that all the aforementioned features appear several times in their respective experimental groups, which suggests that this enhances the possibility of ever finding them at the top of the podium of the feature importance.

It should be noted that the bar graphs in Figures 7 and 8 and the sunburst graphs in Figures 11–14, only show the results of evaluation groups 1 and 3; the time-only and data-fusion feature sets. This election, as stated previously, occurs because these groups are present in Table 9 with the best performance in the classification task, so more significant associations can be derived between features and databases. If we try to detect patterns in the results obtained, we can find certain coincidences with different databases with similar study elements. For example, the hr feature is present with different percentages in DB01, DB03, DB04, DB05, and DB08 in different places of importance. Furthermore, all these databases study gearboxes with helical and spur gears, and the hr feature appears in four of the five sets of features used that achieved the best classification results, as we can see in Table 9. We also have DB02 and CRWU, which have results that share features such as Mean and SFP. Another common trend that we can observe in the features of the top 4 organized by the database they represent is that they usually exchange the position in which they appear within the ranking of the top 4. A high percentage of occurrences in a given top 4 position means that the feature is important and valuable in various fault diagnosis application cases in different types of rotating machinery. In addition, this also indicates that the ranking methods are consistent in identifying said feature as one of the most important. Associating the results shown in Figures 11–14, and Table 9, we can identify that, for a specific database, the ranking method with the best performance places a feature in a high position in the top 4 with a high percentage of appearances, and that the classifiers used in that database achieve a high percentage of classification with some essential features. This result is excellent evidence that this feature can be very informative and may suggest good physical significance for diagnosing faults in that particular database and other similar ones that also capture condition information on rotating machine processes.

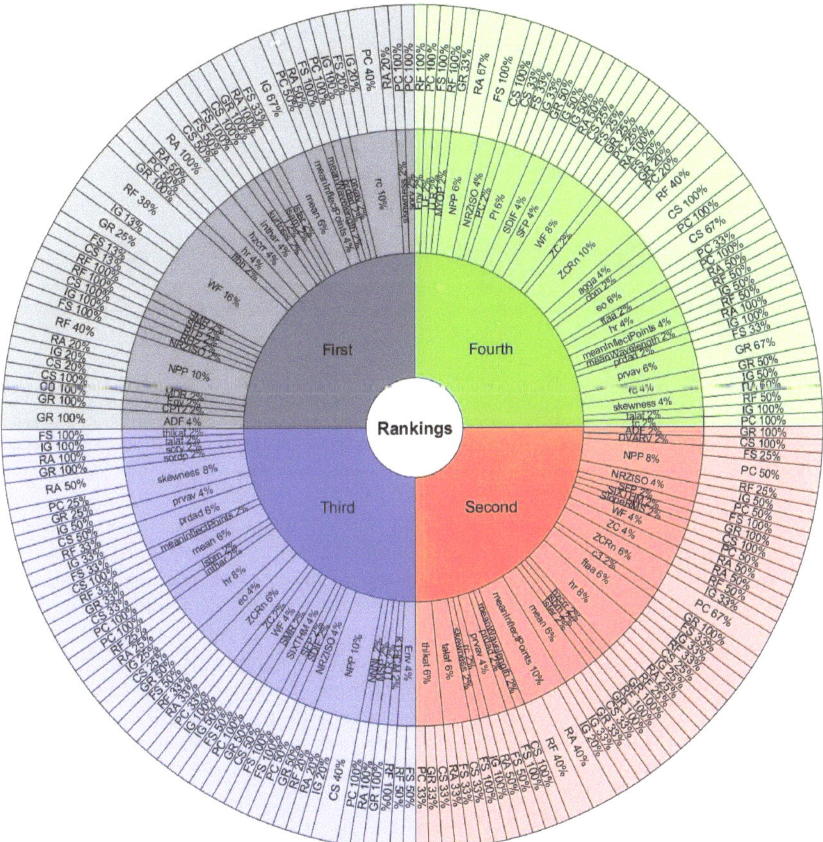

Figure 13. The top four ranked as the most important features and their appearance ratio in those places for all databases for the only time features group. The second ring denotes the percentage of occurrences of the feature in that position, while the third indicates the percentage of times the ranking method places the feature there.

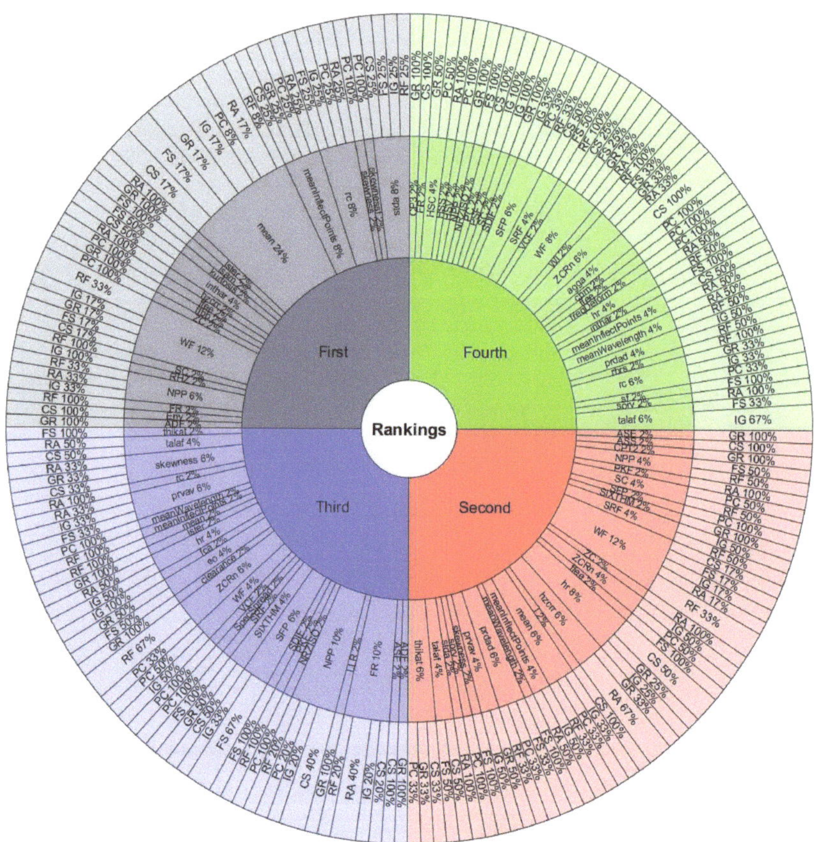

Figure 14. The top four ranked as the most important features and their appearance ratio in those places for all databases for the fusion features group. The second ring denotes the percentage of occurrences of the feature in that position, while the third indicates the percentage of times the ranking method places the feature there.

6. Conclusions

This study presented a comprehensive compilation of 169 hand-crafted features for fault diagnosis in rotating machinery and evaluated their effectiveness using seven databases and various classification models. Our findings demonstrate that hand-crafted features can achieve high classification accuracy, up to 99% in some cases, for various types of rotating machinery faults. Furthermore, the combination of time and frequency domain features often outperformed single-domain feature sets. The advantage of hand-crafted feature extraction is high interpretability, allowing for direct physical insights into fault mechanisms. These features also offer flexibility in application across different types of rotating machinery and show comparable performance to more complex methods, including deep learning approaches.

In this study, a series of formulas and calculations, simple methods, were collected from the literature to perform hand-crafted feature extraction on signals in the time and frequency domains. The proposed methodology for evaluating the performance of hand-crafted features in fault classification tasks has proven to be practical and valuable. Furthermore, the results strongly suggest that it is possible to achieve a good fault diagnosis performance using a combination of different ranking methods and calculated hand-crafted features. The results obtained show the importance of analyzing different domains to achieve a good diagnosis of faults in rotating machinery. The proposed methodology allowed us to objectively compare the performance results of the features, and the ranking methods used allowed us to identify the most relevant features in each database. Some features have a high percentage of occurrences anywhere in the top 4 of importance. They are valuable and viable for various fault diagnosis applications across many types of rotating machinery. In addition, simplicity and low computational cost are some advantages of using hand-crafted features for classical feature extraction processes in condition monitoring signals since, among other things, they allow a whole series of experiments and tests exhaustively without excessive time consumption.

A relative repetitive pattern is seen between relevant features in some databases that have mechanical elements in common. This could suggest a direct physical relation of the phenomenon with the feature values. Despite this, additional analysis and testing are needed to fully confirm and establish the physical significance between the feature and the physical process or mechanical component. The results suggest the importance of a methodical and systematic approach to analyze characteristics and evaluate the performance of features through the classifiers in diagnosing faults in rotating machinery. The classification percentages and the number of features required to achieve that value were informative in determining the effectiveness of the features and identifying those that may be useful for fault identification. These results also suggest that it is crucial to consider interpretability and interpretation of features, as this can significantly impact diagnostic effectiveness. In the same way, the compilation of features carried out and presented in the summary tables is a substantial part of the contribution of this paper. Due to the disparate nature of the literature in terms of the mathematical formalization that the features have, we have found it necessary to unify the nomenclature of the formulas in this compilation. The same summary of nomenclature can be seen in Table 1. A map was built that organizes and specifies in which domains the different types of features are used. The main trends of the features in terms of the information they seek to reveal of the signal were identified. The fact that these diversified trends exist according to the signal representation domain makes sense since each domain has its advantages, disadvantages, and specific objectives of what they want to reveal about the signal; therefore, the features will seek to exploit precisely those strengths, such as waveforms in time and energy distribution in frequency components in the spectrum.

Assuming that the acquired signals have a sufficient sampling rate to effectively capture the phenomenon being measured and are not affected by aliasing or noise, another advantage of hand-crafted features is agnostic to the acquisition rate and time. Whether the signals are large or small, fast or slow, feature calculation on any signal will always be viable, and there is virtually no restriction on the size of the input to the processing since the information extraction stage is separated from the inference part, which does not happen in Deep Learning models [164,165]. This advantage applies particularly to those signals in the time domain. Some exceptions could cause errors, such as counting the number of zero crossings of a time domain signal with no negative side. The result will be 0, which is not a significant feature in this example. On the other hand, the features in the frequency domain could become dependent on the sampling rate due to its direct relationship with the resolution of the spectrum. If the size of the sample corresponding to the signal is reduced, a decrease in the performance of the features in frequency is also expected. Consequently, the precision of the classifiers that use them will decrease. The choice of which features or calculations to use for each case will be left to the judgment and expertise of the subject who performs the analysis. In this work, databases that had signals with different sampling rates and acquisition times were used, achieving very high classification results using hand-crafted features without any problem. In addition, standard methods and models were also used for ranking and classifiers. There was no hyperparameter optimization, so many combinations of rank classifiers could significantly improve their performance if this other process was applied. This election is because the experiment aimed to measure the efficiency of the features, not the power of the classification model. If a group of characteristics serves to classify failures in a rotating machine, then these features are very informative to reflect the status or type of failure of a particular component; therefore, the fault associated with a component of the machine has been characterized. In subsequent actions within a CBM plan, these features and their evolution can be used as markers that provide information on the deterioration of the machine or any of its components. In this way, we would get closer to performing predictive maintenance.

The yields obtained in the classification are comparable with many others based on much more sophisticated processing or with methods based on deep learning with the advantage that they are interpretable and computationally very light. It is necessary to mention that there is no such thing as an optimal group of features that serve all databases. Instead, there is a pool of features from which some can be extracted to achieve acceptable performances in fault classification tasks for rotating machinery. Part of these pools would be in the sunburst of occurrences since the graphs show all those features that were chosen by more than three different ranking methods. Furthermore, this work has found some specific features that work for many datasets that use the same mechanical components. It is also necessary to evaluate other databases of rotating machinery, other methods of dimension reduction in the data, and aspects such as generalization and robustness to variations in the condition monitoring signals to provide further completeness in feature evaluation. It is necessary to continue researching and developing new techniques to achieve an even more precise, effective, and generalizable diagnosis in the future and optimize the feature's interpretability. Although favorable results were obtained, it is important to stress that research is still needed. In future research, it would be interesting to explore the combination of automatic and hand-crafted feature extraction methods that involve other domains or signal transformations or even further expand the feature collection to achieve better performance in rotating machinery fault classification tasks.

Future work will focus on comparing the performance of these hand-crafted features against automated feature extraction methods, particularly deep learning approaches. This comparison should include an expanded range of classification models, incorporating both traditional machine learning and deep learning architectures. Such a comprehensive evaluation would provide valuable insights into the relative strengths and limitations of manual versus automated techniques for fault diagnosis in rotating machinery, potentially revealing new perspectives on feature effectiveness across diverse learning paradigms.

Author Contributions: Conceptualization, R.-V.S.; Methodology, D.C. and M.C.; Software, J.C.M.; Validation, J.C.M., L.-R.O. and F.P.G.M.; Formal analysis, D.C.; Investigation, M.C.; Data curation, J.C.M. and L.-R.O.; Writing—original draft, J.C.M. and L.-R.O.; Writing—review & editing, R.-V.S. and F.P.G.M.; Visualization, R.-V.S.; Project administration, R.-V.S. and L.-R.O. All authors have read and agreed to the published version of the manuscript.

Funding: The authors wish to thank the Universidad Politécnica Salesiana through the GIDTEC research group for their support in this work.

Institutional Review Board Statement: Not applicable.

Informed Consent Statement: Not applicable.

Data Availability Statement: The data presented in this study are available on request from the corresponding author due to privacy policy.

Acknowledgments: The authors would like to express their gratitude to the Universidad Politécnica Salesiana and the GIDTEC research group for their invaluable support in this work. Additionally, we extend our appreciation to the reviewers for their constructive feedback, which has significantly improved the quality of this manuscript.

Conflicts of Interest: The authors declare no conflicts of interest. The funders had no role in the design of the study; in the collection, analyses, or interpretation of data; in the writing of the manuscript; or in the decision to publish the results.

Abbreviations

The following abbreviations are used in this manuscript:

CBM	Condition-Based Maintenance
CPT	Condition Prognostics Technology
DSP	Digital Signal Processing
FFT	Fast Fourier Transform
KNN	k-Nearest Neighbors
ML	Machine Learning
PSD	Power Spectral Density
RCM	Reliability-Centered Maintenance
RF	Random Forest
SVM	Support Vector Machine
TSA	Time Synchronous Averaging

References

1. Kumar, A.; Gandhi, C.P.; Tang, H.; Sun, W.; Xiang, J. Latest innovations in the field of condition-based maintenance of rotatory machinery: A review. *Meas. Sci. Technol.* **2023**, *35*, 022003. [CrossRef]
2. Lei, Y.; Lin, J.; Zuo, M.J.; He, Z. Condition monitoring and fault diagnosis of planetary gearboxes: A review. *Measurement* **2014**, *48*, 292–305. [CrossRef]
3. Guerra, C. Condition Monitoring of Reciprocating Compressor Valves Using Analytical and Data-Driven Methodologies. Ph.D. Thesis, Rochester Institute of Technology, Rochester, NY, USA, 2013.
4. Khadersab, A.; Shivakumar, S. Vibration Analysis Techniques for Rotating Machinery and its effect on Bearing Faults. *Procedia Manuf.* **2018**, *20*, 247–252. [CrossRef]
5. Qiao, W.; Lu, D. A Survey on Wind Turbine Condition Monitoring and Fault Diagnosis—Part I: Components and Subsystems. *IEEE Trans. Ind. Electron.* **2015**, *62*, 6536–6545. [CrossRef]
6. Bevilacqua, M.; Braglia, M. The analytic hierarchy process applied to maintenance strategy selection. *Reliab. Eng. Syst. Saf.* **2000**, *70*, 71–83. [CrossRef]
7. Coria, V.; Maximov, S.; Rivas-Dávalos, F.; Melchor, C.; Guardado, J. Analytical method for optimization of maintenance policy based on available system failure data. *Reliab. Eng. Syst. Saf.* **2015**, *135*, 55–63. [CrossRef]
8. Mitoma, T.; Wang, H.; Chen, P. Fault diagnosis and condition surveillance for plant rotating machinery using partially-linearized neural network. *Comput. Ind. Eng.* **2008**, *55*, 783–794. [CrossRef]
9. Kothamasu, R.; Huang, S.H.; VerDuin, W.H. System health monitoring and prognostics—A review of current paradigms and practices. *Int. J. Adv. Manuf. Technol.* **2006**, *28*, 1012–1024. [CrossRef]
10. Tsui, K.L.; Chen, N.; Zhou, Q.; Hai, Y.; Wang, W. Prognostics and Health Management: A Review on Data Driven Approaches. *Math. Probl. Eng.* **2015**, *2015*, 1–17. [CrossRef]
11. Jardine, A.K.; Lin, D.; Banjevic, D. A review on machinery diagnostics and prognostics implementing condition-based maintenance. *Mech. Syst. Signal Process.* **2006**, *20*, 1483–1510. [CrossRef]
12. Ahmad, R.; Kamaruddin, S. An overview of time-based and condition-based maintenance in industrial application. *Comput. Ind. Eng.* **2012**, *63*, 135–149. [CrossRef]
13. Zhang, L.; Xiong, G.; Liu, L.; Cao, Q. Gearbox health condition identification by neuro-fuzzy ensemble. *J. Mech. Sci. Technol.* **2013**, *27*. [CrossRef]
14. Vachtsevanos, G.; Lewis, F.L.; Roemer, M.; Hess, A.; Wu, B. *Intelligent Fault Diagnosis and Prognosis for Engineering Systems*; John Wiley & Sons: Nashville, TN, USA, 2006.
15. Gawde, S.; Patil, S.; Kumar, S.; Kamat, P.; Kotecha, K.; Abraham, A. Multi-fault diagnosis of Industrial Rotating Machines using Data-driven approach: A review of two decades of research. *Eng. Appl. Artif. Intell.* **2023**, *123*, 106139. [CrossRef]
16. Devendiran, S.; Manivannan, K. Vibration based condition monitoring and fault diagnosis technologies for bearing and gear components—A review. *Int. J. Appl. Eng. Res.* **2016**, *11*, 3966–3975.
17. Al-Arbi, S.; Talbot, C.J.; Wang, T.; Fengshou, G.; Ball, A. Characterization of vibration transmission paths for gearbox condition monitoring. In Proceedings of the Inter Noise 2010 39th International Congress on Noise Control Engineering, Lisbon, Portugal, 3–16 June 2010; p. 10.
18. Kar, C.; Mohanty, A. Monitoring gear vibrations through motor current signature analysis and wavelet transform. *Mech. Syst. Signal Process.* **2006**, *20*, 158–187. [CrossRef]
19. Mohanty, A.; Kar, C. Fault Detection in a Multistage Gearbox by Demodulation of Motor Current Waveform. *IEEE Trans. Ind. Electron.* **2006**, *53*, 1285–1297. [CrossRef]
20. Yasir, H.A.; Salah, M.A.; Roslan, A.R.; Raja Ishak, R.H. Acoustic Emission and Artificial Intelligent Methods in Condition Monitoring of Rotating Machine—A Review. In Proceedings of the National Conference for Postgraduate Research (NCON-PGR 2016), Universiti Malaysia Pahang, Pekan, Malaysia, 24–25 September 2016; pp. 212–219.

21. Ulus, c.; Erkaya, S. An Experimental Study on Gear Diagnosis by Using Acoustic Emission Technique. *Int. J. Acoust. Vib.* **2016**, *21*, 103–111. [CrossRef]
22. Yang, H.; Mathew, J.; Ma, L. Vibration feature extraction techniques for fault diagnosis of rotating machinery: A literature survey. In Proceedings of the Asia-Pacific Vibration Conference, Gold Coast, Australia, 12–14 November 2003.
23. Gao, Z.; Cecati, C.; Ding, S.X. A Survey of Fault Diagnosis and Fault-Tolerant Techniques—Part I: Fault Diagnosis with Model-Based and Signal-Based Approaches. *IEEE Trans. Ind. Electron.* **2015**, *62*, 3757–3767. [CrossRef]
24. Gao, Z.; Cecati, C.; Ding, S. A Survey of Fault Diagnosis and Fault-Tolerant Techniques Part II: Fault Diagnosis with Knowledge-Based and Hybrid/Active Approaches. *IEEE Trans. Ind. Electron.* **2015**, *62*, 3768–3774. [CrossRef]
25. Kan, M.S.; Tan, A.C.; Mathew, J. A review on prognostic techniques for non-stationary and non-linear rotating systems. *Mech. Syst. Signal Process.* **2015**, *62–63*, 1–20. [CrossRef]
26. Vrba, J.; Cejnek, M.; Steinbach, J.; Krbcova, Z. A Machine Learning Approach for Gearbox System Fault Diagnosis. *Entropy* **2021**, *23*, 1130. [CrossRef] [PubMed]
27. Li, C.; Sánchez, R.V.; Zurita, G.; Cerrada, M.; Cabrera, D. Fault Diagnosis for Rotating Machinery Using Vibration Measurement Deep Statistical Feature Learning. *Sensors* **2016**, *16*, 895. [CrossRef] [PubMed]
28. Bayar, N.; Darmoul, S.; Hajri-Gabouj, S.; Pierreval, H. Fault detection, diagnosis and recovery using Artificial Immune Systems: A review. *Eng. Appl. Artif. Intell.* **2015**, *46*, 43–57. [CrossRef]
29. Yin, S.; Ding, S.X.; Xie, X.; Luo, H. A Review on Basic Data-Driven Approaches for Industrial Process Monitoring. *IEEE Trans. Ind. Electron.* **2014**, *61*, 6418–6428. [CrossRef]
30. Xu, Y.; Sun, Y.; Wan, J.; Liu, X.; Song, Z. Industrial Big Data for Fault Diagnosis: Taxonomy, Review, and Applications. *IEEE Access* **2017**, *5*, 17368–17380. [CrossRef]
31. Devijver, P.A.; Kittler, J. *Pattern Recognition: A Statistical Approach*; Prentice Hall: Old Tappan, NJ, USA, 1982.
32. Jankowski, N.; Grabczewski, K. Learning Machines. In *Feature Extraction: Foundations and Applications*; Springer: Berlin/Heidelberg, Germany, 2006; pp. 29–64. [CrossRef]
33. Duda, R.O.; Hart, P.E.; Stork, D.G. *Pattern Classification*, 2nd ed.; A Wiley-Interscience Publication; John Wiley & Sons: Nashville, TN, USA, 2000.
34. Bishop, C. *Neural Networks for Pattern Recognition*; Oxford University Press: Cary, NC, USA, 1995.
35. Sánchez, R.V.; Lucero, P.; Macancela, J.C.; Cerrada, M.; Vásquez, R.E.; Pacheco, F. Multi-fault Diagnosis of Rotating Machinery by Using Feature Ranking Methods and SVM-based Classifiers. In Proceedings of the 2017 International Conference on Sensing, Diagnostics, Prognostics, and Control (SDPC), Shanghai, China, 16–18 August 2017; pp. 105–110. [CrossRef]
36. Duch, W. Filter Methods. In *Feature Extraction*; Springer: Berlin/Heidelberg, Germany, 2006; pp. 89–117. [CrossRef]
37. Kuhn, M.; Johnson, K. *Feature Engineering and Selection*; Chapman & Hall/CRC Data Science Series; CRC Press: London, UK, 2019.
38. Li, Y.; Ang, K.K.; Guan, C. Digital Signal Processing and Machine Learning. In *Brain-Computer Interfaces*; Springer: Berlin/Heidelberg, Germany, 2010; p. 305. [CrossRef]
39. Vetterli, M.; Kovačević, J.; Goyal, V.K. *Foundations of Signal Processing*; Cambridge University Press: Cambridge, UK, 2014.
40. Lee, C.; Landgrebe, D. Feature extraction based on decision boundaries. *IEEE Trans. Pattern Anal. Mach. Intell.* **1993**, *15*, 388–400. [CrossRef]
41. Xu, X.; Liang, T.; Zhu, J.; Zheng, D.; Sun, T. Review of classical dimensionality reduction and sample selection methods for large-scale data processing. *Neurocomputing* **2019**, *328*, 5–15. [CrossRef]
42. Guyon, I.; Elisseeff, A. An Introduction to Feature Extraction. In *Feature Extraction: Foundations and Applications*; Springer: Berlin/Heidelberg, Germany, 2006; pp. 1–25. [CrossRef]
43. Nixon, M. *Feature Extraction & Image Processing*, 2nd ed.; Academic Press: San Diego, CA, USA, 2007.
44. Večeř, P.; Kreidl, M.; Šmíd, R. Condition Indicators for Gearbox Condition Monitoring Systems. *Acta Polytech.* **2005**, *45*. [CrossRef]
45. Qiao, Z.; Lei, Y.; Li, N. Applications of stochastic resonance to machinery fault detection: A review and tutorial. *Mech. Syst. Signal Process.* **2019**, *122*, 502–536. [CrossRef]
46. Bengio, Y. *Deep Learning*; Adaptive Computation and Machine Learning Series; MIT Press: London, UK, 2016.
47. Bengio, Y.; Courville, A.; Vincent, P. Representation Learning: A Review and New Perspectives. *IEEE Trans. Pattern Anal. Mach. Intell.* **2013**, *35*, 1798–1828. [CrossRef]
48. Pin, W.; Hongjie, Y.; Weilie, S.; Yonghua, Z.; Honghao, G. Research on Feature Extraction based on Deep Learning. *Int. J. Hybrid Inf. Technol.* **2015**, *8*, 113–120. [CrossRef]
49. Bengio, Y.; Delalleau, O.; Roux, N.L.; Paiement, J.F.; Vincent, P.; Ouimet, M. Spectral Dimensionality Reduction. In *Feature Extraction*; Springer: Berlin/Heidelberg, Germany, 2006; pp. 519–550. [CrossRef]
50. Bengio, Y. Deep learning of representations for unsupervised and transfer learning. In Proceedings of the 2011 International Conference on Unsupervised and Transfer Learning Workshop, Washington, DC, USA, 2 July 2011; Volume 27, pp. 17–37.
51. Anzanello, M.J. Feature Extraction and Feature Selection: A Survey of Methods in Industrial Applications. In *Wiley Encyclopedia of Operations Research and Management Science*; John Wiley & Sons, Inc.: Hoboken, NJ, USA, 2011. [CrossRef]
52. Li, J.; Zhang, H.; Zhao, J.; Guo, X.; Rihan, W.; Deng, G. Embedded Feature Selection and Machine Learning Methods for Flash Flood Susceptibility-Mapping in the Mainstream Songhua River Basin, China. *Remote Sens.* **2022**, *14*, 5523. [CrossRef]

53. Liu, Q.; Zhang, J.; Liu, J.; Yang, Z. Feature extraction and classification algorithm, which one is more essential? An experimental study on a specific task of vibration signal diagnosis. *Int. J. Mach. Learn. Cybern.* **2022**, *13*, 1685–1696. [CrossRef]
54. Luo, Y.; Daley, S. A comparative study of feature extraction methods for crack detection. *IFAC Proc. Vol.* **2006**, *39*, 1109–1114. [CrossRef]
55. Aggarwal, C.C. (Ed.) *Data Classification*; Chapman & Hall/CRC Data Mining and Knowledge Discovery Series; CRC Press: Boca Raton, FL, USA, 2014.
56. Granitzer, M.; Rath, A.; Kröll, M.; Seifert, C.; Ipsmiller, D.; Devaurs, D.; Weber, N.; Lindstaedt, S. Machine Learning based Work Task Classification. *JDIM* **2009**, *7*, 306–313.
57. Cabrera, D.; Sancho, F.; Sánchez, R.V.; Zurita, G.; Cerrada, M.; Li, C.; Vásquez, R.E. Fault diagnosis of spur gearbox based on random forest and wavelet packet decomposition. *Front. Mech. Eng.* **2015**, *10*, 277–286. [CrossRef]
58. Hastie, T.; Tibshirani, R.; Friedman, J. *The Elements of Statistical Learning*, 2nd ed.; Springer Series in Statistics; Springer: New York, NY, USA, 2009.
59. Lavrač, N.; Podpečan, V.; Robnik-Šikonja, M. *Representation Learning*; Springer International Publishing: Berlin/Heidelberg, Germany, 2021. [CrossRef]
60. Wang, H.; Chen, P. A Feature Extraction Method Based on Information Theory for Fault Diagnosis of Reciprocating Machinery. *Sensors* **2009**, *9*, 2415–2436. [CrossRef]
61. Yan, W.; Qiu, H.; Iyer, N. *Feature Extraction for Bearing Prognostics and Health Management (PHM)—A Survey (Preprint)*; Technical Report AFRL-RX-WP-TP-2008-4309; Air Force Research Laboratory: Wright-Patterson Air Force Base, OH, USA, 2008.
62. Tom, K. *Survey of Diagnostic Techniques for Dynamic Components*; Technical Report ARL-TR-5082; Army Research Laboratory: Adelphi, MD, USA, 2010.
63. Sait, A.S.; Sharaf-Eldeen, Y.I. A Review of Gearbox Condition Monitoring Based on vibration Analysis Techniques Diagnostics and Prognostics. In *Rotating Machinery, Structural Health Monitoring, Shock and Vibration*; Proulx T., Ed.; Conference Proceedings of the Society for Experimental Mechanics Series; Springer: Berlin/Heidelberg, Germany, 2011; pp. 1–2.
64. Han, Z.Y.; Liu, Y.L.; Jin, H.Y.; Fu, H.Y. A Review of Methodologies Used for Fault Diagnosis of Gearbox. *Appl. Mech. Mater.* **2013**, *415*, 510–514. [CrossRef]
65. Zhao, X.; Zuo, M.J.; Liu, Z.; Hoseini, M.R. Diagnosis of artificially created surface damage levels of planet gear teeth using ordinal ranking. *Measurement* **2013**, *46*, 132–144. [CrossRef]
66. Sharma, V.; Parey, A. A Review of Gear Fault Diagnosis Using Various Condition Indicators. *Procedia Eng.* **2016**, *144*, 253–263. [CrossRef]
67. Caesarendra, W.; Tjahjowidodo, T. A Review of Feature Extraction Methods in Vibration-Based Condition Monitoring and Its Application for Degradation Trend Estimation of Low-Speed Slew Bearing. *Machines* **2017**, *5*, 21. [CrossRef]
68. Riaz, S.; Elahi, H.; Javaid, K.; Shahzad, T. Vibration Feature Extraction and Analysis for Fault Diagnosis of Rotating Machinery—A Literature Survey. *Asia Pac. J. Multidiscip. Res.* **2017**, *5*, 103–110.
69. Ogundare, A.; Ojolo, S.; Mba, D.; Duan, F. *Review of Fault Detection Techniques for Health Monitoring of Helicopter Gearbox*; Springer, Cham, Switzerland, 2017; pp. 125–135. [CrossRef]
70. Goyal, D.; Vanraj.; Pabla, B.S.; Dhami, S.S. Condition Monitoring Parameters for Fault Diagnosis of Fixed Axis Gearbox: A Review. *Arch. Comput. Methods Eng.* **2017**, *24*, 543–556. [CrossRef]
71. Wang, T.; Han, Q.; Chu, F.; Feng, Z. Vibration based condition monitoring and fault diagnosis of wind turbine planetary gearbox: A review. *Mech. Syst. Signal Process.* **2019**, *126*, 662–685. [CrossRef]
72. Stetco, A.; Dinmohammadi, F.; Zhao, X.; Robu, V.; Flynn, D.; Barnes, M.; Keane, J.; Nenadic, G. Machine learning methods for wind turbine condition monitoring: A review. *Renew. Energy* **2019**, *133*, 620–635. [CrossRef]
73. Zhang, X.; Zhao, B.; Lin, Y. Machine Learning Based Bearing Fault Diagnosis Using the Case Western Reserve University Data: A Review. *IEEE Access* **2021**, *9*, 155598–155608. [CrossRef]
74. Khan, M.A.; Asad, B.; Kudelina, K.; Vaimann, T.; Kallaste, A. The Bearing Faults Detection Methods for Electrical Machines—The State of the Art. *Energies* **2022**, *16*, 296. [CrossRef]
75. Zhang, S.; Zhou, J.; Wang, E.; Zhang, H.; Gu, M.; Pirttikangas, S. State of the art on vibration signal processing towards data-driven gear fault diagnosis. *IET Collab. Intell. Manuf.* **2022**, *4*, 249–266. [CrossRef]
76. Pandit, R.; Astolfi, D.; Hong, J.; Infield, D.; Santos, M. SCADA data for wind turbine data-driven condition/performance monitoring: A review on state-of-art, challenges and future trends. *Wind Eng.* **2023**, *47*, 422–441. [CrossRef]
77. Romanssini, M.; de Aguirre, P.C.C.; Compassi-Severo, L.; Girardi, A.G. A Review on Vibration Monitoring Techniques for Predictive Maintenance of Rotating Machinery. *Eng* **2023**, *4*, 1797–1817. [CrossRef]
78. Xu, X.; Huang, X.; Bian, H.; Wu, J.; Liang, C.; Cong, F. Total process of fault diagnosis for wind turbine gearbox, from the perspective of combination with feature extraction and machine learning: A review. *Energy AI* **2024**, *15*, 100318. [CrossRef]
79. Zhang, L.; Hu, N. Time Domain Synchronous Moving Average and its Application to Gear Fault Detection. *IEEE Access* **2019**, *7*, 93035–93048. [CrossRef]
80. Bajric, R.; Sprečić, D.; Zuber, N. Review of vibration signal processing techniques towards gear pairs damage identification. *Int. J. Eng. Technol. IJET-IJENS* **2011**, *11*, 124–128.

81. Randall, R.B. *Vibration-Based Condition Monitoring*; Wiley-Blackwell: Hoboken, NJ, USA, 2011.
82. Jayaswal, P.; Wadhwani, A.K.; Mulchandani, K. Machine Fault Signature Analysis. *Int. J. Rotating Mach.* **2008**, *2008*, 1–10.
83. Venugopal, S.; Mcinerny, S. Fusion of Vibration Based Features for Gear Condition Classification. In Proceedings of the 15th International Congress on Sound and Vibration 2008, Daejeon, Republic of Korea, 6–10 July 2008; Volume 2.
84. Bowman, W.F.; Stachowiak, G.W. A review of scuffing models. *Tribol. Lett.* **1996**, *2*, 113–131. [CrossRef]
85. Snidle, R.W.; Evans, H.P.; Alanou, M.P.; Holmes, M.J.A. Understanding scuffing and micropitting of gears. In Proceedings at NATO Research and Technology Organisation Specialists' Meeting on the Control and Reduction of Wear in Military Platforms, Williamsburg, USA, 2003. Available online: https://www.sto.nato.int/publications/STO%20Meeting%20Proceedings/RTO-MP-AVT-109/MP-AVT-109-14.pdf (accessed on 1 February 2024).
86. Martins, R.; Cardoso, N.; Seabra, J. Influence of lubricant type in gear scuffing. *Ind. Lubr. Tribol.* **2008**, *60*, 299–308. [CrossRef]
87. Ninoslav, Z.F.; Rusmir, B.; Cvetkovic, D. Vibration Feature Extraction Methods for Gear Faults Diagnosis—A Review. In Proceedings of the Facta Universitatis, 2015, Working and Living Environmental Protection; pp. 63–72. Available online: http://casopisi.junis.ni.ac.rs/index.php/FUWorkLivEnvProt/article/view/689 (accessed on 1 February 2024).
88. Fu, D.; Liu, J.; Zhong, H.; Zhang, X.; Zhang, F. A novel self-supervised representation learning framework based on time-frequency alignment and interaction for mechanical fault diagnosis. *Knowl.-Based Syst.* **2024**, *295*, 111846. [CrossRef]
89. Haidong, S.; Hongkai, J.; Xingqiu, L.; Shuaipeng, W. Intelligent fault diagnosis of rolling bearing using deep wavelet auto-encoder with extreme learning machine. *Knowl.-Based Syst.* **2018**, *140*, 1–14. [CrossRef]
90. Wang, Z.; Xu, Z.; Cai, C.; Wang, X.; Xu, J.; Shi, K.; Zhong, X.; Liao, Z.; Li, Q. Rolling bearing fault diagnosis method using time-frequency information integration and multi-scale TransFusion network. *Knowl.-Based Syst.* **2024**, *284*, 111344. [CrossRef]
91. Mohammed, A.A.; Neilson, R.D.; Deans, W.F.; MacConnell, P. Crack detection in a rotating shaft using artificial neural networks and PSD characterisation. *Meccanica* **2013**, *49*, 255–266. [CrossRef]
92. Haram, M.; Wang, T.; Gu, F.; Ball, A.D. Electrical Motor Current Signal Analysis using a Modulation Signal Bispectrum for the Fault Diagnosis of a Gearbox Downstream. *J. Phys. Conf. Ser.* **2012**, *364*, 012050. [CrossRef]
93. Lee, S.K.; White, P.R. Two-Stage Adaptive Line Enhancer and Sliced Wigner Trispectrum for the Characterization of Faults from Gear Box Vibration Data. *J. Vib. Acoust.* **1999**, *121*, 488–494. [CrossRef]
94. El Morsy, M.; Achtenová, G. Vehicle gearbox fault diagnosis based on cepstrum analysis. *World Acad. Sci. Eng. Technol. Int. J. Mech. Aerosp. Ind. Mechatron. Eng.* **2014**, *8*, 1533–1539.
95. Lei, Y.; He, Z.; Zi, Y. A new approach to intelligent fault diagnosis of rotating machinery. *Expert Syst. Appl.* **2008**, *35*, 1593–1600. [CrossRef]
96. Yang, Z.; Hoi, W.I.; Zhong, J. Gearbox fault diagnosis based on artificial neural network and genetic algorithms. In Proceedings of the 2011 International Conference on System Science and Engineering, Macau, China, 8–10 June 2011; pp. 37–42. [CrossRef]
97. Nandi, A.K.; Liu, C.; Wong, M.L.D. Intelligent Vibration Signal Processing for Condition Monitoring. In Proceedings of the International Conference Surveillance 7, Institute of Technology of Chartres, Chartres, France, 29–30 October 2013.
98. Yuan, J.; Liu, X. Semi-supervised learning and condition fusion for fault diagnosis. *Mech. Syst. Signal Process.* **2013**, *38*, 615–627. [CrossRef]
99. Bagheri, B.; Ahmadi, H.; Labbafi, R. Application of data mining and feature extraction on intelligent fault diagnosis by Artificial Neural Network and k-nearest neighbor. In Proceedings of the XIX International Conference on Electrical Machines—ICEM 2010, Rome, Italy, 6–8 September 2010; pp. 1–7. [CrossRef]
100. Sreejith, B.; Verma, A.; Srividya, A. Fault diagnosis of rolling element bearing using time-domain features and neural networks. In Proceedings of the 2008 IEEE Region 10 and the Third International Conference on Industrial and Information Systems, Kharagpur, India, 8–10 December 2008; pp. 1–6. [CrossRef]
101. Zhao, X.; Zuo, M.J.; Liu, Z. Diagnosis of pitting damage levels of planet gears based on ordinal ranking. In Proceedings of the 2011 IEEE Conference on Prognostics and Health Management, Denver, CO, USA, 20–23 June 2011; pp. 1–8. [CrossRef]
102. Gupta, K.N. Vibration—A tool for machine diagnostics and condition monitoring. *Sadhana* **1997**, *22*, 393–410. [CrossRef]
103. Gupta, P. Feature Selection by Genetic Programming, and Artificial Neural Network-based Machine Condition Monitoring. *Int. J. Eng. Innov. Technol.* **2012**, *1*, 177–181.
104. Samuel, P.D.; Pines, D.J. A review of vibration-based techniques for helicopter transmission diagnostics. *J. Sound Vib.* **2005**, *282*, 475–508. [CrossRef]
105. Al-Atat, H.; Siegel, D.; Lee, J. A Systematic Methodology for Gearbox Health Assessment and Fault Classification. *Int. J. Progn. Health Manag.* **2011**, *2*, 1–16.
106. Elamvazuthi, I.; Ling, G.A.; Nurhanim, K.A.R.K.; Vasant, P.; Parasuraman, S. Surface electromyography (sEMG) feature extraction based on Daubechies wavelets. In Proceedings of the 2013 IEEE 8th Conference on Industrial Electronics and Applications (ICIEA), Melbourne, Australia, 19–21 June 2013; pp. 1492–1495. [CrossRef]
107. Phinyomark, A.; Quaine, F.; Charbonnier, S.; Serviere, C.; Tarpin-Bernard, F.; Laurillau, Y. Feature extraction of the first difference of EMG time series for EMG pattern recognition. *Comput. Methods Programs Biomed.* **2014**, *117*, 247–256. [CrossRef]
108. Lee, S.W.; Yi, T.; Jung, J.W.; Bien, Z. Design of a Gait Phase Recognition System That Can Cope with EMG Electrode Location Variation. *IEEE Trans. Autom. Sci. Eng.* **2017**, *14*, 1429–1439. [CrossRef]
109. Shen, C.; Wang, D.; Kong, F.; Tse, P.W. Fault diagnosis of rotating machinery based on the statistical parameters of wavelet packet paving and a generic support vector regressive classifier. *Measurement* **2013**, *46*, 1551–1564. [CrossRef]

110. Wang, D.; Tse, P.W.; Guo, W.; Miao, Q. Support vector data description for fusion of multiple health indicators for enhancing gearbox fault diagnosis and prognosis. *Meas. Sci. Technol.* **2011**, *22*, 025102. [CrossRef]
111. Warke, V.; Kumar, S.; Bongale, A.; Kotecha, K. Robust Tool Wear Prediction using Multi-Sensor Fusion and Time-Domain Features for the Milling Process using Instance-based Domain Adaptation. *Knowl.-Based Syst.* **2024**, *288*, 111454. [CrossRef]
112. Ahmed, M.; Abdusslam, S.A.; Baqqar, M.; Gu, F.; Ball, A. *Fault Classification of Reciprocating Compressor Based on Neural Networks and Support Vector Machines*; Chinese Automation and Computing Society: London, UK, 2011.
113. *Rugosidad Superficial*; Technical Report; Universidad Nacional de Mar del Plata: Mar del Plata, Argentina, 2014.
114. Baró Gadea, E. Estudio de la Rugosidad Superficial Mediante Fresado. Master's Thesis, UPC, Escola Tècnica Superior d'Enginyeria Industrial de Barcelona, Departament d'Enginyeria Mecànica, Barcelona, Spain, 2012. Available online: http://hdl.handle.net/2099.1/18775 (accessed on 2 July 2024).
115. Hreha, P.; Radvanska, A.; Knapcikova, L.; Królczyk, G.M.; Legutko, S.; Królczyk, J.B.; Hloch, S.; Monka, P. Roughness Parameters Calculation by Means of On-Line Vibration Monitoring Emerging from AWJ Interaction with Material. *Metrol. Meas. Syst.* **2015**, *22*, 315–326. [CrossRef]
116. Gadelmawla, E.S.; Koura, M.M.; Maksoud, T.M.; Elewa, I.M.; Soliman, H.H. Roughness parameters. *J. Mater. Process. Technol.* **2002**, *123*, 133–145. [CrossRef]
117. Prakash, S.; Palanikumar, K.; Lilly Mercy, J.; Nithyalakshmi, S. Evaluation of Surface Roughness Parameters (Ra, Rz) in Drilling of Mdf Composite Panel Using Box-Behnken Experimental Design (Bbd). *Int. J. Des. Manuf. Technol.* **2011**, *5*, 52–62. [CrossRef]
118. Lei, Y.; Zuo, M.J.; He, Z.; Zi, Y. A multidimensional hybrid intelligent method for gear fault diagnosis. *Expert Syst. Appl.* **2010**, *37*, 1419–1430. [CrossRef]
119. Lei, Y.; He, Z.; Zi, Y.; Chen, X. New clustering algorithm-based fault diagnosis using compensation distance evaluation technique. *Mech. Syst. Signal Process.* **2008**, *22*, 419–435. [CrossRef]
120. Morales, O.C. Time-Frequency Analysis of Mechanic Vibration Signals for Fault Detection in Rotating Machines. Master's Thesis, Universidad Nacional de Colombia, Manizales, Colombia, 2011. Available online: https://repositorio.unal.edu.co/handle/unal/7743 (accessed on 2 July 2024).
121. Jeong, I.K.; Kang, M.; Kim, J.; Kim, J.M.; Ha, J.M.; Choi, B.K. Enhanced DET-Based Fault Signature Analysis for Reliable Diagnosis of Single and Multiple-Combined Bearing Defects. *Shock Vib.* **2015**, *2015*, 814650. [CrossRef]
122. Huang, W.; Kong, F.; Zhao, X. Spur bevel gearbox fault diagnosis using wavelet packet transform and rough set theory. *J. Intell. Manuf.* **2018**, *29*, 1257–1271. [CrossRef]
123. Fuqing, Y.; Kumar, U. Statistical Index Development from Time Domain for Rolling Element Bearings. *Int. J. Perform. Eng.* **2014**, *10*, 313–324.
124. Overview on Extracted Features. Available online: https://tsfresh.readthedocs.io/en/latest/text/list_of_features.html (accessed on 27 February 2024).
125. Zhu, Y.; Jiang, W.; Kong, X.; Zheng, Z.; Hu, H. An Accurate Integral Method for Vibration Signal Based on Feature Information Extraction. *Shock Vib.* **2015**, *2015*, 962793. [CrossRef]
126. Kumar, H.; Pai, S.P.; Sriram, N.; Vijay, G. Rolling element bearing fault diagnostics: Development of health index. *Proc. Inst. Mech. Eng. Part C J. Mech. Eng. Sci.* **2017**, *231*, 3923–3939. [CrossRef]
127. Sharma, A.; Amarnath, M.; Kankar, P.K. Feature extraction and fault severity classification in ball bearings. *J. Vib. Control* **2014**, *22*, 176–192. [CrossRef]
128. Zhang, D.; Entezami, M.; Stewart, E.; Roberts, C.; Yu, D. Adaptive fault feature extraction from wayside acoustic signals from train bearings. *J. Sound Vib.* **2018**, *425*, 221–238. [CrossRef]
129. Kim, H.G.; Moreau, N.; Sikora, T. *MPEG-7 Audio and Beyond: Audio Content Indexing and Retrieval*; John Wiley & Sons, Inc.: Hoboken, NJ, USA, 2005.
130. Lu, L.; Zhang, H.J.; Jiang, H. Content analysis for audio classification and segmentation. *IEEE Trans. Speech Audio Process.* **2002**, *10*, 504–516. [CrossRef]
131. *Algorithms to Measure Audio Programme Loudness and True-Peak Audio Level*; International Telecommunication Union: Geneva, Switzerland, 2011.
132. European Broadcasting Union. *Loudness Normalisation and Permitted Maximum Level of Audio Signals*; European Broadcasting Union: Geneva, Switzerland, 2014.
133. European Broadcasting Union. *Loudness Metering: 'EBU Mode' Metering to Supplement EBU R 128 Loudness Normalization*; European Broadcasting Union: Geneva, Switzerland, 2014.
134. European Broadcasting Union. *Loudness Range: A Measure to Supplement EBU R 128 Loudness Normalization*; European Broadcasting Union: Geneva, Switzerland, 2016.
135. Chowdhury, R.; Reaz, M.; Ali, M.; Bakar, A.; Chellappan, K.; Chang, T. Surface Electromyography Signal Processing and Classification Techniques. *Sensors* **2013**, *13*, 12431–12466. [CrossRef] [PubMed]
136. Mokhlesabadifarahani, B.; Gunjan, V.K. *EMG Signals Characterization in Three States of Contraction by Fuzzy Network and Feature Extraction*; Springer: Singapore, 2015. [CrossRef]
137. Phinyomark, A.; Nuidod, A.; Phukpattaranont, P.; Limsakul, C. Feature Extraction and Reduction of Wavelet Transform Coefficients for EMG Pattern Classification. *Elektron. Elektrotech.* **2012**, *122*, 27–32. [CrossRef]

138. Dhamande, L.S.; Chaudhari, M.B. Compound gear-bearing fault feature extraction using statistical features based on time-frequency method. *Measurement* **2018**, *125*, 63–77. [CrossRef]

139. Giannakopoulos, T.; Pikrakis, A. *Introduction to Audio Analysis: A MATLAB® Approach*; Academic Press: Cambridge, MA, USA, 2014.

140. Salem, A.; Aly, A.; Sassi, S.; Renno, J. Time-Domain Based Quantification of Surface Degradation for Better Monitoring of the Health Condition of Ball Bearings. *Vibration* **2018**, *1*, 172–191. [CrossRef]

141. Chujit, G.; Phinyomark, A.; Hu, H.; Phukpattaranont, P.; Limsakul, C. Evaluation of EMG Feature Extraction for Classification of Exercises in Preventing Falls in the Elderly. In Proceedings of the 10th International PSU Engineering Conference, Songkhla, Thailand, 14–15 May 2012; p. 6.

142. Vakharia, V.; Gupta, V.K.; Kankar, P.K. A comparison of feature ranking techniques for fault diagnosis of ball bearing. *Soft Comput.* **2015**, *20*, 1601–1619. [CrossRef]

143. Tom, K. *A Primer on Vibrational Ball Bearing Feature Generation for Prognostics and Diagnostics Algorithms*; Technical Report ARL-TR-7230; Army Research Laboratory: Adelphi, MD, USA, 2015.

144. Raut, R.K.; Gurjar, A.A. Bio-medical (EMG) Signal Feature Extraction Using Wavelet Transform for Design of Prosthetic Leg. *Int. J. Electron. Commun. Soft Comput. Sci. Eng. (IJECSCSE)* **2015**, *4*, 81–84.

145. Soualhi, M.; Nguyen, K.T.; Soualhi, A.; Medjaher, K.; Hemsas, K.E. Health monitoring of bearing and gear faults by using a new health indicator extracted from current signals. *Measurement* **2019**, *141*, 37–51. [CrossRef]

146. Wijaya, I.G.P.S.; Kencanawati, N. Denoising Acoustic Emission Signal using Wavelet Transforms for Determining the Micro Crack Location Inside of Concrete. *Int. J. Technol.* **2014**, *5*, 259–268. [CrossRef]

147. Cerrada, M.; Zurita, G.; Cabrera, D.; Sánchez, R.V.; Artés, M.; Li, C. Fault diagnosis in spur gears based on genetic algorithm and random forest. *Mech. Syst. Signal Process.* **2016**, *70-71*, 87–103. [CrossRef]

148. Pacheco, F.; de Oliveira, J.V.; Sánchez, R.V.; Cerrada, M.; Cabrera, D.; Li, C.; Zurita, G.; Artés, M. A statistical comparison of neuroclassifiers and feature selection methods for gearbox fault diagnosis under realistic conditions. *Neurocomputing* **2016**, *194*, 192–206. [CrossRef]

149. Li, C.; Sanchez, R.V.; Zurita, G.; Cerrada, M.; Cabrera, D.; Vásquez, R.E. Multimodal deep support vector classification with homologous features and its application to gearbox fault diagnosis. *Neurocomputing* **2015**, *168*, 119–127. [CrossRef]

150. Li, C.; Sanchez, R.V.; Zurita, G.; Cerrada, M.; Cabrera, D.; Vásquez, R.E. Gearbox fault diagnosis based on deep random forest fusion of acoustic and vibratory signals. *Mech. Syst. Signal Process.* **2016**, *76–77*, 283–293. [CrossRef]

151. Pacheco, F.; Drimus, A.; Duggen, L.; Cerrada, M.; Cabrera, D.; Sanchez, R.V. Deep Ensemble-Based Classifier for Transfer Learning in Rotating Machinery Fault Diagnosis. *IEEE Access* **2022**, *10*, 29778–29787. [CrossRef]

152. Cabrera, D.; Sancho, F.; Li, C.; Cerrada, M.; Sánchez, R.V.; Pacheco, F.; de Oliveira, J.V. Automatic feature extraction of time-series applied to fault severity assessment of helical gearbox in stationary and non-stationary speed operation. *Appl. Soft Comput.* **2017**, *58*, 53–64. [CrossRef]

153. Case Western Reserve University. 12k Drive End Bearing Fault Data. Available online: https://engineering.case.edu/bearingdatacenter/12k-drive-end-bearing-fault-data (accessed on 1 February 2024).

154. Urbanowicz, R.J.; Meeker, M.; Cava, W.L.; Olson, R.S.; Moore, J.H. Relief-based feature selection: Introduction and review. *J. Biomed. Inform.* **2018**, *85*, 189–203. [CrossRef] [PubMed]

155. Zhai, Y.; Song, W.; Liu, X.; Liu, L.; Zhao, X. A Chi-Square Statistics Based Feature Selection Method in Text Classification. In Proceedings of the 2018 IEEE 9th International Conference on Software Engineering and Service Science (ICSESS), Beijing, China, 23–25 November 2018; pp. 160–163. [CrossRef]

156. Lee, C.; Lee, G.G. Information gain and divergence-based feature selection for machine learning-based text categorization. *Inf. Process. Manag.* **2006**, *42*, 155–165. [CrossRef]

157. Nasir, I.M.; Khan, M.A.; Yasmin, M.; Shah, J.H.; Gabryel, M.; Scherer, R.; Damaševičius, R. Pearson Correlation-Based Feature Selection for Document Classification Using Balanced Training. *Sensors* **2020**, *20*, 6793. [CrossRef] [PubMed]

158. Gu, Q.; Li, Z.; Han, J. Generalized Fisher score for feature selection. In Proceedings of the Twenty-Seventh Conference on Uncertainty in Artificial Intelligence, Arlington, VA, USA, 14 July 2011; pp. 266–273.

159. Jamalludin, M.D.; Muljono.; Fajar Shidik, G.; Zainul Fanani, A.; Purwanto.; Al Zami, F. Implementation of Feature Selection Using Gain Ratio towards Improved Accuracy of Support Vector Machine (SVM) on Youtube Comment Classification. In Proceedings of the 2021 International Seminar on Application for Technology of Information and Communication (iSemantic), Semarangin, Indonesia, 18–19 September 2021; pp. 28–31. [CrossRef]

160. Rogers, J.; Gunn, S. Identifying Feature Relevance Using a Random Forest. In *Subspace, Latent Structure and Feature Selection*; Springer: Berlin/Heidelberg, Germany, 2006; pp. 173–184. [CrossRef]

161. Evgeniou, T.; Pontil, M. Support Vector Machines: Theory and Applications. In *Machine Learning and Its Applications*; Springer: Berlin/Heidelberg, Germany, 2001; pp. 249–257. [CrossRef]

162. Guo, G.; Wang, H.; Bell, D.; Bi, Y.; Greer, K. KNN Model-Based Approach in Classification. In *On the Move to Meaningful Internet Systems 2003: CoopIS, DOA, and ODBASE*; Springer: Berlin/Heidelberg, Germany, 2003; pp. 986–996. [CrossRef]

163. Breiman, L. Random Forests. *Mach. Learn.* **2001**, *45*, 5–32. [CrossRef]

164. Miao, M.; Sun, Y.; Yu, J. Deep sparse representation network for feature learning of vibration signals and its application in gearbox fault diagnosis. *Knowl.-Based Syst.* **2022**, *240*, 108116. [CrossRef]
165. Yu, J.; Liu, G. Knowledge extraction and insertion to deep belief network for gearbox fault diagnosis. *Knowl.-Based Syst.* **2020**, *197*, 105883. [CrossRef]

Article

Evaluation of the Diagnostic Sensitivity of Digital Vibration Sensors Based on Capacitive MEMS Accelerometers

Marek Fidali [1], Damian Augustyn [1], Jakub Ochmann [2] and Wojciech Uchman [2],*

[1] Department of Fundamentals of Machinery Design, Silesian University of Technology, Konarskiego 18A, 44-100 Gliwice, Poland; marek.fidali@polsl.pl (M.F.); damian.augustyn@polsl.pl (D.A.)
[2] Department of Power Engineering and Turbomachinery, Silesian University of Technology, Konarskiego 18, 44-100 Gliwice, Poland; jakub.ochmann@polsl.pl
* Correspondence: wojciech.uchman@polsl.pl

Abstract: In recent years, there has been an increasing use of digital vibration sensors that are based on capacitive MEMS accelerometers for machine vibration monitoring and diagnostics. These sensors simplify the design of monitoring and diagnostic systems, thus reducing implementation costs. However, it is important to understand how effective these digital sensors are in detecting rolling bearing faults. This article describes a method for determining the diagnostic sensitivity of diagnostic parameters provided by commercially available vibration sensors based on MEMS accelerometers. Experimental tests were conducted in laboratory conditions, during which vibrations from 11 healthy and faulty rolling bearings were measured using two commercial vibration sensors based on MEMS accelerometers and a piezoelectric accelerometer as a reference sensor. The results showed that the diagnostic sensitivity of the parameters depends on the upper-frequency band limit of the sensors, and the parameters most sensitive to the typical fatigue faults of rolling bearings are the peak and peak-to-peak amplitudes of vibration acceleration. Despite having a lower upper-frequency range compared to the piezoelectric accelerometer, the commercial vibration sensors were found to be sensitive to rolling bearing faults and can be successfully used in continuous monitoring and diagnostics systems for machines.

Keywords: MEMS accelerometer; vibration measurement; bearing faults; diagnostics; condition monitoring

Citation: Fidali, M.; Augustyn, D.; Ochmann, J.; Uchman, W. Evaluation of the Diagnostic Sensitivity of Digital Vibration Sensors Based on Capacitive MEMS Accelerometers. *Sensors* **2024**, *24*, 4463. https://doi.org/10.3390/s24144463

Academic Editors: Dong Wang, Shilong Sun and Changqing Shen

Received: 5 June 2024
Revised: 5 July 2024
Accepted: 9 July 2024
Published: 10 July 2024

1. Introduction

Rolling bearings are used in almost every type of rotating machinery. Most machine breakdowns relate to bearing failures; thus, it is very important to diagnose bearing conditions and predict the moment of failure occurrence [1–3]. Many bearings fail prematurely due to contamination, poor lubrication, misalignment, temperature extremes, poor fitting/fits, shaft unbalance, and misalignment [4]. The occurrence of bearing faults leads to an increase in the bearing vibration; therefore, in diagnosing the condition of rolling bearings, measurements and analysis of vibration signals are most often used [3].

1.1. Vibration Symptoms of Bearing Faults

Effective diagnosis of rolling bearings based on vibration measurements first requires an understanding of the relationship between how damage occurs in bearings and the symptoms of that damage visible in vibration signals [5,6]. A bearing consists of rolling elements mounted in a cage and rolling on an inner and outer race. If we take a closer look at the contact area (Figure 1), in an efficient and well-lubricated bearing, the rolling elements are separated from the race surface by a layer of grease such that during the rolling, only the highest peaks of surface roughness will interfere with each other, generating hundreds of small-amplitude short pulses [7]. Due to the random distribution of the roughness, the pulses generated will have the character of random noise. It can therefore be concluded that a healthy bearing is a random noise generator. As the lubrication conditions deteriorate and the lubrication film thickness is gradually reduced, the pulses generated will be more intense, so the level of perceived noise will be higher.

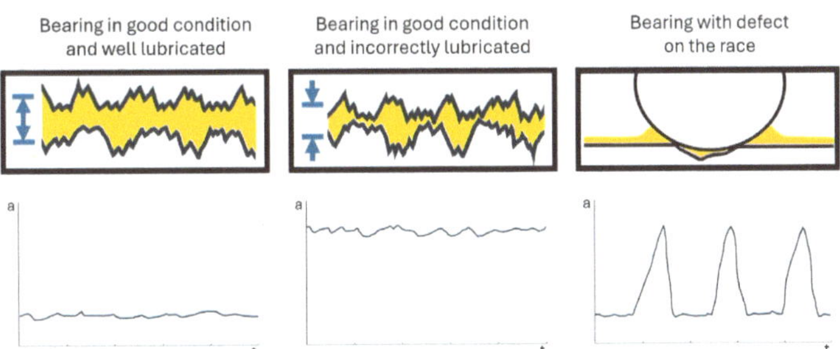

Figure 1. The influence of the surface conditions of interacting bearing elements on the number and intensity of generated vibration signals.

In an operating bearing, as a result of various wear mechanisms, including, but not limited to, fatigue wear, a small loss of material may develop on the surface of one of the raceways (Figure 2), causing each of the rolling elements to come into collision with the damage systematically, while generating cyclic pulses with an amplitude exceeding the noise level by up to 1000 times.

Figure 2. Material loss on the inner race of the bearing.

The frequency and intensity of the pulses will strongly depend on the bearing geometry (number and diameter of rolling elements and bearing race diameters) and shaft speed. The characteristic frequencies of the pulses arising from the various bearing events can be calculated from the analytical relationships shown below. The rolling element pass frequency over the single defect on the outer-race BPFO (ball pass frequency, outer race) is defined as follows:

$$BPFO = \frac{nf_r}{2}\left\{1 - \frac{d}{D}\cos\phi\right\}, \tag{1}$$

where n is the number of rolling elements, f_r is the rotational frequency, d is the rolling element diameter, D is the pitch diameter, and ϕ is the angle of load. The ball pass frequency over the single defect on the inner race BPFI is defined as follows:

$$BPFI = \frac{nf_r}{2}\left\{1 + \frac{d}{D}\cos\phi\right\}. \tag{2}$$

The frequency related to the cage speed FTF (fundamental train frequency) is defined as follows:

$$\text{FTF} = \frac{f_r}{2}\left\{1 - \frac{d}{D}\cos\phi\right\}.$$ (3)

The rolling element spin frequency BSF(RSF) is defined as follows:

$$\text{BSF(RSF)} = \frac{D}{2d}\left\{1 - \left(\frac{d}{D}\cos\phi\right)^2\right\}.$$ (4)

The impulses generated when rolling elements collide with a defect on one of the races are called shock or impact impulses and, due to the high stiffness of the elements involved in their generation, are characterized by a very short duration of a few to tens of microseconds. The pulses generate elastic waves in the material, which propagate at a speed of around 5000 m/s in steel. Furthermore, due to the short pulse duration, in the spectrum, the pulse energy is distributed over a very broad frequency band beyond 40 kHz [8].

The bearing is not an isolated component but cooperates with the shaft and, additionally, carries loads from, among other things, forces generated by the residual imbalance of the rotor and/or shaft misalignment. Consequently, the shock pulses generated during the initial stage of bearing degradation are very weak in relation to the signal components generated by inertia forces. It follows that in order to detect bearing damage at an early stage, it is useful to measure vibrations over a wide frequency band, covering the ultrasonic range, and in order to extract weak pulses caused by mechanical damage from the broadband signal, it is necessary to use appropriate methods for vibration signal processing and analysis.

1.2. Methods of Bearing Diagnostics

The condition of rolling bearings can be assessed by the results of diagnostic tests using temperature measurements, lubricant tests, thrust torque measurements, ultrasonic measurements, and vibration and noise measurements [2].

Due to the high availability of test equipment, vibration measurement and analysis is one of the more frequently used methods for diagnosing rolling element bearings. In the field of vibration signal analysis for rolling element bearing condition assessment, more or less sophisticated methods of signal analysis and evaluation based on the Hilbert transform and analysis of the vibration acceleration envelope signals in the time and frequency domains are used [1]. Over the years, a number of commercial solutions have been developed for rolling bearing diagnostics, such as the SPM (shock pulse method) and SPM HD (shock pulse method (higher definition)) from the SPM Instrument AB, the spike energy spectrum (gSE) from Rockwell Automation/ENTEK and PeakVue from CSI/Emerson, SEE (spectral emitted energy) and AEE (acoustic emission enveloping), ENV Acc and HFD from SK, and the BCU (bearing condition unit) from Schenck, among others. As industrial practice shows, the condition assessment of rolling element bearings is often based on basic numeric estimators of vibration acceleration signals after prior high-pass filtering and subsequent observation of time series as a function of operating time and trend analysis. The most commonly determined vibration acceleration signal amplitude estimators (signal features) are the peak value (aPeak) and/or the rms value (aRMS). Using numerical parameters, it is also possible to refer to limit values defined in the standards or practical diagnostic recommendations developed by diagnosticians or some companies [9]. An example of a standard which defines the criteria for evaluating the bearing condition based on point features determined from a broadband vibration acceleration signal is ISO 13373-3 [10].

The values of rms and peak amplitudes can also be used to determine the dimensionless parameter, like crest factor (Equation (5)). This represents the ratio of the peak value of the vibration signal to its RMS value in a given vibration frequency range. If the crest factor increases, the rolling bearing deteriorates. However, in the last phase of damage, the value of the peak factor may decrease. Therefore, this ratio should be used from the beginning of the bearing's life.

$$C_f = \frac{|x_{Peak}|}{x_{RMS}} \tag{5}$$

where x_{Peak} is the peak amplitude and x_{RMS} is the effective amplitude.

Another parameter used for bearing defect detection is kurtosis. This statistical parameter (Equation (6)) describes the flatness of a Gaussian distribution, and, for a strictly random signal, its value equals 3.0. Because bearings in good condition theoretically, it should generate random noise, and kurtosis serves as an indicator of a healthy bearing. If mechanical degradation in the bearing begins, the kurtosis increases, and for deteriorated bearings, it can be higher than 10 or 15.

$$K = \frac{\frac{1}{n}\Sigma_{i=1}^{n}(x_i - \mu)^4}{\sigma^4} - 3 \tag{6}$$

where x_i is the i-th value of the feature, μ is the population mean, σ is the population standard deviation, and n is the sample size.

The usefulness of using numerical features in assessing the bearing condition is reflected in the VDI 3832 standard [11], which defines, among other things, a diagnostic parameter helpful for assessing the condition of rolling element bearings $K(t)$, also called the Sturm diagnostic coefficient. It is calculated according to Equation (7) from the product of the peak and rms values of the vibration accelerations in the frequency range from 1 to 10 kHz, which is related to the product of the reference rms and peak values of the vibration accelerations measured at the beginning of the bearing's operation.

$$K(t) = \frac{aRMS(0){\cdot}aPeak(0)}{aRMS(t){\cdot}aPeak(t)}, \tag{7}$$

where $aRMS(0)$ is the RMS for the start point in time, $aPeak(0)$ is the maximum value for start, $aRMS(t)$ is the current RMS, and $aPeak(t)$ is the current maximum value.

The value of the parameter $K(t)$ decreases with the deterioration of the bearing condition, making it possible to define the ranges of the limit values for the parameter and to relate them to the expected bearing condition. The limit values of the $K(t)$ parameter are shown in Table 1.

Table 1. Classification of bearing condition according to the ranges of the parameter $K(t)$.

$K(t)$	Classification of Bearing Condition
>1	Condition improvement
1.0–0.5	Standard bearing condition
0.5–0.2	Deteriorating bearing condition
0.2–0.02	Advanced damage
<0.02	Failure

Rolling bearing vibration diagnostic methods based on numerical parameters allow for the damage to be detected early enough that bearing replacement can be planned and carried out at the most convenient time for the production process and before potential failure.

1.3. Measuremnt of Bearing Vibration

Piezoelectric accelerometers (IEPE, ICP) have been used for many years in rolling bearing diagnostics, which, when connected to portable vibration meters or continuous monitoring systems, allow for the effective detection and identification of bearing damage at a very early stage [12]. Capacitive accelerometers, so-called MEMS accelerometers, have been on the market for a long time, alongside piezoelectric accelerometers [6,13,14].

Landi et al. [15] presented a prototype MEMS sensor accelerometer for monitoring vibrations over a wide frequency range. The research presented included a sensor calibration procedure and was carried out on an in-house test stand. Staszewski et al. [16] presented a MEMS vibration sensor design with a wide frequency range up to 10 kHz, which can replace traditional sensors due to high sensitivity, low noise, and lower costs. The sensor prototype tested on a rig with a faulty rolling element bearing demonstrated effectiveness in fault detection and comparable performance to a piezoelectric accelerometer. Zusman [17] presented a comparison of traditional piezoelectric and modern MEMS-based vibration sensors used in machinery condition monitoring and fault diagnostics. Experimental data and detailed comparisons of output noise level and spectrum density for several popular piezoelectric and MEMS vibration sensors are presented.

Rossi et al. [18] focused on demonstrating the sufficient accuracy of MEMS-based data monitoring compared to a reference, a conventional mini-integrated circuit piezoelectric (ICP). Investigating the vibration of turbofan engine fan blades, the MEMS was shown to have a satisfactory level of measurement accuracy of $\pm 5\%$ deviation with respect to the ICP at the angular velocity tested from 0 to 300 rpm. Varanis et al. [19] presented the use of MEMS sensors for measuring mechanical vibrations and a broad literature review on their use in various applications. Two experiments were also performed comparing the amplitudes and frequencies of oscillations measured by MEMS sensors and piezoelectric accelerometers in the time and frequency domains. Augustyn et al. [20] presented the results of research on the identification of the frequency characteristics of a digital MEMS accelerometer dedicated to monitoring the condition of machines. The specified characteristics indicate the possibility of using the sensor for basic machine diagnostics in accordance with the ISO 10816 [21] and ISO 20816 standards [22]; however, non-linearities at the limits of the measurement band may limit its use in precise scientific measurements. Anslow [23] presented the design of a mechanical housing for a MEMS accelerometer, which ensures high-quality vibration data for machine condition monitoring (CbM). This paper presents modal analysis, vibration sensor design guidelines, and housing design examples for single-axis and three-axis MEMS accelerometers, highlighting the importance of avoiding resonance and ensuring the appropriate housing natural frequency. However, Albarbar et al. [24] have shown experimentally that the selection of a suitable MEMS sensor is crucial for adequate monitoring of the desired quantity. In their study, they compared data obtained with sensors dedicated to measuring different types of signals: sinusoidal, random, and impulsive. They showed apparent differences between the results and suggested using one of the sensors for purposes other than monitoring the condition of the machine.

1.4. Contemporary Digital Vibration Sensors and Its Usefulness to Bearing Faults Diagnostics

With the advent of Industry 4.0 and IIoT technology, many automation companies are using MEMS accelerometers to build vibration sensors; they allow for vibration measurement and direct evaluation of vibration signals thanks to an integrated ADC and microcontroller [25–27].

These types of sensors have a digital output; thus, they can be called digital vibration sensors.

A digital vibration sensor can be considered as such if it has at least one of the following features:

- ADC converter and microcontroller.
- The ability to process and analyse measured signals.
- Ability to linearization of processing characteristics.
- Digital two-way communication interface.
- Self-test and auto calibration unit.
- The ability to learn and make independent decisions.

In order to be able to implement these functions, integrated in the sensor are a measuring transducer, a conditioning system, a microprocessor, and a communication interface which provides an estimation of vibration parameters and transfers it to the sensor registers;

it can be read using popular data exchange protocols in automation, such as Modus RTU or IO Link. The use of this type of solution simplifies the design and implementation of continuous monitoring and diagnostic systems and facilitates data transfer to predictive and cloud systems. (Figure 3).

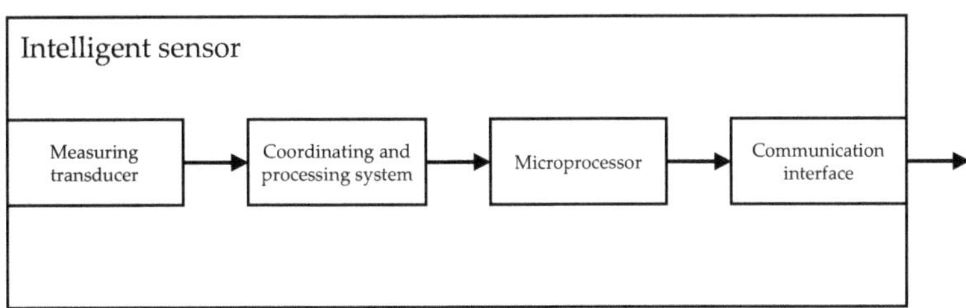

Figure 3. A block diagram of the digital accelerometer.

Commercial vibration sensors based on MEMS accelerometers have been available on the market for some time now, allowing for the measurement of a whole range of numerical parameters useful in machine diagnostics. Table 2 provides a comparison of the parameters of exemplary digital vibration sensors, while Table 3 summarizes the diagnostic parameters determined by the sensors and provided by the digital interface [25–27].

Table 2. Comparison of parameters of exemplary digital vibration sensors based on MEMS accelerometers.

	Balluff BCM0002	**Banner QM30VT2**	**Sick MPB10**
Number of axes	3	2	3
Measuring range	±16 g	N/A	±8 g
Measuring range vRMS	N/A	0–46 mm/s	0–100 mm/s (at 88 Hz)
Frequency range	2–2500 Hz	10–4000 Hz	0.78–3200 Hz
Accuracy	±10% (2–1800 Hz) ±3 dB (2–2500 Hz)	±10% (at 25 °C)	±6%
Interface	IO-Link 1.1	RS-485 (Modbus RTU)	IO-Link 1.1
Operating temperature	−25 to +70 °C	−40 to +105 °C	−40 to +80 °C

Table 3. Comparison of diagnostic parameters estimated on vibration signals by digital vibration sensors.

Balluff BCM0003	**Banner QM30VT2**		**Sick MPB10**
	10–1000 Hz	**1000–4000 Hz**	
RMS Peak to Peak Max Kurtosis Crest Factor Skewness	vRMS (mm/s) vPeak (mm/s) aRMS (G) vPeak Component Frequency (Hz) Simplified Order Spectrum	aRMS (G) aPeak (G) Kurtosis Crest Factor	aRMS vRMS Variance Skewness Peak to Peak Shape factor Crest factor Impulse factor FFT spectrum analysis

It is noticeable that the sensors make available the classic vibration parameters for assessing the overall condition of the machines based on the, e.g., ISO 20816 [28] standard. These include rms vibration velocity amplitudes (vRMS) measured in the 10–1000 Hz band. There are also standard parameters used in evaluating the condition of bearings, such as rms amplitudes (aRMS), peak (aPeak), and peak-to-peak amplitude (aPP) of vibration accelerations, which, depending on the sensor, are determined in the full available frequency band or can be determined in the high-frequency band above 1000 kHz. When analyzing the parameters of the above-presented vibration sensors, the frequency band does not exceed 4 kHz. Considering how bearing faults occur and how vibration signals are emitted, this range may not be sufficient in some cases.

Despite the low price of the aforementioned sensors and the simplicity of their implementation, the question arises as to how the vibration parameters determined by MEMS accelerometers are sensitive to bearing damage at different levels of severity. This paper attempts to answer this question by presenting the results of diagnostic sensitivity estimation for rolling element bearing measurements performed with the use of the two commercially available sensors with embedded MEMS accelerometers.

1.5. Novelty of This Research

The novelty of this study lies in its precise evaluation of the diagnostic sensitivity of digital vibration sensors based on MEMS accelerometers in detecting rolling bearing faults. Compared to previous studies, this approach stands out by directly comparing these sensors with traditional piezoelectric accelerometers in controlled laboratory conditions. A key finding is the effectiveness of MEMS sensors in detecting typical fatigue faults in bearings despite their lower-frequency bandwidth, making them a cost-effective alternative to piezoelectric vibration sensors.

2. Materials and Methods

2.1. The Test Bench and Experiment Description

In order to assess the sensitivity of the diagnostic parameters determined by modern MEMS-based digital vibration sensors to bearing faults of different intensities and under different operating conditions, a series of active diagnostic experiments were planned and carried out on a test rig located at the Department of Fundamentals of Machinery Design of the Silesian University of Technology in Gliwice. The test stand consisted of a drive motor and a motor controller, allowing for rotational speed change; a bearing housing for mounting the tested bearings; and a loading system, allowing for radial load application to the tested bearing. The test rig was equipped with measurement systems to measure bearing housing vibrations using two commercial digital vibration sensors (SE1 and SE2) (see Table 2, items 1 and 2) and a piezoelectric accelerometer connected to an industrial programmable signal processing module. A PCB T352C34 (PCB Piezotronics, Inc., Depew, NY, USA) miniature piezoelectric accelerometer with a sensitivity of 100 mV/g and frequency range 0.5–10,000 Hz was used. A piezoelectric accelerometer was applied to collect reference data, which were used for comparison with data from the digital sensors being evaluated. The first tested sensor (SE1) was connected to the manufacturer's dedicated measurement and data acquisition module, interfacing with the PC via a web browser. The second tested digital vibration sensor (SE2) was connected to a PC using a dedicated RS485-to-USB serial transmission converter. A script written in the MATLAB R2020b environment was used to acquire data from SE2. The piezoelectric vibration sensor was interfaced with processing electronics, also connected to a PC, which was equipped with dedicated software. All the sensors were mounted using a magnet holder. Figure 4 presents a diagram of the laboratory stand.

Figure 5 presents the experimental setup used for the research.

The tests were carried out on a set of 11 deep-groove ball bearings with polymer cage type 6303, 6 of which were brand new bearings that were considered to be in perfect condition. The new bearings were assigned identifiers N1–N6. The technical conditions of the remaining 5 bearings are characterized in Table 4.

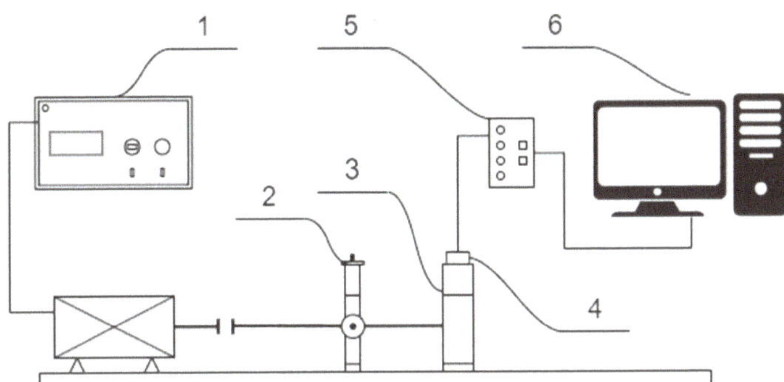

Figure 4. The test bench: 1—engine rotation speed controller; 2—radial load; 3—bearing housing; 4—vibration sensor with magnet holder; 5—processing electronics; 6—PC.

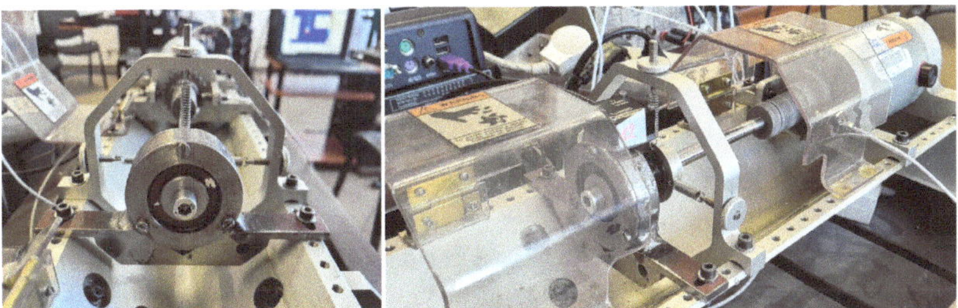

Figure 5. Experimental setup.

Table 4. Condition classification of investigated bearings.

Bearing ID	Condition Characterization
D1	Damaged outer race
D2	Damaged inner race
D3	Damaged rolling element
D4	Bearing cage damaged (1 crack)
D4_2	Bearing cage damaged (2 cracks)
D4_3	Bearing cage damaged (3 cracks)
D4_4	Bearing cage missing
D5	No bearing lubricant

For bearings D1–D4, the damage was introduced manually. In the case of bearing D4, progressive cage damage was simulated between measurements by cutting through the cage at selected points to finally remove it completely. For each bearing, vibration measurements were taken at three shaft speeds, 600 rpm, 1500 rpm, and 3000 rpm, and each was loaded with the same radial force.

For the piezoelectric sensor, the raw acceleration signal was recorded at a sampling rate of 100 kSamples/s for a period of 10 s in the full frequency range of 2–10,000 Hz. The collected signals were subjected to processing and analysis. Processing consisted of bandpass filtering in bands 10–10,000 Hz and 1000–10,000 Hz. The processed acceleration signals were segmented into time sub-realizations of 1 s duration and then analyzed to determine diagnostic parameters corresponding to those determined by the digital vibration sensors tested. Processing and analysis of the acceleration signals from the piezoelectric sensor were carried out in the Python computational environment.

2.2. Method of Evaluation of the Diagnostic Sensitivity of Investigated Digital Vibration Sensors

For the purposes of the described research, diagnostic sensitivity can be defined as a quantitative measurement of the relative change in the value of a diagnostic signal feature due to a small change in the technical condition of the diagnosed object [29–31]. It can be assumed that if a small change in the state causes a significant relative change in the value of the diagnostic parameter, we can speak of the parameter's sensitivity to change. It can be assumed that the technical condition against which the changes of condition will be determined will be some reference condition; e.g., in the case of bearings, this is the good condition, characterized by the bearing at the beginning of operation. Therefore, assuming that the value of a specific characteristic of the diagnostic signal will be a measure of the current condition, the sensitivity measure S_p can be defined according to the following equation:

$$S_p(c_i, t_j) = \left| 1 - \frac{p(c_i, t_j)}{p(c_0, t_0)} \right| \cdot 100\%, \qquad (8)$$

where c_i is the value of the characteristic technical condition at the moment of time t_j, c_0 is the reference value of the characteristic technical condition, and t_0 is the beginning of the object's operation. In the considerations, the moment of time t_j should be considered in the sense of operating time counted in hours, days, or months. In this case, the feature value can be estimated from the vibration signal over a short integration time interval, counted in milliseconds or seconds. It can be considered that if the sensitivity value meets the following condition, i.e., $S_p \geq 25\%$, the change in the value of the characteristic is significant.

Due to the fact that two digital sensors and one piezoelectric sensor were used for the tests, we decided to compare the diagnostic sensitivities of the sensors in such a way that the differences between the sensitivities between the piezoelectric sensor and each of the digital sensors could be determined. For this purpose, a measurement of differential sensitivity (DS_{SE}), defined as follows, was introduced:

$$DS_{SE} = S_{P_n}^{SE_{ref}} - S_{P_n}^{SE_i}, \qquad (9)$$

where S_{P_n} is the diagnostic sensitivity of a given signal characteristic, SE_{ref} is the reference sensor, and SE_i is the sensor under test. It can be assumed that if the value of DS_{SE} meets the following condition, i.e., $DS_{SE} \geq 25\%$, then the difference in the way the value of the characteristic is estimated by the sensor under test is significant.

The data collected during the experiments was processed, ordered, and analyzed. As the feature values for bearings in good condition will serve as reference values, the quality of the collected feature values of the vibration signals for bearings in good condition was assessed first. The coefficient of variation (CV), determined as follows, was used to assess the quality of the data:

$$CV = \frac{\sigma}{\mu} \cdot 100\%, \qquad (10)$$

where σ is standard deviation and μ is a mean value of signal feature value.

3. Results

The Test Bench and Experiment Description

The CV values determined by Equation (8) are shown in Table 5.

The coefficient of variation for the new bearings in the case of the piezoelectric sensor took values lower than or close to the coefficients of variation of the digital sensors. A deviation can be seen for the kurtosis and CF parameters, where the CV index values were higher for the piezoelectric sensor at rotational speeds of 1500 and 3000. This behavior should be explained by the higher standard deviations resulting from the wider frequency range for which the signal features were determined.

Table 5. Comparison of CV values for the considered vibration signal features estimated for new bearings in perfect condition.

Speed	600			1500			3000		
Feature Name	**SEref**	**SE1**	**SE2**	**SEref**	**SE1**	**SE2**	**SEref**	**SE1**	**SE2**
aRMS	3.25	4.66	-	3.51	7.52	-	2.7	6.22	-
aRMS HF	3.18	-	8.05	3.43	-	7.61	2.82	-	9.18
aPeak HF	14.19	-	19.96	15.56	-	14.6	15.42	-	15.86
aPP	14.48	11.15	-	14.50	13.02	-	16.34	9.12	-
K	9.11	12.51	19.92	7.59	7.53	9.48	17.49	5.33	6.39
CF	12.83	10.85	15.16	13.74	9.17	12.32	14.07	8.48	10.88

For each of the determined features of the vibration signals, sensitivity values were determined, which are summarized in Figure 6 for sensor SE1 and Figure 7 for sensor SE2.

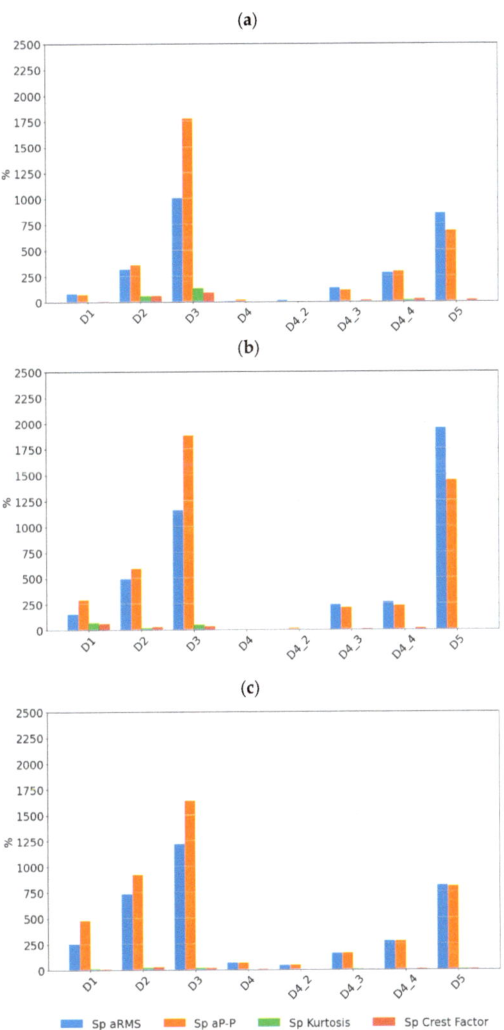

Figure 6. Sensitivity value of features for sensor SE1 at speeds of (**a**) 600 rpm, (**b**) 1500 rpm, and (**c**) 3000 rpm.

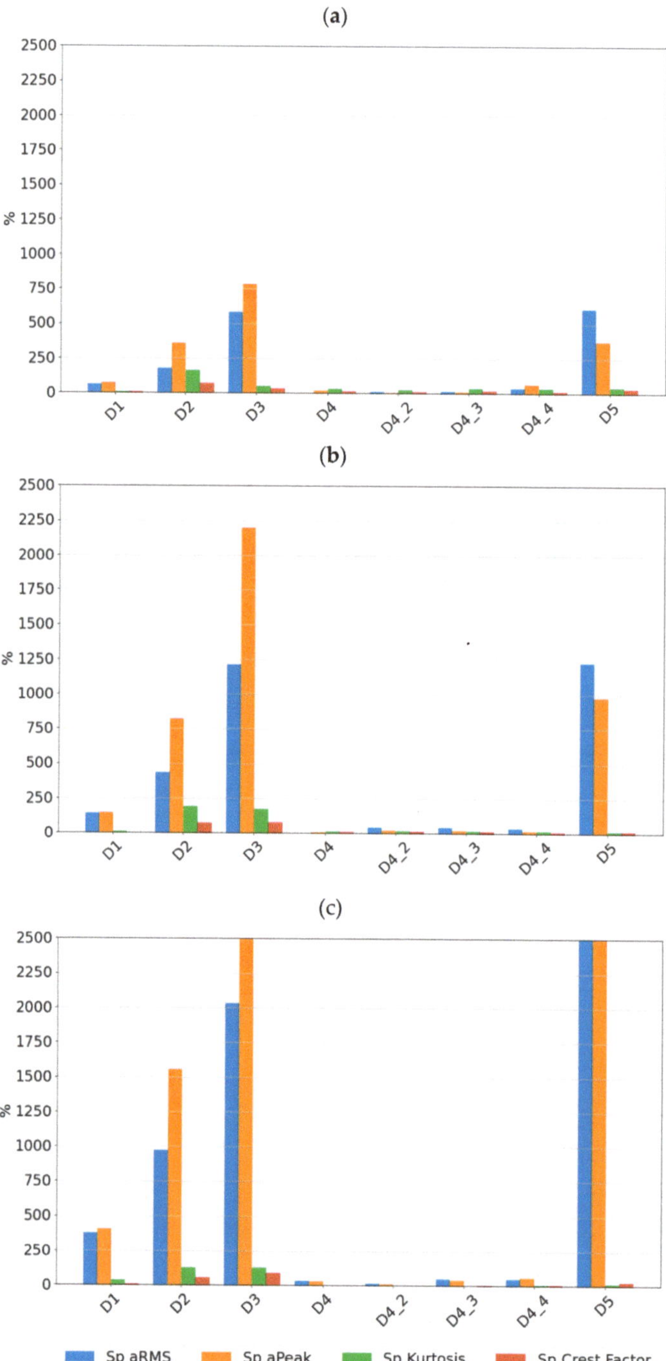

Figure 7. Sensitivity value of features for sensor SE2 at speeds of (**a**) 600 rpm, (**b**) 1500 rpm, and (**c**) 3000 rpm.

For the piezoelectric accelerometer, the sensitivity values are presented in Figure 8.

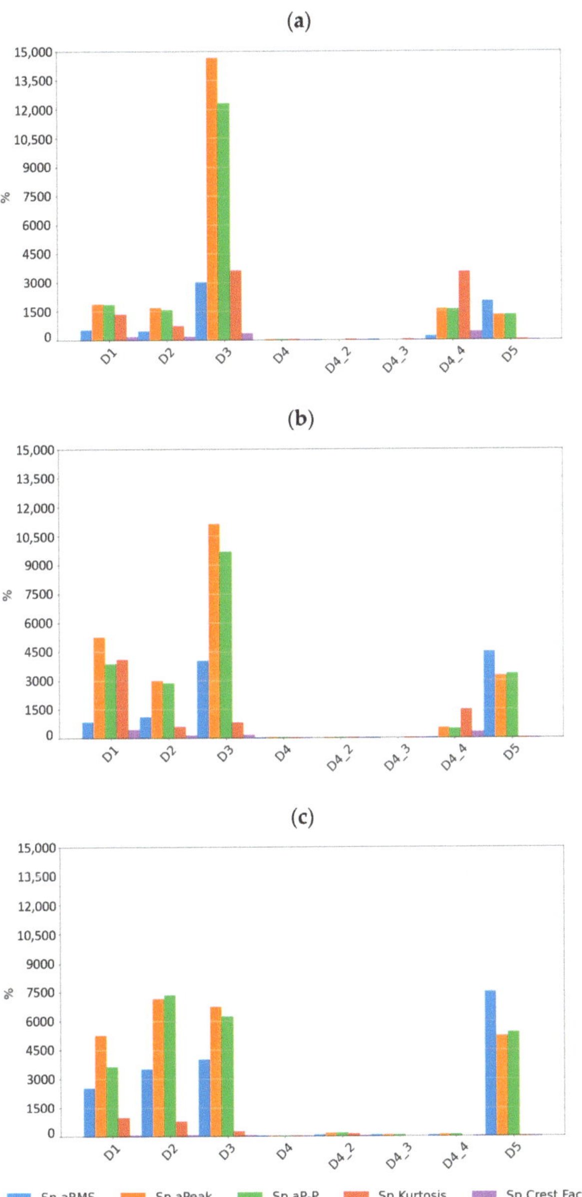

Figure 8. Sensitivity value of features for sensor SEref at speeds of (**a**) 600 rpm, (**b**) 1500 rpm, and (**c**) 3000 rpm.

By analyzing the sensitivity values determined, it can be seen that the tested sensors SE1 and SE2 show very high sensitivity to typical bearing damages. Both sensors showed the highest sensitivity to the rolling element defect, followed by damage to the inner race, damage to the outer race, and lack of lubrication. From a diagnostic parameters point of view, peak and peak-to-peak values showed the highest sensitivity, although, in the absence of lubrication, the rms amplitude showed the highest sensitivity. It is worth noting that in the non-lubricated case, the sensitivity increased with rotational speed and the highest values were achieved for the features determined from the high-pass filtered signal.

For all features for the sensors considered, the lowest sensitivity was observed for cage damage. This is a specific kind of damage manifesting itself with different symptoms depending on its intensity, the design of the cage, and the material from which it is made. From a spectrum analysis perspective, the cage damage manifests itself with a frequency component equal to $0.4f_n$, where f_n is the rotational speed frequency. This makes it possible to see that the high-pass filtering of the signal in this case can make the detection of this damage more difficult. This phenomenon is apparent if we compare the feature values of the SE1 sensor, which were determined for the full sensor frequency range, and the feature values of the SE2 sensor, where were determined for the 1000–4000 Hz range.

Figures 9–11 show the sensitivity plots of the feature values for the different stages of cage damage considered.

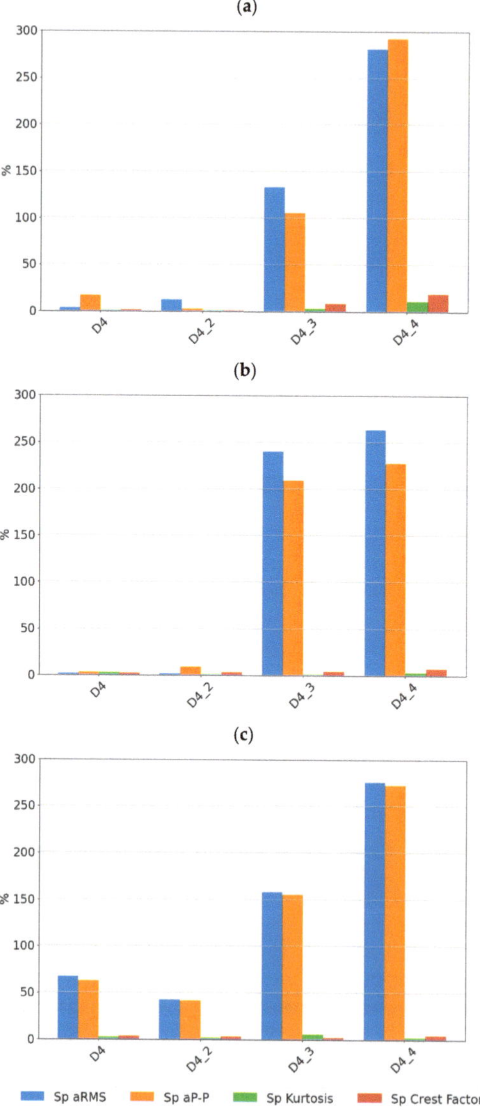

Figure 9. Sensitivity values for different stages of cage damage for sensor SE1 at speeds of (**a**) 600 rpm, (**b**) 1500 rpm, and (**c**) 3000 rpm.

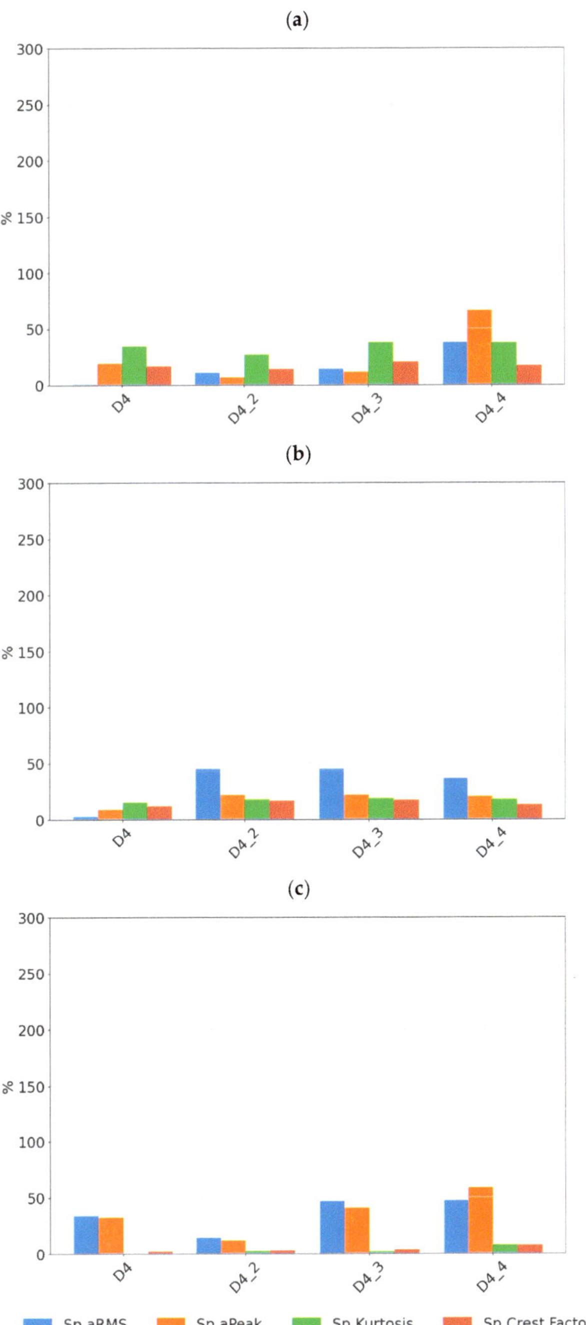

Figure 10. Sensitivity values for different stages of cage damage for sensor SE2 at speeds of (**a**) 600 rpm, (**b**) 1500 rpm, and (**c**) 3000 rpm.

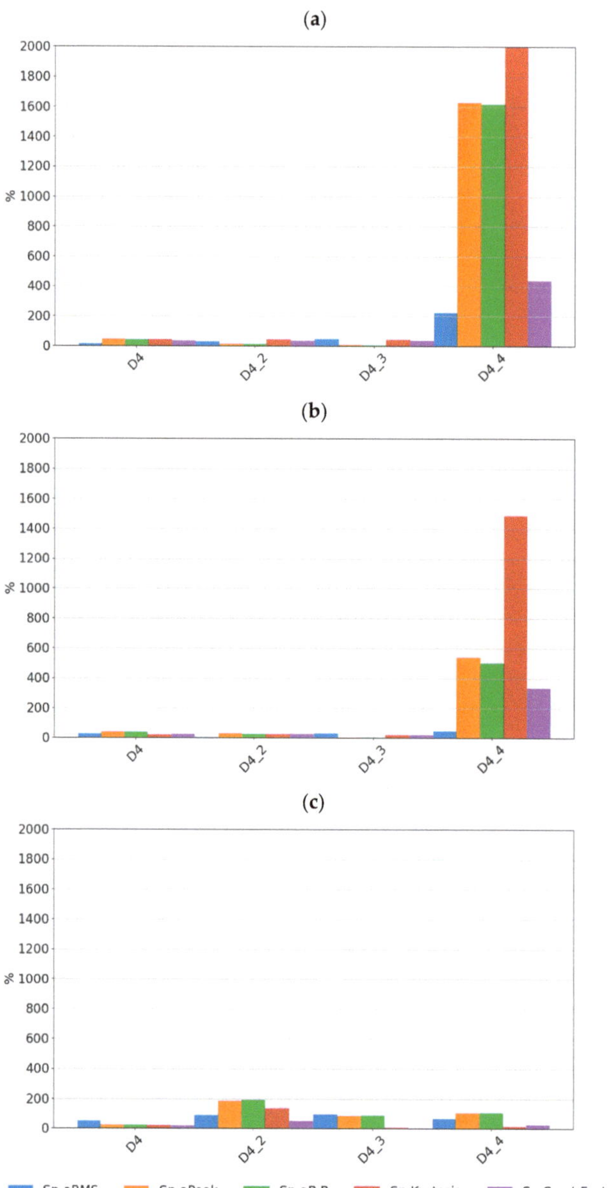

Figure 11. Sensitivity values for different stages of cage damage for sensor SEref at speeds of (**a**) 600 rpm, (**b**) 1500 rpm, and (**c**) 3000 rpm.

It can be seen that for low-intensity cage damage (D4—single break; D4_2—two breaks), the sensitivity for both sensors does not exceed 100%. In contrast, high-intensity cage damage is best detected on the basis of peak and rms amplitudes determined over the full frequency range. The peak and peak-to-peak values of the accelerations are also a diagnostic parameter that characterizes this type of damage well.

From the sensitivity values point of view, it was observed that the sensitivity increases with increasing speed, which is the expected effect for bearings, but in the case of cage damage, no significant increase in sensitivity with speed was observed.

If one relates the values of the sensitivity of the features of the vibration signals of the tested sensors SE1 and SE2 to the features calculated for signals from the piezoelectric sensor by analyzing the values of the differential sensitivity SD presented in Figures 12 and 13, it is easy to see that in most cases, the values are positive, and for typical bearing damage, the values are very high, which indicates that in the case of rolling element bearings, the determination of the signal features in a wide frequency range allows for the early detection of typical defects related to the material fatigue of the races and rolling elements. In the case of a lack of lubrication and cage damage, the results do not clearly indicate an advantage for the piezo sensor; in which case, for example, sensor SE2 had better sensitivity to a lack of lubrication and sensor SE1 greater sensitivity to intensive cage damage at 3000 rpm.

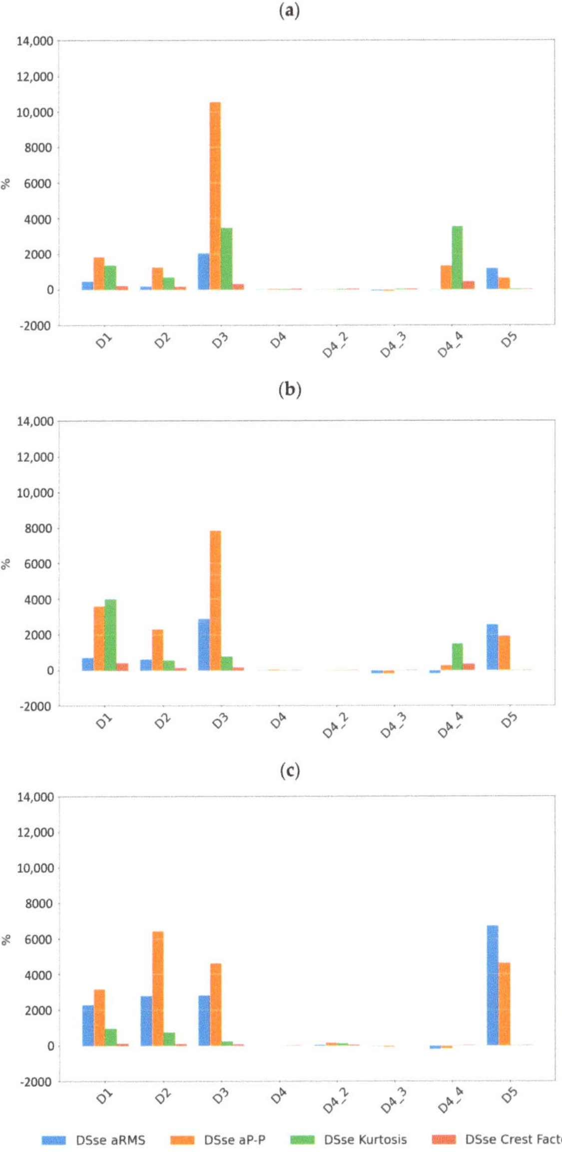

Figure 12. Differential sensitivity value for sensor SE1 at speeds of (**a**) 600 rpm, (**b**) 1500 rpm, and (**c**) 3000 rpm.

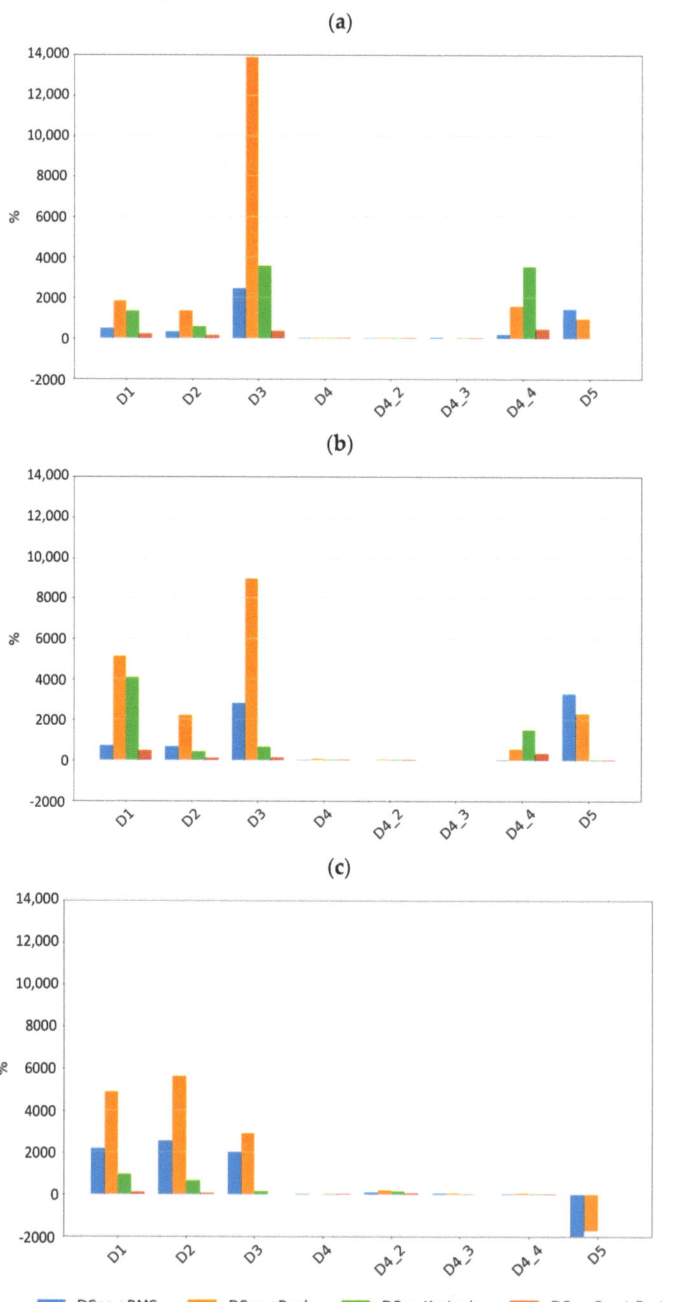

Figure 13. Differential sensitivity value for sensor SE2 at speeds of (**a**) 600 rpm, (**b**) 1500 rpm, and (**c**) 3000 rpm.

When comparing the differences in the sensitivity of the signal features between the piezoelectric sensor and the digital sensors, it was noted that for the piezoelectric sensor, kurtosis was a much more sensitive parameter. In the case of typical bearing failures, kurtosis did not show high sensitivity in the case of the SE1 and SE2 sensors, which may be due to the way in which it is estimated, which is not completely known as far as the SE1 and SE2 sensors are concerned.

4. Conclusions

On the basis of the research carried out, it can be concluded that commercial digital vibration sensors are sensitive to basic rolling bearing damage of medium and high intensity, which makes it possible to detect bearing faults and prevent unexpected machine failures, provided that the correct warning and alarm thresholds are set in condition monitoring systems and that maintenance services respond correctly to the emergence of alarm signals. The sensitivity of sensor bearing fault detection is dependent on the frequency band. The higher the sensor's processing capabilities in the higher frequency range, the higher the sensitivity to even low-intensity damage increases, as can be seen from a comparison of the two sensors SE1 and SE2, in which the diagnostic parameters of the vibration signals were determined in the bands 2–3500 Hz and 1000–4000 Hz, respectively. This fact is also confirmed by comparing the sensitivity of the signal features of digital sensors with a piezoelectric sensor, whose upper frequency of the measurement range was 10,000 Hz.

It can also be argued that the high diagnostic sensitivity values for the piezoelectric sensor are due to the nature of the sensor's operation; however, at this stage of the research, it is not possible to state unequivocally that the use of a piezoelectric transducer increases the diagnostic sensitivity to a decisive degree compared to MEMS capacitive accelerometers. This would require a comparison of accelerometers in similar processing bands, which will be the subject of the authors' further research.

Research shows that diagnostic sensitivity depends on the frequency band as well as the type of damage. It can be assumed that these two factors determine the ability of sensors to detect various bearing defects. To investigate these relationships, it is necessary to conduct broader studies on a wider statistical sample and an expanded number of vibration sensors. This will be the subject of further research by the authors.

Author Contributions: Conceptualization, M.F.; methodology, M.F.; software, D.A.; validation, M.F., D.A., J.O. and W.U.; formal analysis, M.F. and D.A.; investigation, D.A.; resources, M.F. and D.A.; data curation, M.F. and D.A.; writing—original draft preparation, M.F.; writing—review and editing, M.F., D.A. and J.O.; visualization, D.A.; supervision, M.F.; project administration, M.F. and D.A.; funding acquisition, M.F. All authors have read and agreed to the published version of the manuscript.

Funding: The research was financed by Silesian University of Technology statutory research funds.

Institutional Review Board Statement: Not applicable.

Informed Consent Statement: Not applicable.

Data Availability Statement: The data presented in this study are available on request from the corresponding author.

Conflicts of Interest: The authors declare no conflicts of interest.

References

1. Howard, I. *A Review of Rolling Element Bearing Vibration: Detection, Diagnosis and Prognosis*; Department of Defense: Sydney, Australia, 1994.
2. Randall, R.B.; Antoni, J. Rolling Element Bearing Diagnostics—A Tutorial. *Mech. Syst. Signal Process.* **2011**, *25*, 485–520. [CrossRef]
3. Mathew, J.; Alfredson, R.J. The Condition Monitoring of Rolling Element Bearings Using Vibration Analysis. *J. Vib. Acoust.* **1984**, *106*, 447–453. [CrossRef]
4. Ghazaly, N.; Samy, A.; Mousa, M.O. Bearing Fault Detection Techniques—A Review. *Turk. J. Eng. Sci. Technol.* **2015**, *3*, 1–18.
5. *ISO 15243:2017*; Rolling Bearings. Damage and Failures. Terms, Characteristics and Causes. International Standards Organization: Geneva, Switzerland, 2017.
6. Niu, W.; Fang, L.; Xu, L.; Li, X.; Huo, R.; Guo, D.; Qi, Z. Summary of Research Status and Application of MEMS Accelerometers. *J. Comput. Commun.* **2018**, *6*, 215–221. [CrossRef]
7. SPM Instrument AB. Bearing Condition Monitoring: Case Studies. Available online: https://www.spminstrument.com/ (accessed on 5 June 2024).
8. Sundström, T. *An Introduction to the SPM HD Method, R&D*; SPM Instrument AB: Strängnäs, Sweden, 2010.

9.   Adash. DDS Software Manual. 2023. Available online: https://adash.com/ (accessed on 5 June 2024).

10. *ISO 13373-3:2015*; Condition Monitoring and Diagnostics of Machines Vibration Condition Monitoring Part 3: Guidelines for Vibration Diagnosis. International Standards Organization: Geneva, Switzerland, 2015.

11. *VDI 3832:2013*; Measurement of Structure-Borne Sound of Rolling Element Bearings in Machines and Plants for Evaluation of Condition. Verein Deutscher Ingenieure: Düsseldorf, Germany, 2013.

12. Tandon, N.; Choudhury, A. A Review of Vibration and Acoustic Measurement Methods for the Detection of Defects in Rolling Element Bearings. *Tribol. Int.* **1999**, *32*, 469–480. [CrossRef]

13. Murphy, C. Choosing the Most Suitable Predictive Maintenance Sensor. Analog Devices. 2020. Available online: https://www.analog.com/en/resources/technical-articles/choosing-the-most-suitable-predictive (accessed on 5 June 2024).

14. Murphy, C. Why MEMS Accelerometers Are Becoming the Designer's Best Choice for CbM Applications. Analog Devices. 2021. Available online: https://www.analog.com/en/resources/technical-articles/why-memes-acceler-are-best-choice-for-cbm-apps.html (accessed on 5 June 2024).

15. Landi, E.; Prato, A.; Fort, A.; Mugnaini, M.; Vignoli, V.; Facello, A.; Mazzoleni, F.; Murgia, M.; Schiavi, A. Highly Reliable Multicomponent MEMS Sensor for Predictive Maintenance Management of Rolling Bearings. *Micromachines* **2023**, *14*, 376. [CrossRef] [PubMed]

16. Staszewski, W.; Jablonski, A.; Dziedziech, K.; Barszcz, T. Modelling and testing of MEMS based vibration sensor for rolling element bearing fault detection. In Proceedings of the Diagnostyka Maszyn: XLVII Ogólnopolskie Sympozjum, Wisła, Poland, 1–5 March 2020.

17. Zusman, G. The Comparison of Piezoelectric and MEM Sensors Intended for Vibration Condition Machinery Monitoring. In Proceedings of the Latin American Workshop on Structural Health Monitoring, Cartagena de Indias, Colombia, 5–7 December 2023; CIMNE: Barcelona, Spain, 2024.

18. Rossi, A.; Bocchetta, G.; Botta, F.; Scorza, A. Accuracy Characterization of a MEMS Accelerometer for Vibration Monitoring in a Rotating Framework. *Appl. Sci.* **2023**, *13*, 5070. [CrossRef]

19. Varanis, M.; Silva, A.; Mereles, A.; Pederiva, R. MEMS Accelerometers for Mechanical Vibrations Analysis: A Comprehensive Review with Applications. *J. Braz. Soc. Mech. Sci. Eng.* **2018**, *40*, 527. [CrossRef]

20. Augustyn, D.; Fidali, M. Identification and Evaluation of the Characteristics of a Selected Commercial Mems Based Vibration Sensor for the Machine Condition Monitoring. *Diagnostyka* **2023**, *24*, 1–8. [CrossRef]

21. *ISO 10816-8:2014*; Mechanical Vibration — Evaluation of machine vibration by measurements on non-rotating parts. International Standards Organization: Geneva, Switzerland.

22. *ISO 20816-1:2016*; Mechanical Vibration—Measurement and Evaluation of Machine Vibration. International Standards Organization: Geneva, Switzerland.

23. Anslow, R. *How to Design a Good Vibration Sensor Enclosure Using Modal Analysis*; Analog Devices: Wilmington, MA, USA, 2022.

24. Albarbar, A.; Mekid, S.; Starr, A.; Pietruszkiewicz, R. Suitability of MEMS Accelerometers for Condition Monitoring: An Experimental Study. *Sensors* **2008**, *8*, 784–799. [CrossRef]

25. Sick AG. Condition Monitoring Sensors for Vibration, Shock, and Temperature Monitoring (Technical Documentation). Available online: https://www.sick.com/ (accessed on 5 June 2024).

26. Balluff GmbH. Condition Monitoring Sensor with Integrated Data Preprocessing, (Technical Documentation). Available online: https://www.balluff.com/ (accessed on 5 June 2024).

27. Banner Engineering Corp. Sure Cross® QM30VT2 Vibration and Temperature Sensor (Technical Documentation). Available online: https://www.bannerengineering.com/ (accessed on 5 June 2024).

28. *ISO 20816-3:2022*; Mechanical Vibration—Measurement and Evaluation of Machine Vibration Part 3: Industrial Machinery with a Power Rating above 15 kW and Operating Speeds between 120 r/Min and 30 000 r/Min. International Standards Organization: Geneva, Switzerland, 2022.

29. Wachla, D. Badanie Wrażliwości Diagnostycznej Cech Sygnałów Wibroakustycznych Przekładni Zębatych w Dziedzinie Czasu. In *Methods of Artificial Intelligence in Mechanics and Mechanical Engineering*; Silesian University of Technology: Gliwice, Poland, 2000; pp. 319–324.

30. Wojtusik, J. Wrażliwość Diagnostyczna Wzajemnych Cech Par Sygnałów Wibroakustycznych w Ujęciu Metod Klasyfikacji i Grupowania. In *Methods of Artificial Intelligence in Mechanics and Mechanical Engineering*; Silesian University of Technology: Gliwice, Poland, 2000; pp. 363–366.

31. Cholewa, W.; Cholewa, W.; Kiciński, J. *Diagnostyka Techniczna: Metody Odwracania Nieliniowych Modeli Obiektów*; Silesian University of Technology: Gliwice, Poland, 2001; ISBN 978-83-906533-8-9.

Article

A Deep Learning Method for Bearing Cross-Domain Fault Diagnostics Based on the Standard Envelope Spectrum

Lubin Zhai [1], Xiufeng Wang [1,*], Zeyiwen Si [2] and Zedong Wang [1]

[1] College of Mechanical Engineering, Xi'an Jiaotong University, Xi'an 710049, China; zlb_5200@stu.xjtu.edu.cn (L.Z.); 2860245534@qm sjhvej@163.com (Z.W.)

[2] School of Mathematics, University of Bristol, Bristol BS8 1QU, UK; iy22624@bristol.ac.uk

* Correspondence: wangxiufeng@xjtu.edu.cn; Tel.: +86-13259779741

Abstract: Intelligent fault diagnostics based on deep learning provides a favorable guarantee for the reliable operation of equipment, but a trained deep learning model generally has low prediction accuracy in cross-domain diagnostics. To solve this problem, a deep learning fault diagnosis method based on the reconstructed envelope spectrum is proposed to improve the ability of rolling bearing cross-domain fault diagnostics in this paper. First, based on the envelope spectrum morphology of rolling bearing failures, a standard envelope spectrum is constructed that reveals the unique characteristics of different bearing health states and eliminates the differences between domains due to different bearing speeds and bearing models. Then, a fault diagnosis model was constructed using a convolutional neural network to learn features and complete fault classification. Finally, using two publicly available bearing data sets and one bearing data set obtained by self-experimentation, the proposed method is applied to the data of the fault diagnostics of rolling bearings under different rotational speeds and different bearing types. The experimental results show that, compared with some popular feature extraction methods, the proposed method can achieve high diagnostic accuracy with data at different rotational speeds and different bearing types, and it is an effective method for solving the problem with cross-domain fault diagnostics for rolling bearings.

Keywords: rolling bearings; cross-domain fault diagnostics; standardized envelope spectrum; convolutional neural networks

Citation: Zhai, L.; Wang, X.; Si, Z.; Wang, Z. A Deep Learning Method for Bearing Cross-Domain Fault Diagnostics Based on the Standard Envelope Spectrum. *Sensors* **2024**, *24*, 3500. https://doi.org/10.3390/s24113500

Academic Editor: Steven Chatterton

Received: 11 April 2024
Revised: 15 May 2024
Accepted: 23 May 2024
Published: 29 May 2024

1. Introduction

Rolling bearings are an important part of rotating machinery and equipment, and they are easily damaged under long time operation and bad working conditions. Sudden failure will affect the normal operation of the equipment, resulting in economic losses and even casualties [1]. Therefore, it is of great importance to monitor and diagnose the operating condition of rolling bearings.

Signal processing and intelligent diagnostic methods have been successively applied to rolling bearing fault diagnostics, which has attracted the attention of a large number of scholars due to the fact that intelligent diagnostic methods do not require specialized technicians. Intelligent diagnostic methods usually include two steps: feature extraction and fault classification. Feature extraction is a signal processing method based on the time domain (TD), frequency domain (FD), and time–frequency domain (TFD) [2,3] to extract feature indicators that can characterize the health state. Traditional machine learning methods, such as support vector machines (SVM) [4], principal component analysis (PCA) [5], and artificial neural networks (ANN) [6], have been widely used for fault classification in intelligent diagnostics. However, due to their shallow architectures, they have difficulty learning effective features from raw signals, and the diagnostic performance relies on expert a priori knowledge and signal analysis tools.

Compared with traditional machine learning, deep learning models can automatically discover deep features in the original signals, and have received extensive attention and research in recent years. Deep learning algorithms, such as the convolutional neural networks (CNN) [7], the recurrent neural networks (RNN) [8], th deep autoencoder (DAE) [9] and the deep belief network (DBN) [10], have been successfully applied to bearing fault diagnostics.

CNN is a typical deep learning algorithm with a strong local feature extraction capability and performs well in image recognition and classification tasks. Hoang et al. [11] converted one-dimensional time-domain vibration signals into two-dimensional vibration images and then used CNN for vibration image classification to identify bearing faults. Zhang et al. [12] proposed an improved CNN model using time–frequency images as input for bearing fault diagnosis that is highly adaptable to workload variations. Hasan et al. [13] fused the multi-domain information of raw bearing vibration signals into a two-dimensional composite image, and then fed the composite image into a multi-task learning (MTL)-based CNN model for fault diagnostics, which was able to accurately detect faults in the presence of simultaneous changes in speed and health conditions. Chen et al. [14] used cyclic spectral analysis to construct a frequency domain graph as an input to CNN to reveal the hidden periodic behavior of each fault type in bearing vibration signals, which reduces the difficulties with feature learning in deep diagnostic models. Sobie et al. [15] sequentially performed envelope extraction, simultaneous angular domain averaging, and normalization of the bearing vibration signals before feeding them into a CNN, which allows the fault classification of experimental data with different shaft speeds and bearing geometries.

Although the CNN-based rolling bearing intelligent diagnostic methods have achieved remarkable results, there are still some problems:

(1) Although some feature extraction methods based on the time domain, frequency domain, and time–frequency domain have been applied to intelligent diagnostics, these feature extraction methods do not fully consider the a priori knowledge of the local fault characteristics of the bearings, have poor robustness, and are ineffective in cross-domain diagnostics.

(2) Many existing feature extraction methods do not take into account the differences between source and target domains caused by changes in bearing speeds and bearing models, making it difficult for deep learning models to learn common features between domains when performing cross-domain diagnostics, and degrading the diagnostic performance.

To solve the above problems, a standard envelope spectrum is constructed based on the envelope spectrum morphology of rolling bearing faults. The standard envelope spectrum reveals the signal characteristics of rolling bearings that do not vary with rotational speed and model, but only with changes in health state. That is, the difference between the source and target domains due to different speeds and models is eliminated by taking into account the a priori knowledge of experts. A standard sample library reflecting the health state of rolling bearings is established from the existing bearing data set, and fault diagnostics without target domain samples are realized. A convolutional neural network model is constructed to learn the common features between the source and target domains and perform fault classification by taking the standard envelope spectrogram as input.

The main contributions of this paper can be summarized as follows:

(1) A standard envelope spectrum is constructed that reveals the unique characteristics of different bearing health states.

(2) An intelligent diagnostic method based on vibration signals for the SES-CNN of rolling bearings is proposed, which is effective in the cross-domain diagnostics of bearing data with different rotational speeds and different models.

(3) The proposed method focuses on effective feature representation for the cross-domain fault diagnostics of rolling bearings, establishes a standard sample library reflecting the health state of the rolling bearings, and the diagnostic process does not require target domain samples and models of high complexity.

The rest of the paper is organized as follows: in the next section, the relevant theoretical background is introduced. Section 3 analyzes and presents the proposed framework for rolling bearing fault diagnostics. Section 4 describes the datasets used and provides a comprehensive evaluation and comparison of the methods. Finally, some conclusions are given in Section 5.

2. Theoretical Basis

2.1. Envelope Spectrum

When the rolling bearing element surface produces local defects, in the rolling body and the inner and outer ring, mutual motion processes will produce periodic impact vibrations, and the frequency of vibration is called failure characteristic frequency. The failure characteristic frequency depends on the shaft speed and bearing type (geometry). Different component failures correspond to different failure characteristic frequencies. The formulas for the outer ring failure characteristic frequency, inner ring failure characteristic frequency, rolling body failure characteristic frequency, and cage failure characteristic frequency calculation are as follows:

$$f_o = 0.5 z f_r \left(1 - \frac{d}{D} \cos \alpha \right) \tag{1}$$

$$f_i = 0.5 z f_r \left(1 + \frac{d}{D} \cos \alpha \right) \tag{2}$$

$$f_b = \frac{1}{2} \frac{D}{d} f_r \left[1 - \left(\frac{d}{D}\right)^2 \cos^2 \alpha \right] \tag{3}$$

$$f_c = 0.5 f_r \left(1 - \frac{d}{D} \cos \alpha \right) \tag{4}$$

where z is the number of balls, f_r is the rotation frequency, d is the ball diameter, D is the raceway pitch diameter, and α is the contact angle.

Envelope spectrum analysis is an effective method for rolling bearing fault diagnostics [16]. Usually, the original signal undergoes Hilbert demodulation to obtain the envelope signal, and then the envelope signal undergoes a Fourier transform to obtain the envelope spectrum.

Let $x(t)$ represent a vibration signal whose analytical signal is expressed by the Hilbert transform as:

$$\tilde{x}(t) = x(t) + jH\{x(t)\} = x(t) + j\frac{1}{\pi} \int_{-\infty}^{+\infty} \frac{x(\tau)}{t - \tau} d\tau \tag{5}$$

where $H\{\cdot\}$ denotes the Hilbert transform and j is the imaginary unit.

The envelope signal is then obtained from the following equation:

$$Env(t) = |a(t)| = \sqrt{(x(t))^2 + (H\{x(t)\})^2} \tag{6}$$

The envelope spectrum is obtained by applying a Fourier transform to the envelope signal, which is given by the following equation:

$$Es(f) = F\{|a(t)|\} = \int_{-\infty}^{+\infty} |a(t)| e^{-j2\pi f t} dt \tag{7}$$

where $F\{\cdot\}$ denotes the fast Fourier transform (FFT).

The envelope spectrum can effectively reveal the failure characteristic frequency of rolling bearings, and the expected envelope spectrum patterns of different components of rolling bearings when failures occur are shown in Figure 1 [17]. For the outer ring fault, the main frequency components in the envelope spectrum are the outer ring fault characteristic frequency and harmonics; there is no sideband. For the inner ring fault, the main frequency components in the envelope spectrum are the inner ring fault eigenfrequency and harmonics, and the sideband interval is the rotating frequency sideband. For rolling element faults, the main frequency components in the envelope spectrum are the rolling element fault eigenfrequency and harmonics, and the sideband interval is the sideband of the cage fault eigenfrequency.

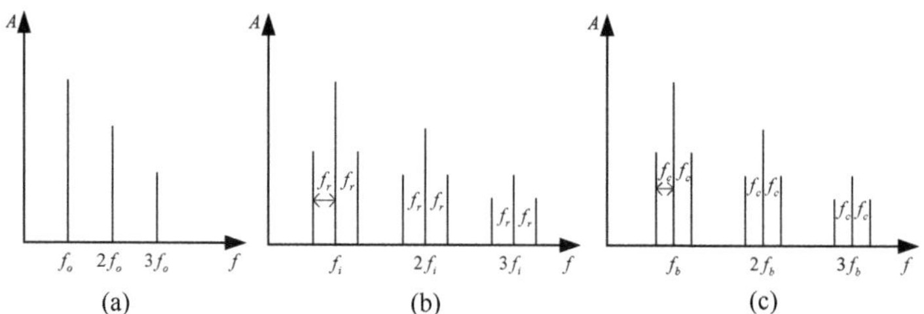

Figure 1. Bearing fault envelope spectrum morphology: (**a**) Outer ring fault; (**b**) Inner ring fault; (**c**) Rolling element fault.

2.2. Convolutional Neural Networks

A convolutional neural network (CNN) is a deep learning model that performs well at visual recognition tasks, such as image classification and target detection. A typical CNN architecture consists of multiple layers, such as convolutional, pooling, and fully connected layers, which perform different functions and are primarily used to extract features from input data and perform classification [18].

The basic block of the CNN feature extraction part is the convolutional layer. The convolutional layer applies a set of convolutional kernels (also known as learnable filters) to the input data, with each kernel extracting specific features. The efficiency of feature extraction is significantly affected by the filter size, the convolutional step size, and the number of filters used. The convolutional layer captures patterns and structures by sliding these filters over the input data. This operation allows the network to learn a hierarchical representation, starting with simple features, such as edges and corners, and gradually evolving to more complex features. The convolution formula is:

$$y_{l,j}^{conv} = \sum_{i=1}^{k} w_{i,j}^{l} * y_{l-1,i}^{pool} + b_{j}^{l} \tag{8}$$

where $y_{l,j}^{conv}$ represents the convolutional value of the jth channel in convolutional layer l, $y_{l-1,i}^{pool}$ represents the ith channel output in pooling layer $l-1$, $w_{i,j}^{l}$ represents the kernel of convolutional layer l, b_{j}^{l} represents the bias of the jth channel in the convolutional layer l, and $*$ represents the convolutional operation.

After the convolution operation is completed, whether or not the neurons in the convolution layer are awakened depends on the activation function. Activation functions such as ReLU (Rectified Linear Units) can introduce nonlinearities, promote sparsity, improve gradient propagation, and enable the network to capture complex patterns, which ultimately improves the overall performance and learning ability of convolutional neural networks:

$$y_{l,j}^{Relu} = f(y_{l,j}^{conv}) = \max[0, y_{l,j}^{conv}] \tag{9}$$

where $y_{l,j}^{Relu}$ represents the jth channel output in the convolutional layer l, and $f(\cdot)$ represents the activation function.

In the feature extraction part, several convolutional layers are always followed by a pooling layer, i.e., a sampling layer. The main purpose of the pooling layer is to reduce the dimensionality of the feature maps and to reduce the number of feature maps, thus reducing the computational complexity. Maximum pooling and average pooling are commonly used pooling operations. The maximum pooling function is as follows:

$$y_{l,j}^{pool} = \max(w(s_1, s_2) \cap y_{l-1,j}^{Relu}) \tag{10}$$

where $w(s_1, s_2)$ represents the pooling window, which can slide with a certain step, s_1 and s_2 correspond to the dimension of the pooling window, $y_{l-1,j}^{Relu}$ represents the jth channel output in the convolutional layer $l - 1$, and $\cap$ represents the overlap between the pooling window and the channel output.

The final part of a CNN usually consists of fully connected layers. These layers act as classifiers, taking the high-level representations extracted from the previous layer and mapping them to the target classes. The output layer is usually a softmax layer that generates class probabilities. The formula for the fully connected layer is as follows:

$$y = \sigma((w_f)^T s_m + b_f) \tag{11}$$

where w_f represents the weight matrix used to connect the two fully connected layers, b_f represents the bias, s_m represents the input data of the fully connected layer, and $\sigma(\cdot)$ represents the activation function in the fully connected layer.

3. The Proposed SES-CNN Model

3.1. Construction of Standard Envelope Spectrum (SES)

The envelope spectra of rolling bearings are characterized differently when localized defects occur in different elements of the bearing. However, the distribution of spectral lines in the envelope spectrogram changes due to the fact that different types of bearings have different failure characteristic frequencies. Similarly, changes in rotational speed change the distribution of spectral lines in the envelope spectra. Therefore, a standard envelope spectrum is constructed, and the spectral lines of the relevant fault characteristic frequencies in the envelope spectrum are fixed at the specified positions, so that the spectrum not only retains the characteristic differences of different component faults, but also eliminates the differences in the distribution of the characteristic spectral lines caused by changes in rotational speed and bearing models, and has the characteristic that it does not change with changes in bearing rotational speed and models, but only changes due to changes in health. The process of constructing the standard envelope spectrum is shown in Figure 2 with the following steps:

Step 1: To improve the signal-to-noise ratio, a fast kurtogram method is used to select the optimal frequency band of the signal, then a bandpass filter is applied to the optimal frequency band to obtain the filtered signal.

Step 2: A Hilbert transform is applied to the filtered signal to obtain the envelope signal.

Step 3: A Fourier transform is applied to the envelope signal to obtain the envelope spectrum.

Step 4: The fault characteristic frequency of each component of the bearing is calculated using Equations (1)–(4). First, take the outer ring fault characteristic frequency and its harmonic nf_o, the inner ring fault characteristic frequency and its harmonic nf_i, and the rolling element fault characteristic frequency and its harmonic nf_b as the search center frequency. Next, set the frequency search range according to the form of $[F(1 - \alpha), F(1 + \alpha)]$, and extract the point with the largest amplitude in the search range of the search center frequency in the envelope spectrum as the spectral peak point. Then, the frequency corresponding to the spectral peak point at the inner ring fault characteristic frequency and its harmonic nf_i is taken as the actual inner ring fault characteristic frequency and its harmonic, and these frequencies are taken as the new search center frequency. The frequency search range is set in the form of $[F - f_r(1 + \alpha), F - f_r(1 - \alpha)]$ and $[F + f_r(1 - \alpha), F + f_r(1 + \alpha)]$, and the point with the largest amplitude in the search range of the search center frequency in the envelope spectrum is extracted as the spectral peak point. Similarly, the frequencies corresponding to the spectral peaks at the characteristic frequencies and harmonics of the rolling body ring faults and their harmonics nf_b are taken as the actual rolling body fault characteristic frequencies and their harmonics. These frequencies are taken as the new search center frequencies, and the frequency search ranges are set in the form of $[F - f_c(1 + \alpha), F - f_c(1 - \alpha)]$ and $[F + f_c(1 - \alpha), F + f_c(1 + \alpha)]$, and the point with the largest amplitude in the search center frequency search range of the envelope spectrum is extracted as the spectral peak point. After the above search, a total of 21 spectral peaks (F_i, a_i) are obtained where $1 \leq n \leq 3$ and $1 \leq i \leq 21$, and both n and i are integers. That is, a total of 21 frequency components are considered, including the bearing outer ring fault, the inner ring fault, the rolling element fault characteristic frequency of the first 3 harmonics, as well as the inner ring fault, and the rolling element fault characteristic frequency of the first 3 harmonics around the first-order modulation sidebands. Due to the influence of factors such as bearing slippage, the bearing fault characteristic frequency calculated by kinematics theory and the actual fault frequency often have errors, so this paper considers the error coefficient of $\alpha = 0.015$.

Step 5: Normalize the amplitude of the previously searched peak points (F_i, a_i) of the spectrum using the following formula:

$$A_i = \frac{a_i}{a_{i\max}} \tag{12}$$

where $1 \leq i \leq 21$ and $a_{i\max}$ is the maximum value in a_i.

Step 6: Establish a new coordinate system where the vertical coordinate range is 0–1 and the horizontal coordinates, from left to right, are the outer ring fault (OF), inner ring fault (IF), rolling element fault (BF), being the three fault characteristics of the region. The normalization of the spectral peak point (F_i, A_i) is placed in the corresponding fault characteristics of the region, in accordance with the order of F_i from small to large, arranged in the corresponding horizontal coordinate position, to generate a new spectral line graph known as the standard envelope spectrum.

Step 7: Save the resulting standard envelope spectrum as a grayscale image in a jpg format with a resolution of $64 \times 64 \ px$.

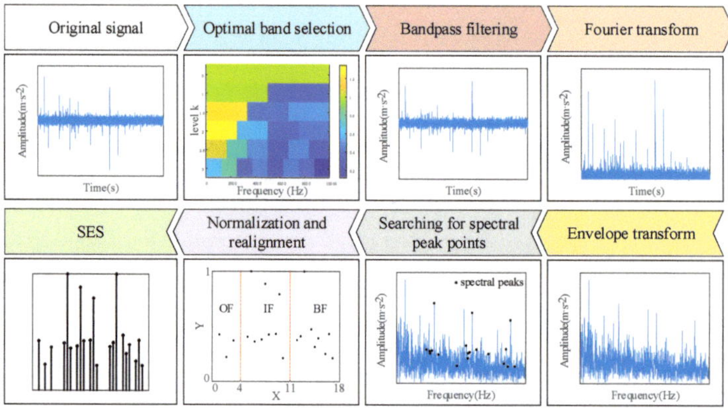

Figure 2. The process of constructing a standard envelope spectrum.

3.2. The Architecture of the Proposed CNN

The CNN model constructed in this paper consists of two convolutional layers, two batch normalization layers, two pooling layers, one spreading layer, and two fully connected layers, where the convolutional layers are used for feature extraction, the pooling layer is used to reduce the spatial dimensionality of the feature maps, the batch normalization layer is used to prevent overfitting and speed up convergence, the spreading layer is used to spread the multidimensional feature maps into one-dimensional vectors, and the fully connected layer is used for classification and output prediction. The size of the convolutional layer filter is (3, 3), the step size is (1, 1), the same filling method is chosen, ReLU is chosen as the activation function, the number of the first convolutional layer filter is 16, and the number of the second convolutional layer filter is 32. Both convolutional layers are followed by a batch normalization layer and a pooling layer where the type of the pooling layer is the maximum pooling layer, the size of the filter is (2, 2), the step size is (2, 2), the size of the filter is (2, 2), and the same padding method is chosen. The output is then flattened by a spreading layer and mapped to the output categories by two fully connected layers, where the number of neurons in the first fully connected layer is 128 and ReLU is chosen as the activation function, and the number of neurons in the second fully connected layer is four and softmax is chosen as the activation function. The model is trained using the ADAM optimizer, the cross-entropy loss is used as the loss function, and the accuracy and loss value are used as metrics to evaluate the training performance of the network model. The details of the layer type and the parameters used are shown in Table 1.

Table 1. Parameters of the proposed CNN architecture.

Layer	Layer Types	Kernel	Number of Filters	Filter Size	Stride	Output Size	Activation Function
1	Input					$64 \times 64 \times 1$	
2	Conv	Kernel	16	3×3	(1,1)	$64 \times 64 \times 16$	ReLU
3	BN					$64 \times 64 \times 16$	
4	MaxPool	Pooling size	16	2×2	(2,2)	$32 \times 32 \times 16$	ReLU
5	Conv	Kernel	32	3×3	(1,1)	$32 \times 32 \times 32$	
6	BN					$32 \times 32 \times 32$	ReLU
7	MaxPool	Pooling size	32	2×2	(2,2)	$16 \times 16 \times 32$	
8	Flatten					8192	
9	FC					128	ReLU
10	FC					4	Softmax

3.3. Fault Diagnosis Framework Based on SES and CNN

In this paper, a deep learning method for bearing cross-domain fault diagnostics based on a reconstructed envelope spectrum is proposed, and the flow chart of the entire method is shown in Figure 3, with the main steps summarized as follows:

Step 1: Acquisition. Bearing vibration data accelerometers are used to acquire raw vibration signals from the bearings in various health states.

Step 2: Standardization processing. Standardize the original vibration signal to obtain the standard envelope spectrum of the signal.

Step 3: Standard sample library construction. A one-to-one correspondence between the standard envelope spectrum and the healthy rotational state of the bearing is used to construct a standard sample library.

Step 4: Model training. The CNN model is constructed, the standard sample library is divided into the training set, validation set and test set in the ratio of 7:2:1 to train the model, and the network model is optimally updated according to the training results to obtain the rolling bearing fault identification model.

Step 5: Fault type identification. The vibration data under different speeds and different bearing types are fed into the trained model as a test set to obtain the fault classification and visualization results.

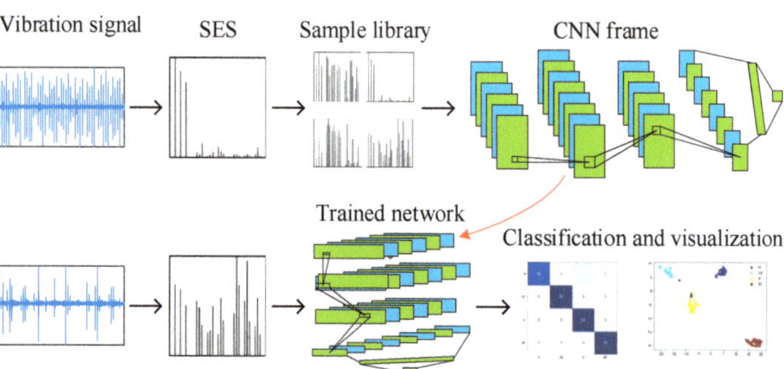

Figure 3. Proposed method frame.

4. Experimental Validation

In this section, a series of cross-domain tasks are designed, using two publicly available datasets and a dataset obtained from self-experimentation, to validate the effectiveness of the method proposed in this paper and to compare its performance with that of some popular signal preprocessing methods in terms of classification accuracy and generalization ability.

4.1. Experiment Setup and Data Description

In this paper, vibration signals from three devices were collected for analysis. The bearing parameters of each device are given in Table 2, and a detailed description of the data used is given in Table 3.

Table 2. Data set bearing parameters.

Data Set	Type	Number of Balls (z)	Roller Diameter d (mm)	Pitch Diameter D (mm)	Contact Angle α (°)
A	6203	8	6.75	29.05	0
B	Rexnord ZA-2115	16	8.4	71.5	15.17
C	6312/C3	8	22	94	0

Table 3. Data set specification.

Data Set	Set	Data Number	Rotational Speed	Sample Number	Class Label
PU Bearings	A1	H: K001, K002, K003, K004, K005, K006	1500 rpm	480	0
		OF: KA01, KA09,KA04,KA16		320	1
		IF: KI01, KIO3, KI04, KI16, KI17, KI18		480	2
	A2	H:K001, K002, K003, K004, K005, K006	900 rpm	480	0
		OF:KA01, KA09, KA04, KA16		320	1
		IF:KI01, KIO3, KI04, KI16, KI17, KI18		480	2
IMS Bearings	B	H: Data set 2, Bearing 1, Filess 1–200	2000 rpm	200	0
		OF: Data set 2, Bearing 1, Files 513–712		200	1
		IF: Data set 1, Bearing 3, Files 2056–2155		100	2
		BF: Data set 1, Bearing 4, Files 1757–1956		200	3
Experimental bearings	C	H: Bearing 1, Files 1–300	3120 rpm	300	0
		OF: Bearing 4, Files 501–800		300	1
		BF: Bearing 1, Files 3001–3300		300	3

A brief description of the dataset is given as follows:

(1) Data set A was obtained from the University of Paderborn, Germany [19], and the test rig included an electric motor, a torque measurement shaft, a rolling bearing test module, a flywheel, and a load motor, as shown in Figure 4. The data set includes normal

bearing data and defective bearing data, in which the defective bearings are obtained by both manual machining damage and accelerated life test damage, including three fault states: an outer ring fault, an inner ring fault, and a compound fault. Each bearing is carried out under four working conditions, there are 20 data points for each working condition, each data acquisition time was 4 s, and the sampling frequency was 64 kHz. The bearing data was under a torque of 0.7 Nm, a radial load force of 1000 N, and rotational speeds of 1500 rpm and 900 rpm were selected for this validation.

(2) Data set B is from the Intelligent Maintenance System (IMS) [20]. The test rig is shown in Figure 5, with four Rexnord ZA-2115 double row bearings mounted on a shaft. The data is the full life data of the bearings and consists of three data sets, each containing the vibration data of the four bearings. The speed of the bearings was 2000 rpm, the sampling frequency was 20 kHz, the sampling interval was 10 min, the sampling time was 1 s, and a data file with 20,480 sampling points was generated. Data set 1 contains 2156 files, where bearing 3 ran until the inner ring was damaged and bearing 4 ran until the rolling element was damaged. Data set 2 contains 984 files, where bearing 1 ran until the outer ring was damaged.

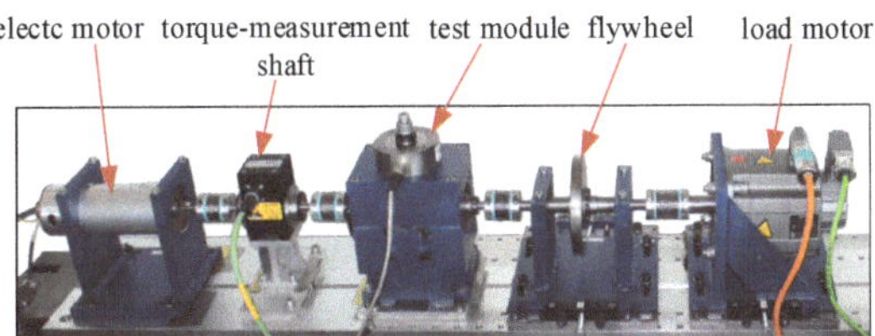

Figure 4. Modular test rig.

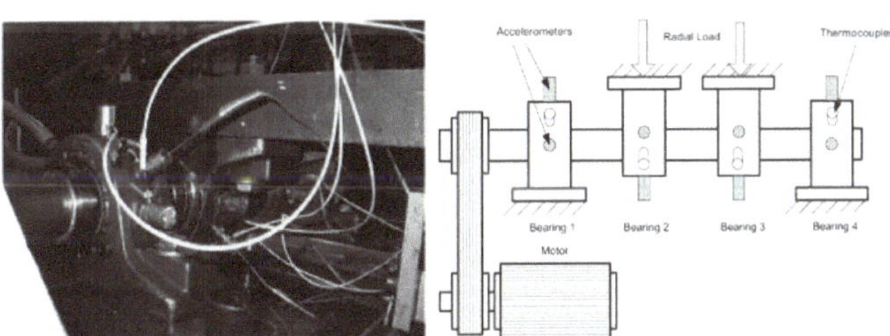

Figure 5. Bearing test rig.

(3) Data set C is the real failure data of the bearings obtained by conducting full life tests of the bearings. A total of four sets of bearings were installed on the test rig, and acceleration sensors were used to collect vibration signals. The arrangement of the test rig and measurement points is shown in Figure 6. The test speed was 3120 rpm, the sampling frequency was 12,800 Hz, the sampling interval was 1 min, the sampling time was 1.28 s, and 1 data file was generated. As shown in Figure 7a, bearing 1 ran until the rolling element was damaged, resulting in 5570 data files. As shown in Figure 7b, bearing 4 ran until the outer ring was damaged, resulting in 2473 data files.

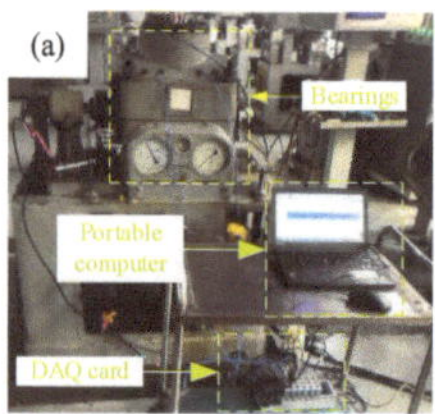
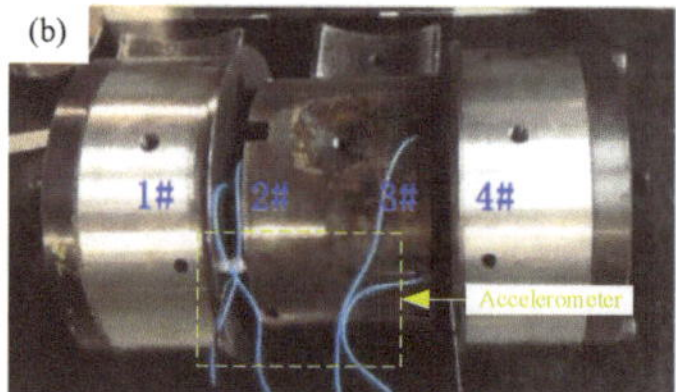

Figure 6. Bearing life test device: (**a**) Bearing test rig; (**b**) Bearing arrangement.

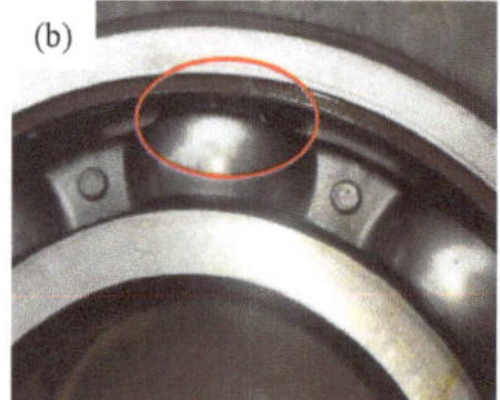

Figure 7. Bearing fault: (**a**) Bearing with outer race fault; (**b**) Bearing with roller fault.

To ensure that the envelope spectrum has sufficient frequency resolution, the acquisition time for each of the three data sets above was assumed to be 1 s for each sample. Some of the raw time domain waveforms for the three datasets are shown in Figure 8, indicating that the same health states in the three datasets have nothing in common in the time domain.

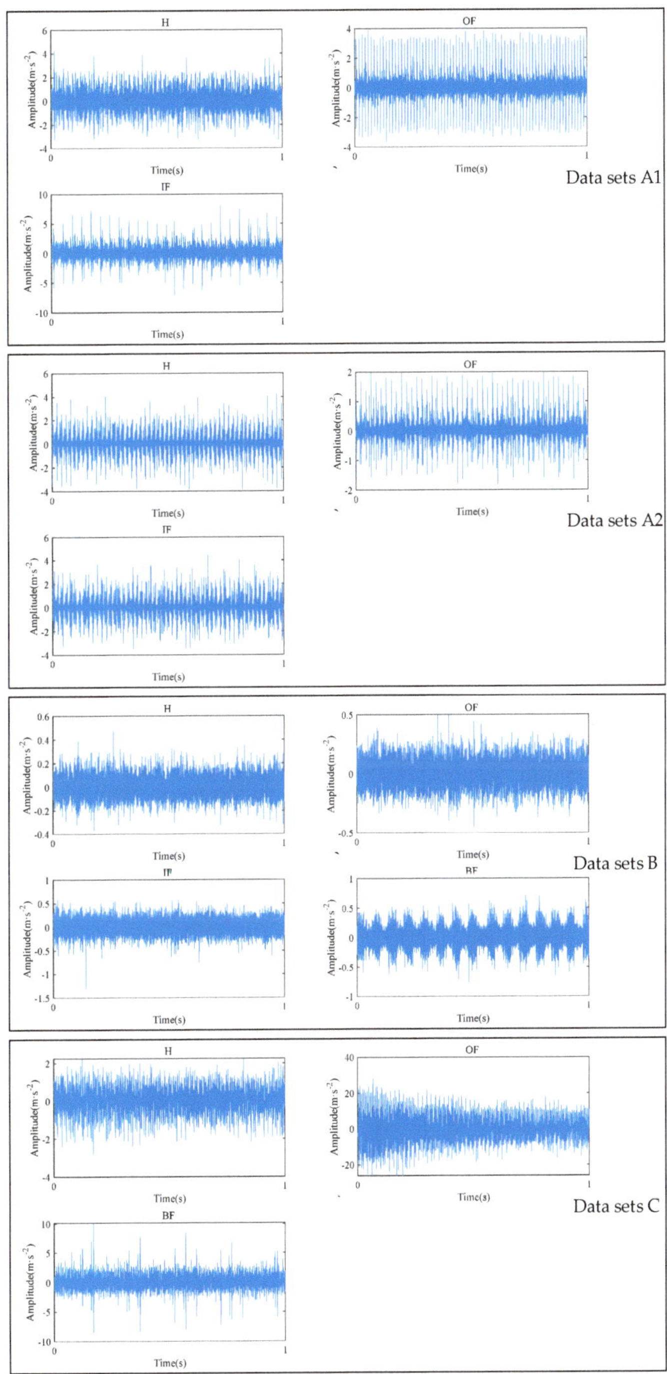

Figure 8. The time domain waveforms of the bearing datasets.

4.2. Analysis of Standard Envelope Spectrum (SES)

Data sets A1 and A2 are data from the same bearing at different speeds, and the standardized envelope spectra of some of their different health data are shown in Figure 9.

Data sets B and C are from two different types of bearings, and their standardized envelope spectra of some different health data are shown in Figure 10. It can be clearly seen that the SES plot shows unique features for a given state of health.

The standard envelope spectrum of the same bearing at the same speed for different health states is shown in Figure 10a–d. As shown in Figure 10a, for the normal condition, the spectral lines in the three characteristic regions are prominent. This is because the normal bearing does not have cyclic shocks caused by defects, does not have prominent peaks in the envelope spectra, and there is not much difference in the magnitude of the spectral lines after normalization. As shown in Figure 10b, for the outer ring fault, the amplitude of the spectral lines in the characteristic region of the outer ring fault is prominent, which is because the amplitude at the characteristic frequency of the outer ring fault and its harmonics in the envelope spectrum are prominent. As shown in Figure 10c, for the inner-circle fault, the amplitude of the spectral lines in the characteristic region of the inner-circle fault is prominent because the amplitude at the characteristic frequency of the inner-circle fault and its harmonics and side frequencies are prominent in the envelope spectrum. As shown in Figure 10d, for the rolling body fault, the amplitude of the spectral lines in the characteristic region of the rolling body fault is prominent because the amplitude at the characteristic frequency of the rolling body fault and its harmonics and side frequencies are prominent in the envelope spectrum.

As shown in Figure 9, for data from the same bearing at different speeds, the standard envelope spectrum for the same flaw type has a high degree of similarity, and both have more prominent spectral line amplitudes in the characteristic region of the corresponding flaw type. Similarly, as shown in Figure 10, the standard envelope spectrum of the same fault type for the data from two different types of bearings also has a high degree of similarity. This indicates that the standard envelope spectrum effectively reveals the signal characteristics of rolling bearings that do not change with speed and model, but only due to changes in health.

In fact, the envelope spectrum patterns of the bearings are different, for example, some faults have prominent amplitudes for the first three harmonics of the fault characteristic frequency, some have prominent amplitudes for only one harmonic of the fault characteristic frequency, some inner ring/rolling element faults have no sidebands, and so on. At this point, it is necessary for the CNN to synthesize the spectral line conditions of the three fault characteristic regions of the standard envelope spectrum to learn and summarize the common characteristics of the corresponding health states. It is also necessary to utilize multiple bearing fault data sets as much as possible to construct a standard sample library, enrich the standard envelope spectrum form of the standard sample library, so that when it is used for new diagnostic data, the most similar standard envelope spectrum from the sample library can be easily found, and its corresponding health state is the diagnostic result.

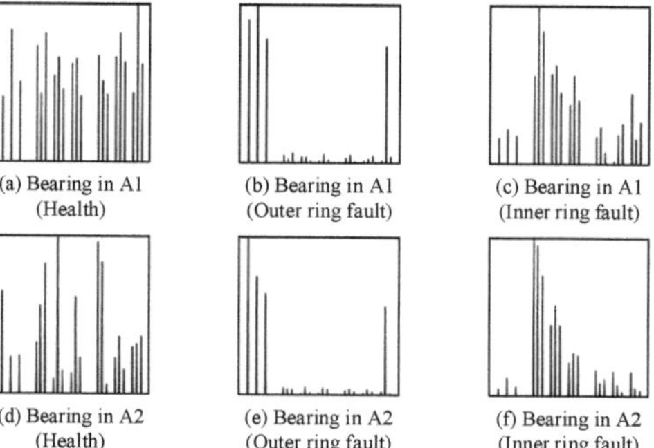

Figure 9. Standardized envelope spectra for different health conditions for data sets A1 in (**a–c**) and A2 in (**d–f**).

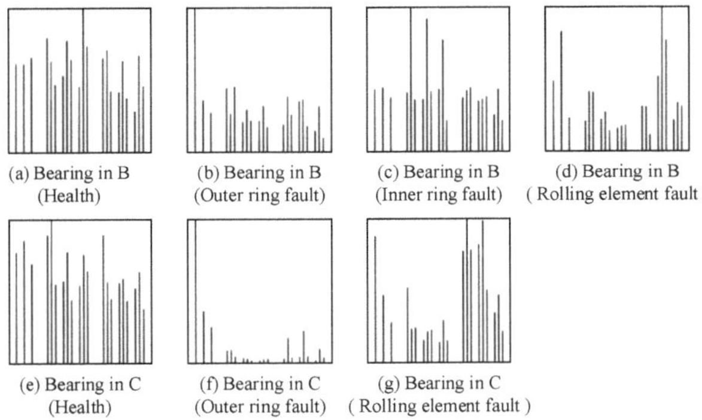

Figure 10. Standardized envelope spectra for different health conditions for data sets B in (**a–d**) and C in (**e–g**).

4.3. Fault Diagnosis under Different Domains

In this section, five cross-domain diagnostic tasks are designed to verify the cross-domain diagnostic performance of the method proposed in this paper. Tasks 1 and 2 are to use the data of one rotational speed as the training set (source domain) and the data of another rotational speed as the test set (target domain) on the data of the same type of bearing. Tasks 3–6 are training and test sets of data from different bearing models, where Task 6 uses data from two different bearing models as the training set. For example, B → C means that data set B is used for training and data set C is used for testing. For comparison, the dataset used for testing uses 50 samples for each health state. The dataset details are shown in Table 2.

4.3.1. Comparison with Time–Frequency Analysis Methods

Time–frequency analysis describes the correlation between the time and frequency domains of a signal and is an effective tool for dealing with non-stationary and transient signals [21]. The time–frequency transformed time–frequency map as a two-dimensional image is often used as input to CNN-based diagnostic models. Therefore, this paper compares the diagnostic effectiveness of STFT spectrograms, CWT spectrograms [22], VMD-HT spectrograms [23], MDFVI spectrograms [13], CSCoh spectrograms [14], and the proposed SES spectrograms as inputs. Each task is repeated 10 times, and the average diagnostic accuracy using the proposed SES-CNN method is shown in Table 3, and the results using other methods are also shown in Table 4. Task 6 is run once and its confusion matrix is shown in Figure 11, where the green font represents the number of correctly classified samples.

Table 4. Comparison results of different input features under different cross-domain tasks.

No.	Source → Target	STFT	CWT	VMD-HT	MDFVI	CSCoh	SES
1	A1 → A2	67.33%	67.33%	34.67%	71.67%	65.33%	98.67%
2	A2 → A1	67.33%	66.67%	33.33%	88.00%	70.00%	92.00%
3	B → A1	34.0%	34.67%	33.33%	66.67%	34.00%	94.00%
4	B → A2	33.33%	33.33%	34.00%	54.67%	33.33%	95.33%
5	B → C	34.00%	33.33%	34.67%	49.33%	33.33%	90.00%
6	A + C → B	26.00%	25.00%	25.50%	36.50%	31.50%	87.00%

It can be seen that the classification accuracy of the methods used for comparison is low. This is as expected, due to the large difference between the domains caused by the different bearing speeds and bearing models, the feature extraction methods used for comparison cannot reduce this difference, and it is difficult for deep learning to learn the common features between the source and target domains without the target domain samples used for training. Compared with these methods, the method proposed in this paper shows higher classification accuracy in experiments, which indicates the effectiveness of its feature extraction, as shown in Figure 11, where most of the samples in the target domain are correctly classified.

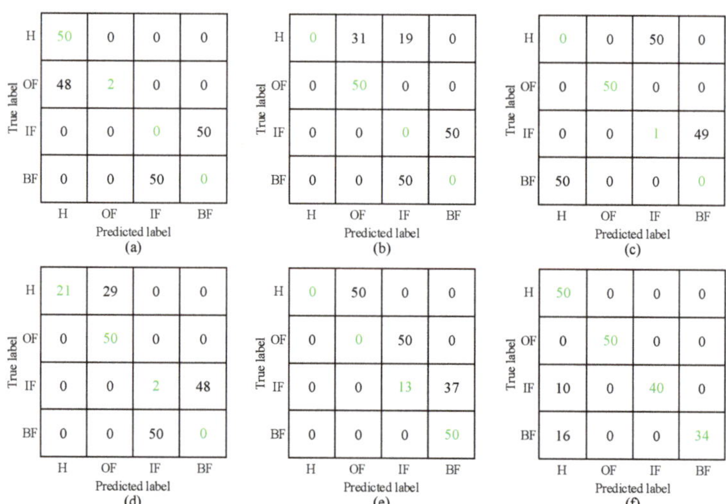

Figure 11. Confusion matrix of five methods in Task 6: (**a**) STFT; (**b**) WT; (**c**) VMD-HT; (**d**) MDFVI; (**e**) CSCoh; (**f**) SES.

4.3.2. Feature Visualization Analysis

To further illustrate the feature learning effectiveness of the proposed method for cross-domain diagnostics, the performance of the proposed SES method on Task 6 is feature visualized using the t-distributed stochastic neighborhood embedding (t-SNE) technique [24], which maps high-level outputs from 128 to 2 dimensions in Layer 10. The results are shown in Figure 12. The four different colors indicate four different categories. It is clear from the visualization results that data points with different health states are well separated and data points with the same health state are clustered together. In addition, a small number of data points overlap between H and IF and between H and BF, suggesting that a small number of IF and BF samples are incorrectly classified as H samples, which would lead to reduced diagnostic accuracy. This observation is consistent with the results shown in Figure 11. Overall, for the untrained data of different bearing models in Task 6, the proposed method can clearly distinguish different categories. Therefore, it can be concluded that the proposed method is able to learn better discriminative features and shows strong cross-domain diagnostic capability.

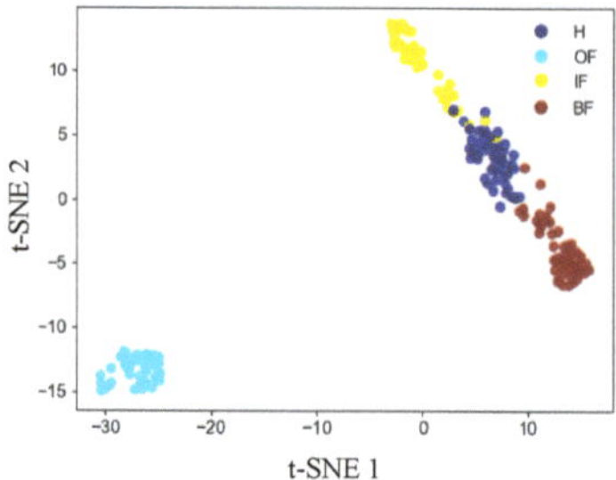

Figure 12. Features visualization based on t-SNE.

5. Conclusions

In order to improve the cross-domain fault diagnostic capability of rolling bearings, this paper proposes a new deep learning-based fault diagnostics framework that combines SES and CNN. Cross-domain diagnostic experiments are carried out on three different sets of equipment, and the following conclusions can be drawn from the analyses of the experimental results:

(1) The constructed SES, as a pre-processing step, reveals the signal characteristics of rolling bearings that do not change with rotational speed and model number, but only due to changes in health status.

(2) The SES eliminates the differences between the source and target domains due to the different rotational speeds and models, which greatly reduces the difficulty the CNN has in learning the common features between the two different domains and improves the diagnostic performance.

(3) In the cross-domain task between six different rotational speeds and between different bearing models, the proposed method shows high classification accuracy in cross-domain diagnostics compared with some popular pre-processing methods.

Author Contributions: Conceptualization, X.W.; methodology, L.Z. and Z.W.; software, L.Z., X.W. and Z.S.; validation, Z.S.; formal analysis, L.Z.; investigation, Z.S.; resources, X.W.; data curation, L.Z. and Z.S.; writing—original draft preparation, Z.W.; writing—review and editing, L.Z. and Z.S.; visualization, L.Z.; supervision, X.W.; project administration, X.W.; funding acquisition, X.W. All authors have read and agreed to the published version of the manuscript.

Funding: This research was funded by the Joint Funds of the National Natural Science Foundation of China (Grant No. U2267206).

Institutional Review Board Statement: Not applicable.

Informed Consent Statement: Not applicable.

Data Availability Statement: Data set A was obtained from the University of Paderborn, Germany, original data download link: https://blog.csdn.net/ynn4818172/article/details/122755894 (accessed on 10 April 2024). Data set B is from the Intelligent Maintenance System (IMS), original data download link: https://ti.arc.nasa.gov/tech/dash/groups/pcoe/prognostic-data-repository/#bearing (accessed on 10 April 2024).

Conflicts of Interest: The authors declare no conflicts of interest.

References

1. Hou, W.; Zhang, C.; Jiang, Y.; Cai, K.; Wang, Y.; Li, N. A new bearing fault diagnosis method via simulation data driving transfer learning without target fault data. *Measurement* **2023**, *215*, 112879. [CrossRef]
2. Li, Y.; Ding, K.; He, G.; Jiao, X. Non-stationary vibration feature extraction method based on sparse decomposition and order tracking for gearbox fault diagnosis. *Measurement* **2018**, *124*, 453–469. [CrossRef]
3. Verstraete, D.; Ferrada, A.; Droguett, E.L.; Meruane, V.; Modarres, M. Deep learning enabled fault diagnosis using time-frequency image analysis of rolling element bearings. *Shock Vib.* **2017**, *2017*, 5067651. [CrossRef]
4. Abbasion, S.; Rafsanjani, A.; Farshidianfar, A.; Irani, N. Rolling element bearings multi-fault classification based on the wavelet denoising and support vector machine. *Mech. Syst. Signal Process.* **2007**, *21*, 2933–2945. [CrossRef]
5. Malhi, A.; Gao, R.X. PCA-based feature selection scheme for machine defect classification. *IEEE Trans. Instrum. Meas.* **2004**, *53*, 1517–1525. [CrossRef]
6. Samanta, B.; Al-Balushi, K.R. Artificial neural network based fault diagnostics of rolling element bearings using time-domain features. *Mech. Syst. Signal Process.* **2003**, *17*, 317–328. [CrossRef]
7. Zhao, B.; Zhang, X.; Li, H.; Yang, Z. Intelligent fault diagnosis of rolling bearings based on normalized CNN considering data imbalance and variable working conditions. *Knowl.-Based Syst.* **2020**, *199*, 105971.
8. Liu, H.; Zhou, J.; Zheng, Y.; Jiang, W.; Zhang, Y. Fault diagnosis of rolling bearings with recurrent neural network-based autoencoders. *ISA Trans.* **2018**, *77*, 167–178. [CrossRef] [PubMed]
9. Kong, X.; Mao, G.; Wang, Q.; Ma, H.; Yang, W. A multi-ensemble method based on deep auto-encoders for fault diagnosis of rolling bearings. *Measurement* **2020**, *151*, 107132. [CrossRef]
10. Jin, Z.; He, D.; Wei, Z. Intelligent fault diagnosis of train axle box bearing based on parameter optimization VMD improved DBN. *Eng. Appl. Artif. Intell.* **2022**, *110*, 104713. [CrossRef]

11. Hoang, D.T.; Kang, H.J. *Rolling Element Bearing Fault Diagnosis Using Convolutional Neural Network and Vibration Image*; Elsevier: Amsterdam, The Netherlands, 2019.
12. Zhang, Y.; Xing, K.; Bai, R.; Sun, D.; Meng, Z. An enhanced convolutional neural network for bearing fault diagnosis based on time–frequency image. *Measurement* **2020**, *157*, 107667. [CrossRef]
13. Hasan, M.J.; Islam, M.M.M.; Kim, J.M. Bearing fault diagnosis using multidomain fusion-based vibration imaging and multitask learning. *Sensors* **2021**, *22*, 56. [CrossRef] [PubMed]
14. Chen, Z.; Mauricio, A.; Li, W.; Gryllias, K. A deep learning method for bearing fault diagnosis based on cyclic spectral coherence and convolutional neural networks. *Mech. Syst. Signal Process.* **2020**, *140*, 106683. [CrossRef]
15. Sobie, C.; Freitas, C.; Nicolai, M. Simulation-driven machine learning: Bearing fault classification. *Mech. Syst. Signal Process.* **2018**, *99*, 403–419. [CrossRef]
16. Chen, B.; Zhang, W.; Gu, J.X.; Song, D.; Cheng, Y.; Zhou, Z.; Gu, F.; Ball, A.D. Product envelope spectrum optimization-gram: An enhanced envelope analysis for rolling bearing fault diagnosis. *Mech. Syst. Signal Process.* **2023**, *193*, 110270. [CrossRef]
17. McFadden, P.D.; Smith, J.D. Model for the vibration produced by a single point defect in a rolling element bearing. *J. Sound Vib.* **1984**, *96*, 69–82. [CrossRef]
18. Guo, Z.; Yang, M.; Huang, X. Bearing fault diagnosis based on speed signal and CNN model. *Energy Rep.* **2022**, *8*, 904–913. [CrossRef]
19. Lessmeier, C.; Kimotho, J.K.; Zimmer, D.; Sextro, W. Condition monitoring of bearing damage in electromechanical drive systems by using motor current signals of electric motors: A benchmark data set for data-driven classification. In Proceedings of the PHM Society European Conference, Bilbao, Spain, 5–8 July 2016; Volume 3.
20. Qiu, H.; Lee, J.; Lin, J.; Yu, G. Wavelet filter-based weak signature detection method and its application on rolling element bearing prognostics. *J. Sound Vib.* **2006**, *289*, 1066–1090. [CrossRef]
21. Dong, G.; Chen, J. Noise resistant time frequency analysis and application in fault diagnosis of rolling element bearings. *Mech. Syst. Signal Process.* **2012**, *33*, 212–236. [CrossRef]
22. Gu, J.; Peng, Y.; Lu, H.; Chang, X.; Chen, G. A novel fault diagnosis method of rotating machinery via VMD, CWT improved CNN. *Measurement* **2022**, *200*, 111635. [CrossRef]
23. Dragomiretskiy, K.; Zosso, D. Variational mode decomposition. *IEEE Trans. Signal Process.* **2013**, *62*, 531–544. [CrossRef]
24. Van der Maaten, L.; Hinton, G. Visualizing data using t-SNE. *J. Mach. Learn. Res.* **2008**, *9*, 2579–2605.

Article

Prediction of Pre-Loading Relaxation of Bolt Structure of Complex Equipment under Tangential Cyclic Load

Xiaohan Lu [1], Min Zhu [1,*], Chao Li [1,*], Shengnan Li [2], Shengao Wang [1] and Ziwei Li [1]

1   College of Nuclear Science and Technology, Naval University of Engineering, Wuhan 430033, China
2   College of Power Engineering, Naval University of Engineering, Wuhan 430033, China
*   Correspondence: min0zhu@163.com (M.Z.); lcandzyn@163.com (C.L.)

Abstract: Bolts have the advantages of simple installation and easy removal. They are widely applied in aerospace and high-speed railway traffic. However, the loosening of bolts under mixed loads can lead to nonlinear decreases in pre-loading. This affects the safety performance of the structure and may lead to catastrophic consequences. Existing techniques cannot be used to monitor the bolt performance status in time. This has caused significant problems with the safety and reliability of equipment. In order to study the relaxation law of bolt pre-loading, this paper carries out an experimental analysis for 8.8-grade hexagonal bolts and calibrates the torque coefficient. We also studied different loading waveforms, nickel steel plate surface roughnesses, tangential displacement frequencies, four different strengths and bolt head contact areas of the bolt, the initial pre-loading, and the effects of tangential cyclic displacement on pre-loading relaxation. This was done in order to accurately predict the degree of bolt pre-loading loosening under external loads. The laws are described using the allometric model function and the nine-stage polynomial function. The least squares method is used to identify the parameters in the function. The results show that bolts with a smooth surface of the connected structure nickel steel flat plate, high-frequency working conditions, half-sine wave, and a high-strength have better anti-loosening properties. Taking 5–10 cycles of cyclic loading as a boundary, the pre-loading relaxation is divided into two stages. The first stage is a stage of rapid decrease in bolt pre-loading, and the second stage is the slow decrease process. The performance prediction study shows that the allometric model function is the worst fitted, at 71.7% for the small displacement condition. Other than that, the allometric model function and the nine-stage polynomial function can predict more than 85.5% and 90.4%, which require the use of least squares to identify two and ten unknown parameters, respectively. The complexity of the two is different, but both can by better indicators than the pre-loading relaxation law under specific conditions. It helps to improve the monitoring of bolt loosening and the system use cycle, and it can provide theoretical support for complex equipment working for a long time.

Keywords: pre-loading; M8 bolt; nickel steel flat plate; tangential force; performance prediction

Citation: Lu, X.; Zhu, M.; Li, C.; Li, S.; Wang, S.; Li, Z. Prediction of Pre-Loading Relaxation of Bolt Structure of Complex Equipment under Tangential Cyclic Load. *Sensors* **2024**, *24*, 3306. https://doi.org/10.3390/s24113306

Academic Editors: Dong Wang, Shilong Sun and Changqing Shen

Received: 28 April 2024
Revised: 18 May 2024
Accepted: 20 May 2024
Published: 22 May 2024

1. Introduction

There are a large number of bolt structures in large and complex equipment [1,2], and the operational life of the bolt greatly affects the safety and complexity of the equipment. The bolt structure prevents the connecting structure from sticking and slipping after being subjected to a tangential load in the form of pre-loading. A small pre-loading is likely to cause the bolt to loosen, leading to safety problems in large and complex equipment and affecting the overall safety of the reactor. Excessive pre-loading can easily lead to losses in bolt strength and stiffness [3].

Pre-loading directly affects the reliability of the bolted structure of complex equipment. The magnitude of pre-loading is affected by factors such as the bolt material, diameter, bearing speed, lubricant temperature, and friction coefficient [4,5]. Yang [6] compared the loosening degree of different bolts under a tangential force, and the performance of the anti-loosening bolts was 9.43% lower than that of ordinary bolts. Hu [7] studied the effects of the bolt diameter, pitch, and friction coefficient on the size of pre-loading based on a finite element analysis based on the secondary development using Python and ABAQUS. The results showed that at the same torque, the size of the bolt diameter is inversely proportional to the pre-loading; the pitch has a small effect on the pre-loading; and the friction coefficient has a significant effect. Reference [8] points out that at lower speeds, the bolt pre-loading needs to be increased to maintain stability. In the case of mechanical finishing, the bolt pre-loading should be reduced to ensure that the stiffness is within a reasonable range. Under the coupling of multiple influencing factors, different tangential cyclic displacements cause irregular pre-loading relaxation. As a result, it is difficult to accurately evaluate the overall performance of large and complex equipment.

Pre-loading relaxation is prone to lead to the functional failure of the bolt structure. It affects safety and reliability, and researchers have conducted many studies on the relaxation of pre-loading [9–12]. Yang [13] assumed that the spring stiffness is proportional to the magnitude of the force and constructed a model that can assess the pre-loading relaxation effect. Wang [10] used the cooling method to set the bolt pre-loading and investigated the effects of factors such as the friction coefficient and excitation replication on self-relaxation. Shen [14] established a wear model based on the secondary development of ABAQUS 2016 software, and the results showed that the pre-loading relaxation phenomenon can be predicted by fatigue damage theory and the wear model. At present, scholars have carried out more studies on the mechanisms, calculations, monitoring, and prevention of pre-loading relaxation [15–19], especially for analysis based on finite element software. However, there are fewer studies on the complex working conditions of high-precision instruments such as complex equipment. In the process of storage, transportation, and work of complex equipment, complex equipment may have impact, slip, and so on. These situations can generate large tangential displacements on the equipment, as shown in Figure 1. Therefore, it is necessary to predict the bolt pre-loading in this process. The existing methods for monitoring pre-loading mainly include the strain gauge sensor method [20,21], piezoelectric impedance method [22,23], ultrasonic method [24,25], and image recognition method [26–28]. These four methods are widely used in the fabrication of a wide range of physical field sensors. In particular, they play an important role in the fields of force measurement, weighing, and pressure sensors. However, such methods are complicated to implement and cannot monitor the change of pre-loading values in real time. Therefore, there is a need to build mathematical functions that can characterize the degree of pre-loading relaxation to solve the problem of unpredictable pre-loading decay during bolt operation.

In this paper, high-strength alloy nickel steel material is used as the basis from a safety point of view for large and complex equipment. The pre-loading experiment was designed and carried out. Nickel steel plates with a high strength, high hardness, and corrosion resistance are widely used in a variety of important complex equipment such as aircraft engines. However, due to its low filling factor, this material is less frequently used in existing common equipment, and there are fewer studies on nickel steel materials in the field at present. Therefore, this paper focuses on nickel steel structures in complex equipment, with a Z-shaped flat plate used to simplify the complexity of the equipment. The relationships between the mechanical properties of fasteners based on the ISO 898-2:2022 specifications are studied [29]. Conducting experiments at ambient temperatures from 10 °C to 35 °C ensures the physical properties of the screw and nut. The relationship between the pre-loading and maximum external tightening torque of M8 external hexagonal bolts was studied, and different pre-loading measurement methods and bolt tightening methods were analyzed. Pre-loading relaxation experiments were designed, and torque coefficient calibration between different bolts was carried out. The effects of different waveforms, surface roughnesses of nickel steel plates, tangential displacement frequencies, bolts, torques, and tangential cyclic displacements on pre-loading relaxation were studied. Parameter identification was carried out for different tangential displacements and initial tightening. An allometric model function and nine-stage polynomial function expression that could describe the degree of pre-loading relaxation were constructed. These can provide theoretical support for the safety and reliability of large and complex equipment.

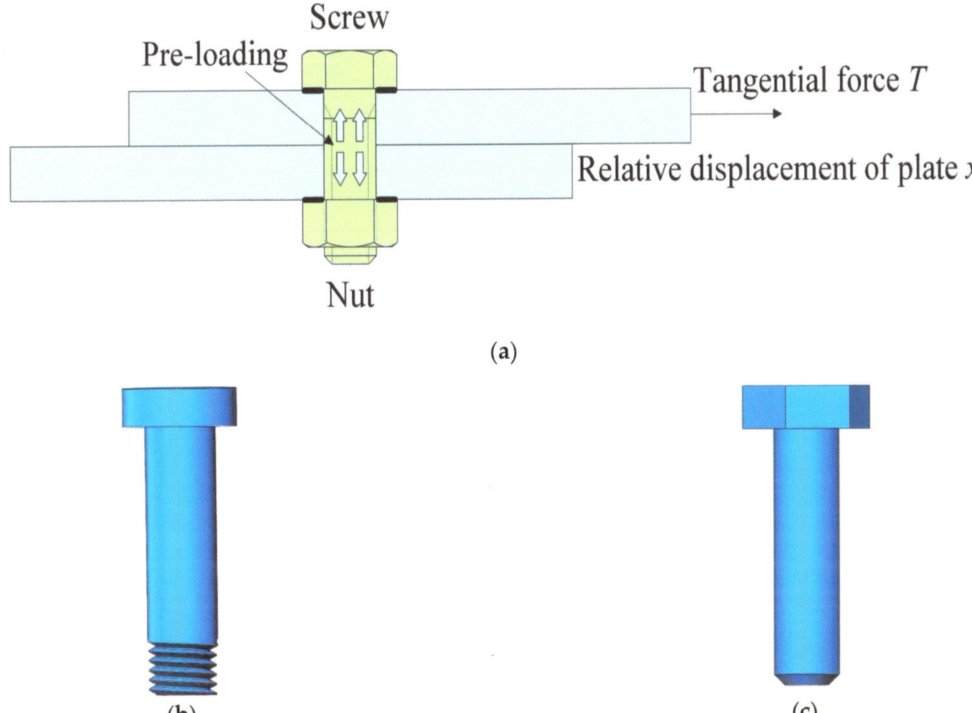

Figure 1. Structure of bolted joints under mixed loads. (**a**) Simplified bolt assembly, (**b**) with lifting angle bolts, (**c**) without lifting angle bolts.

2. Methods

2.1. Theoretical Formula for Calculating Pre-Loading

The total torque acting on the bolt structure is the sum of the friction torque produced by the threaded pair and the friction torque produced between the nut, which is as follows [30]:

$$N = N_1 + N_2 \tag{1}$$

where N_1 and N_2 are calculated using the following formula:

$$N_1 = \frac{1}{2} \times F_y \times d_2 \times \tan(\alpha + \beta) \tag{2}$$

$$N_2 = \mu_w \times F_y \times r_d \tag{3}$$

where F_y is the bolt pre-loading, d_2 is the thread center diameter, α is the thread rise angle, β is the equivalent friction angle, μ_w is the friction coefficient of the inner surface of the nut, and r_d is the equivalent friction radius.

The angle of thread rise is small in relation to the equivalent friction angle and can therefore be approximated as follows:

$$\tan(\alpha + \beta) = \tan\alpha + \tan\beta \tag{4}$$

The equivalent friction radius can be expressed as follows:

$$r_d = \frac{1}{3} \times \frac{(d_1{}^3 - d_0{}^3)}{d_1{}^2 - d_0{}^2} \tag{5}$$

where d_1 is the maximum outer diameter of the nut and d_0 is the diameter of the bolt hole.

Therefore, the relationship between the tightening torque and pre-loading can be expressed as follows:

$$N = \frac{1}{2} \times F_y \times d_2 \times \tan(\alpha + \beta) + \frac{\mu_w \times F_y}{3} \times \frac{(d_1{}^3 - d_0{}^3)}{d_1{}^2 - d_0{}^2} \tag{6}$$

The equation can be simplified as follows:

$$N = K_2 F_y d \tag{7}$$

where K_2 is the torque coefficient to be checked and d is the diameter of the bolt.

When the bolt is tightened, pre-loading should be controlled within a reasonable range. Too much pre-loading can lead to material fracture, failure, and yielding. Too little pre-loading can lead to reliability and safety problems in the equipment. The maximum tightening torque within the elastic range should be taken into account when assembling the bolt to the object to be connected to prevent equipment problems caused by excessive torque. Pre-tightening stress includes tensile stress σ and shear stress τ, which are calculated as follows:

$$\sigma = \frac{F_y}{A_s} = \frac{4F_y}{\pi d_c{}^2} \tag{8}$$

$$\tau = \frac{16N_1}{\pi d_c{}^3} \tag{9}$$

where $A_s = \frac{\pi}{4}(d - 0.9382P)^2$ is the stress cross-section area, P is the pitch, and $d_c = \sqrt{\frac{4A_s}{\pi}}$ is the equivalent diameter.

Therefore, the relationship between the tangential and tensile stresses in the bolt can be expressed as follows:

$$\frac{\tau}{\sigma} = \frac{2d_2 \tan(\alpha + \beta)}{d_c} \tag{10}$$

According to the empirical formula, when the diameter of the bolt is less than 64 mm, it can be approximated as $d_2 \approx 1.12d_c$. When the angle of thread rise and the equivalent friction angle is small, it can be approximated as $\tan(\alpha + \beta) = 0.19$. Therefore, it can be concluded that $\frac{\tau}{\sigma} \approx 0.43$.

The following is known from the theory of the fourth strength:

$$\sigma_{\max} = \sqrt{\sigma^2 + (3\tau)^2} \tag{11}$$

This can be solved as $\sigma_{\max} \leq 0.79\sigma_s$, where σ_s is the yield strength. In the elastic range, the theoretical maximum bolt pre-loading can be expressed as follows:

$$F_{y\max} = \sigma_{\max} A_s = 0.79\sigma_s \times \frac{\pi}{4}d_c{}^2 = 0.1975\sigma_s \pi d_c{}^2 \tag{12}$$

The theoretical maximum tightening torque can be found by bringing the formula into Equation (6) as follows:

$$N_{\max} \approx 0.1975 \times \pi d_c{}^2 \sigma_s \left[\frac{d_2(\tan(\alpha + \beta))}{2} + \mu_w r_d \right] \tag{13}$$

A GB/T 5783-2016 standard [31] 8.8 grade M8 external hexagonal bolt is used to calculate the maximum pre-loading and maximum torsion gauge. In this paper, the friction coefficient of the nickel steel flat plate is taken as 0.2, and its basic parameters are shown in Table 1. According to Equations (12) and (13), its theoretical maximum pre-loading and maximum torque can be calculated as 18,881 N, 34.3 Nm. It conforms to the design specifications and provides theoretical support for the selection of initial torque in the following.

Table 1. Bolt basic parameters.

Parameter	Numerical Value	Parameter	Numerical Value
Thread center diameter d_2	7.2 mm	Stress cross-sectional area A_S	39.167 mm^2
Diameter of the bolt hole d_0	8.8 mm	Equivalent diameter d_c	7.06 mm
Maximum outer diameter of the nut $d_1 \approx 1.5d$	13.5 mm	Pitch P	1 mm
$\tan(\alpha + \beta)$	0.19	Yield strength σ_s	640 MPa

2.2. Experimental Design of Pre-Loading Relaxation

2.2.1. Main Experimental Equipment

We used the 50 kN electro-hydraulic servo fatigue testing machine produced by Xi'an Tongsheng Instrument Manufacturing Co. (Xi'an, China). The machine can be applied to all kinds of components, materials, and dynamic and static mechanical properties tests. The main technical indicators are shown in Table 2. We used EVOTest 2.1.1.0 software to control the various performance parameters of the universal testing machine. The software is an external control system that can be close-loop controlled. Its test machine contains a variety of control methods, which can provide different kinds of stresses, strains, speeds, displacements, forces, and other external load application methods. The machine uses a fuzzy proportion integration differentiation (PID) control algorithm to regulate the loading process, which is able to obtain a high control accuracy. The universal experimental system is shown in Figure 2.

Table 2. Main technical indexes of the universal testing machine.

Performance Indicator	Parameter Range	Performance Indicators	Parameter Range
Maximum test force	±60 kN	Dynamic fluctuation	±2%
Maximum dynamic test force	±50 kN	Displacement range	±100 mm
Measuring range of test force	2–100% FS	Test frequency range	0.01–50 Hz
Test force accuracy	±1% FS	Rack strength	3.3×10^8 N/m
Two-column main frame	Hard chrome plated		

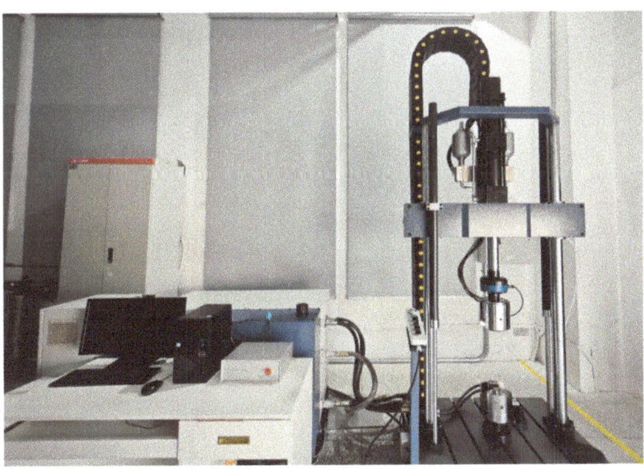

Figure 2. Starring experimental machine.

Due to the long-term operation of the experimental machine, the hydraulic oil is at a high temperature for a long time, and the rated temperature of the universal experimental machine is 55 °C, which needs to be water-cooled. This experiment uses an in-line water cooler, which is rated at 22 °C, and it uses softened water to circulate the hydraulic oil for cooling. The oil pump of the universal experimental machine is started using the EVOTest software. The locking cylinder is loosened, and the upper beam is adjusted to a suitable position. The cylinder is locked, the nickel steel flat plate is clamped in the upper and lower fixtures, and the experimental parameters are adjusted using the software.

2.2.2. Main Measuring Devices

(1)   Ring force sensor

In this paper, the strain gauge sensor method is selected to implement the monitoring of pre-loading recession values. Bengbu Jinnuo Sensor Co., Ltd. (Bengbu, China) Produced the JHBM-4 ring force sensor, and its specific parameters shown in Table 3.

Table 3. JHBM-4 ring force sensor main indicators.

Performance Indicator	Parameter Range
Range	0–20,000 N
Sensitivity	1.0~2.0 ± 0.1 mv/V
Combined accuracy	0.2% F·S
Operating temperature range	−20~+70 °C
Allowable overload	120% F·S
Excitation voltage	5~12 VDC
Creep	±0.1% F·S/30 min

The ring force sensor is subject to bolt-induced pressure on the hub tab, as shown in Figure 3. In the experiment, it was placed in the middle surface of the bolt and the nickel steel plate to ensure that the convex surface was facing the bolt to enable it to accurately detect the bolt pre-loading.

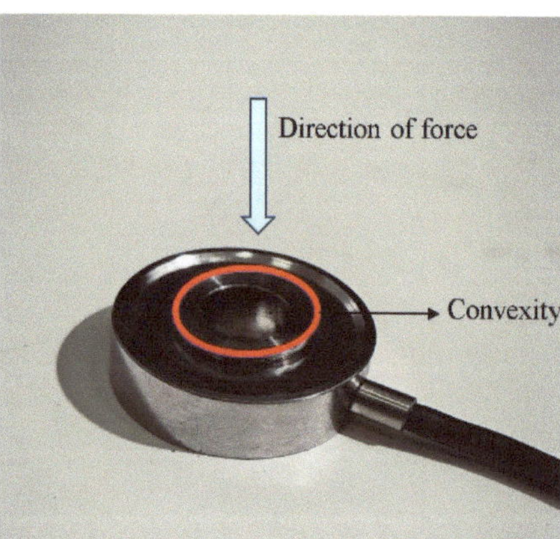

Figure 3. Working principle of the ring force sensor.

(2)   Intelligent display instrument

The ring force measuring sensor was connected to MCK-Z-I intelligent instrument as shown in Figure 4. It received signals from the sensor and generated electrical signals, and its specific parameters are shown in Table 4.

Figure 4. MCK-Z-I intelligent meter.

Table 4. Main parameters.

Performance Indicator	Parameter Range	Performance Indicator	Parameter Range
Input method	mV signals, standard variable signals or frequency signal	Baud rate	2400, 4800, 9600, 19,200 bps
Sampling speed	10 times/s, 80 times/s	Power consumption	Less than 5 VA
Precision	±0.02% F·S	Operating temperature	−20~50 °C
Communication interface	Standard serial RS-232/485 interfaces	Power supply	220 VAC/24 VDC

Before the experiment, the instrument needs to be calibrated for zero and display. Zero calibration requires entering the calibration mode and setting the value without preloading as the zero point, while display calibration needs to be based on the following calibration formula:

$$k_1 = \frac{X_1}{X_2} \times k_2 \tag{14}$$

where k_1 is the calibration coefficient to be set, which ranges from 0.0010 to 9.9999; k_2 is the initial calibration coefficient rated at 1; X_1 is the value to be displayed; and X_2 is the currently displayed value.

Five-kilogram and ten-kilogram weights were placed on top of the ring force transducer tabs. The steady-state value was recorded in the intelligent display meter. The calibration coefficients were calculated to be set by the formula, and the coefficients were re-entered into the meter to complete the calibration. The calibrated meter was connected to a computer via the conversion cable, and the digital transmitter communication software was used to monitor and record the pre-loading recession value in real time.

(3)   High-precision digital torque spanner

The tightening process of the bolt structure is divided into three stages. First, the bolt is not in contact with the plate, and its pre-tightening force is close to 0. Secondly, when the bolt head and nut are close to the plate, the pre-loading increases continuously. Finally, when the bolt reaches the yield limit and continues to tighten, pre-loading and torque will decrease, which may lead to the bolt fracturing. In large, complex equipment, bolt tightening methods are usually the torque method [32,33] and torque-angle method [34]. In this paper, the torque method is used to apply torque to obtain pre-loading, and the high-precision digital display torque wrench produced by Idema Company is selected, as shown in Figure 5. When tightening, the plate should be fixed, and the bolt head should be fixed at the same time to prevent experimental errors caused by the two. The torque wrench used in this experiment has a measuring range of 0~60 Nm, and its working life can reach 10,000 times. The accuracy can reach ±2% when tightening a bolt and ±2.5% when unloading a bolt. The target torque value can be preset, and the alarm for reaching the preset torque value improves the precision of the applied torque.

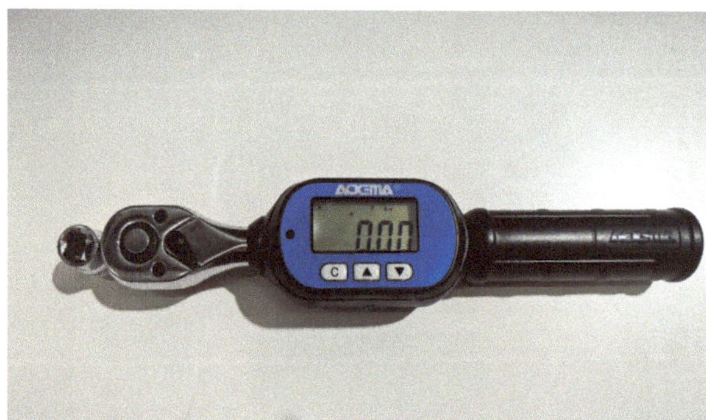

Figure 5. Torque spanner.

2.2.3. Preparation of Laboratory Supplies

We selected national standard 8.8-grade M8 external hexagonal bolts and internal hexagonal bolts and 12.9-grade M8 external hexagonal bolts and internal hexagonal bolts as the specimens. Bolts with obvious defects were excluded before the test. To prevent damage to the threads, each bolt was loaded and unloaded only once during the experiment. Complex equipment in the complex structure is not convenient for the study of the pre-loading. Therefore, this paper simplifies the nickel steel high-strength material into a Z-shaped plate. This structure allows the bolt to be stressed at the center during pre-loading work, eliminating the effects of torque on the bolt, as shown in Figure 6. The simplified nickel steel flat plate can accurately simulate the force on complex equipment when the bolt is acting [35].

(a)

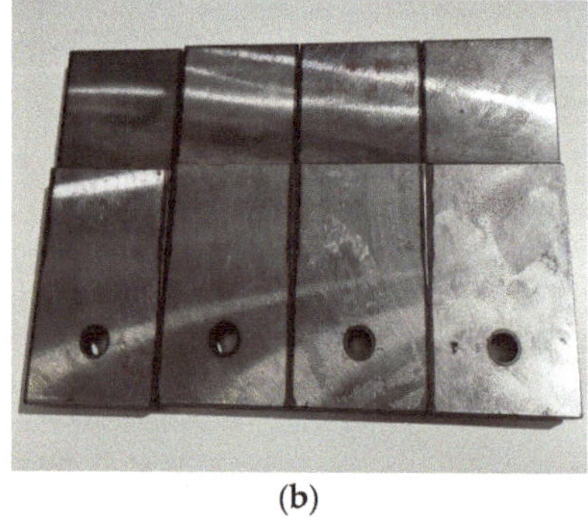

(b)

Figure 6. Experimental consumables. (**a**) Bolts, (**b**) nickel steel plate.

The nickel steel flat plate was fixed in the fixture in the universal testing machine, and the overall assembly is shown in Figure 7. The ring force transducer was placed between the screw and the nickel steel flat plate, and the predetermined torque was applied using a torque spanner. Before the experiment, the tangential force and displacement in the universal testing machine were adjusted to zero. The ring force sensor was placed between the nut and the nickel steel flat plate. The pre-loading change value was read through the intelligent display meter connected to the sensor. The tangential load was applied using the EVOTest software on the computer side. The tangential cyclic load can be applied with a sine wave and half-sine wave. The computer and the intelligent display meter can record the real-time change value.

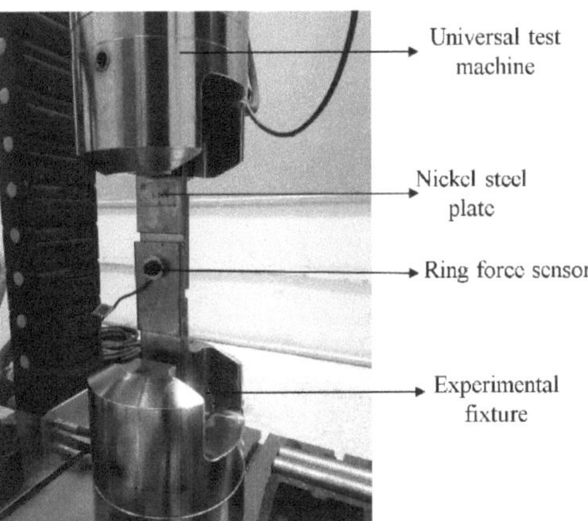

Figure 7. Overall assembly of the experiment.

Before the pre-loading relaxation experiment started, the torque spanner and intelligent display instrument needed to be calibrated to verify the torque coefficient of the bolt to ensure the accuracy and feasibility of the experimental instrument. In this paper, the torque coefficient was calibrated for 8.8-grade M8 external hexagonal bolts and internal hexagonal bolts and 12.9-grade M8 external hexagonal bolts and internal hexagonal bolts. The maximum torque was calculated to be 34.3 Nm for 8.8-grade M8 bolts and 44 Nm for 12.9-grade bolts. A smaller torque leads to the bolts not reaching the pre-tensioning effect, and a larger torque leads to the bolts failing, which affects the subsequent experimental tests. Therefore, this paper selected 4–28 Nm torque. In the form of incremental selection of five torque values, can make the torque cover the range of the standard.

The nickel steel flat plate was fixed in the universal experimental machine, and a spanner was used to fix the nut to prevent the nut from loosening during tightening. The required torque was set using a torque spanner, and the pre-loading was applied in the form of a tightening screw. Five torques of 4 Nm, 10 Nm, 16 Nm, 22 Nm, and 28 Nm were applied to each set of bolts in turn. Each bolt was applied only once for loading and unloading to avoid the problem of reduced bolt pre-loading due to repeated tightening. Each set of experiments was repeated three times, and the average value was taken to circumvent the effects of experimental chance. The experimental results are shown in Figure 8, and the linear expression of the torque coefficients is shown in Table 5. Since the slope of the fitted curve is large, fluctuations at smaller slopes can affect the intercept to produce larger fluctuations. The value of Pearson' s r ranges from −1 to 1. When the value is greater than 0.8, it can indicate that the original data are strongly correlated with the fitted curve. The closer that the R-squared and adjusted R-squared values are to 1, the better the fit. The 8.8 external hexagon bolts fit best, and 8.8 internal hexagon bolts fit worse. However, all of them reached more than 98.5%, indicating that the overall fitting effect of the curves was good. Its slope was stable within a certain range.

The torque coefficient is affected by the tightening speed of the nut, the presence or absence of shims, the thickness of shims, the presence or absence of lubrication, the material, and the ambient temperature. Existing studies have proved [36] that the torque coefficients of the above four types of bolts range from 0.2 to 0.45 with the installation of shims based on the calculation of the torque using Equation (7). Under the roughness Ra1.6 of nickel steel plate specimens, the torque coefficient of 8.8-grade M8 external hexagonal bolts ranges from 0.2360 to 0.2525, and the torque coefficient of 8.8-grade M8 internal hexagonal bolts ranges from 0.2312 to 0.2615. The torque coefficient of 12.9-grade M8 external hexagonal bolts ranges from 0.2374 to 0.2576, and the torque coefficient of 12.9-grade M8 internal hexagonal bolts ranges from 0.2504 to 0.2722. The torque coefficient of 8.8-grade M8 external hexagonal bolts is small, and the torque coefficient of 12.9-grade M8 internal hexagonal bolts is the largest. Their torque coefficients are all within a reasonable range, proving the accuracy of the experimental equipment and measurement methods in this paper.

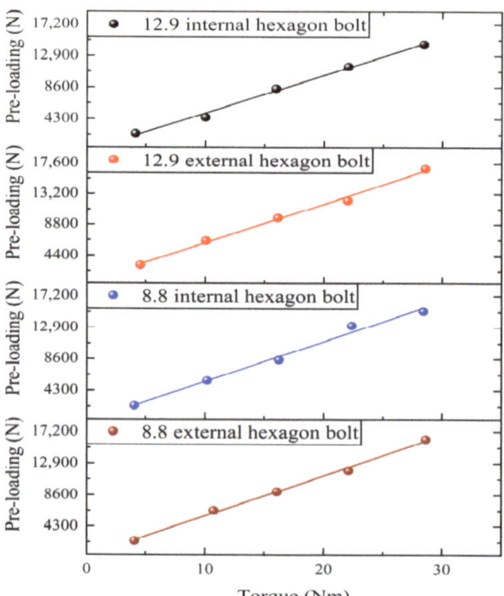

Figure 8. Torque coefficient calibration.

Table 5. Calculation of pre-loading slope.

Equation	$y = a + b{\cdot}x$			
Intercept	149.4 ± 319.5	-14.6 ± 576.4	496.6 ± 380.2	-61.5 ± 366.0
Slope	512.3 ± 17.3	509.3 ± 31.3	505.9 ± 20.7	479.2 ± 20.0
Pearson' s r	0.99829	0.99436	0.99749	0.99739
R-squared (COD)	0.99658	0.98876	0.99499	0.99478
Adjusted R-squared	0.99543	0.98501	0.99332	0.99304

2.3. Bolt Performance Prediction Methods

The bolts were subjected to mixed loads during operation, which led to a relaxation of the pre-loading. However, it was not possible to monitor the pre-loading again using instruments. Therefore, this paper proposes to characterize the pre-loading relaxation law using a mathematical function. The variation of pre-loading under different tangential cyclic loads is predicted. Among the fitting methods are the allometric model function [37], high-order polynomial function [38], and Gaussian function [39].

(1)   Allometric model function

The allometric model function is a method for obtaining optimal parameter estimates in the form of power functions. Built on the basis of a "hierarchy of universes and developments", it allows for quantitative access to models and behaviours. The function is used to describe a curve that grows or decays in the form of a power exponent. The basic formula is as follows:

$$y = a \cdot x^b \tag{15}$$

where x and y are the independent and dependent variables, respectively, and a and b are the coefficients given by the model.

When x is small, the dependent variable changes more significantly, and its slope changes more. When x is large, the dependent variable and slope are regionally stable. It is in line with the law of pre-loading relaxation, so it can be used to predict the performance assessment of pre-loading relaxation.

(2)   Higher-order polynomial function

The higher-order polynomial fitting algorithm is a method for approximating data points using a polynomial function that minimizes the error between the fitted function and the actual data points. The basic idea is to improve the accuracy of the fit to the data by increasing the order of the polynomial. The general form of high-order polynomial fitting is as follows:

$$y = a_0 + a_1 x + a_2 x^2 + \ldots + a_n x^n \tag{16}$$

where a_0, a_1, a_2, ..., a_n denote the polynomial coefficients and n denotes the order of the polynomial. The optimal values of the polynomial coefficients can be solved using mathematical statistical methods such as least squares to obtain an optimal fitting function.

It is important to note that high-order polynomial fitting is prone to overfitting problems. The fitting function is too complex and too sensitive to the data, resulting in a poor fit. To avoid the overfitting problem, regularization methods can be used to optimize the high-order polynomial fitting. In practice, it is important to choose the appropriate polynomial order according to the complexity of the data. Low-order polynomials can be chosen for simple data, while high-order polynomials are required for complex data. At the same time, the fitting results must be evaluated and tested to ensure the validity and reliability of the fitted function. Higher-order polynomials can be fitted directly without a specific physical model. As the number of times increases, it will show the phenomenon that the degree of fit also becomes higher. However, when the number of times is high to a certain degree and then continues to increase, the number of times will exhibit the overfitting phenomenon. After pre-calculation and deduction, the polynomial is the highest degree of fit when the polynomial is 9 times.

$$y = a_0 + a_1 x + a_2 x^2 + a_3 x^3 + a_4 x^4 + a_5 x^5 + a_6 x^6 + a_7 x^7 + a_8 x^8 + a_9 x^9 \tag{17}$$

The specific steps of the fitting process are as follows: use the readtable function to read the data in the file and set the fitting order n from 1 to 9 in a loop. Use the polyfit function to fit the read data, then solve the parameters to judge the degree of fit.

(3)   Gaussian function

Gaussian functions are widely used in statistics to express the normal distribution. The function approximates the set of data points for prediction. The function is characterized as a bell-shaped curve with a multinomial Gaussian function using the following formula:

$$y = a_1 e^{-\left(\frac{x-b_1}{c_1}\right)^2} + a_2 e^{-\left(\frac{x-b_2}{c_2}\right)^2} + a_3 e^{-\left(\frac{x-b_3}{c_3}\right)^2} + a_4 e^{-\left(\frac{x-b_4}{c_4}\right)^2} + a_5 e^{-\left(\frac{x-b_5}{c_5}\right)^2} \tag{18}$$

where a is the height of the peak of the curve, b is the center of the peak region, and c characterizes the width of the bell curve.

The Gaussian function makes predictions based on historical statistics. This creates a predictive model that describes subsequent developments. The above formula is only a common Gaussian function curve fitting form. The specific application may be based on the needs of the problem and the characteristics of the data to choose the appropriate form of the function. In addition, in order to obtain the best fitting effect, it is usually necessary to optimize the parameters using the least squares method after a large number of analyses of the pre-loading data.

In order to verify the prediction effect of the three functions, this paper carried out relaxation experiments on 8.8-grade M8 external hexagonal bolts and internal hexagonal bolts, as well as 12.9-grade M8 external hexagonal bolts and internal hexagonal bolts. The three functions were fitted to the obtained data, and the fitting parameters are shown in Table 6.

Table 6. Fitting accuracy and parameters.

Experimental Category	Fitting Accuracy			Parameter Count		
	Allometric Model Function	Nine-Stage Polynomial Function	Gaussian Function	Allometric Model Function	Nine-Stage Polynomial Function	Gaussian Function
8.8 external hexagonal bolt	0.99794	0.99477	0.99818	2	10	15
8.8 internal hexagonal bolt	0.99919	0.99346	0.99876	2	10	15
12.9 external hexagonal bolt	0.99562	0.98914	0.99814	2	10	15
12.9 internal hexagonal bolt	0.98687	0.98566	0.99917	2	10	15

The accuracy of most of the three fitting functions was above 0.98, which indicated a high fitting effect. However, the Gaussian function had 15 unknown parameters and the most complicated structure. Therefore, for the sake of the simplicity of the subsequent calculations, the Gaussian function is discarded in this paper. The allometric model function, a the nine-stage polynomial function, is used to predict the performance of the bolt.

3. Results and Discussion

3.1. Multiple Factor Impact Analysis

3.1.1. Different Surface Roughness Analysis

Different surface roughnesses of nickel steel flat plates affect the pre-loading of the bolt connection structure. To further analyze the effects of roughness on the pre-loading relaxation of bolted joint structures, the controlled variable method was used to study 8.8-grade hexagonal bolts acting in nickel steel flat plates with Ra0.8 and Ra6.3 roughnesses. Experimentally, a tightening torque of 22 Nm was applied, and a tangential cyclic displacement control of 0.5 mm with a half sine wave was used. Two hundred groups of repetitive experiments were carried out, and the value of pre-tightening force in the MCK-Z-I intelligent meter was recorded every 10 groups. The experimental results are shown in Figure 9. The greater the roughness of the nickel steel plate, the more the pre-loading decayed. Its decay rate was approximately the same, which was the same as the results of the literature [40], proving the correctness of the experimental design of this paper.

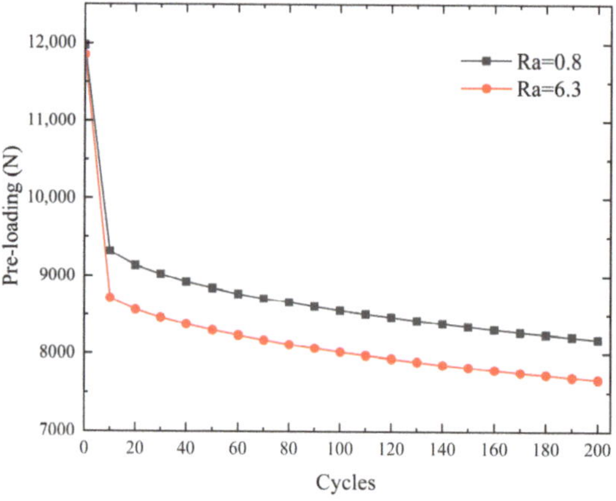

Figure 9. Pre-loading relaxation at different surface roughnesses.

3.1.2. Multiple Cycle Frequency Loading

In order to investigate the pre-loading loosening behavior of the bolted joints with half sine wave displacement control at different loading frequencies, three (0.1 Hz, 1.0 Hz, 5.0 Hz) cyclic frequency experimental conditions were selected. A nickel steel plate with Ra1.6 roughness was used to control the tangential cyclic displacement of 1 mm, and 100 sets of loosening experiments were carried out. The value of pre-loading in the oscilloscope was recorded every 5 groups of cycles, and the experimental results are shown in Figure 10. The most loosening of the bolt pre-loading occurred under lower-frequency cyclic tangential loading. With higher-frequency cyclic tangential loads, the bolt pre-loading loosening was less. The contact time between the threaded sub and the nickel steel flat plate screw hole was longer under low-frequency loading. The connection interface was stressed for a longer period of time, resulting in more loosening of the bolt pre-loading.

As the number of cycles increases, the bolt gradually begins to loosen, as shown in Figure 11a. The force required to reach a tangential displacement of 1 mm gradually decreases. This leads to the gradual loosening of the bolt, which affects the safety of the equipment. The upper peak of displacement can reach 1 mm stably, but there is some fluctuation in the lower peak. The existence of a gap between the nickel steel plate and the bolt leads to instability in the displacement when the plate returns to the initial position, as shown in Figure 11b. This fluctuation is unavoidable because the gap between the two cannot be eliminated.

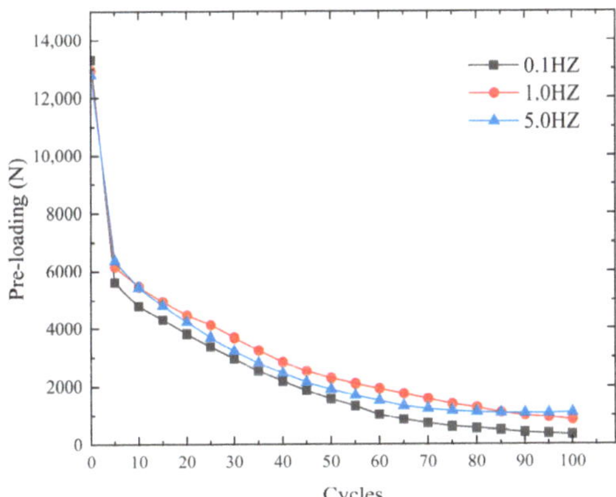

Figure 10. Comparison of three cycle frequencies.

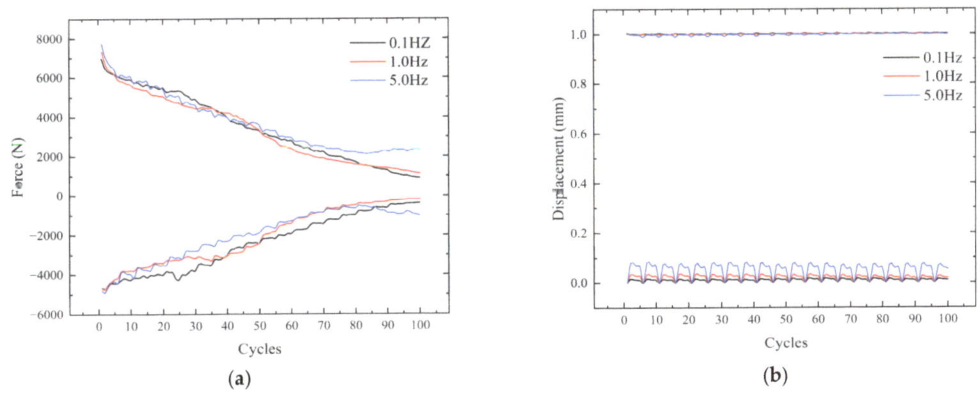

(a)

(b)

Figure 11. Variation curves at different frequencies. (**a**) Force change curve. (**b**) Displacement change curve.

3.1.3. Differential Study of Waveform Control

The universal experimental machine can be equipped with sine and half-sine waves with different tangential cyclic displacements, as shown in Figure 12. Under waveform control, the rising phase of the curve is the universal testing machine stretching phase. When the displacement value reaches the set value, the curve begins to fall, which is the universal laboratory machine compression stage.

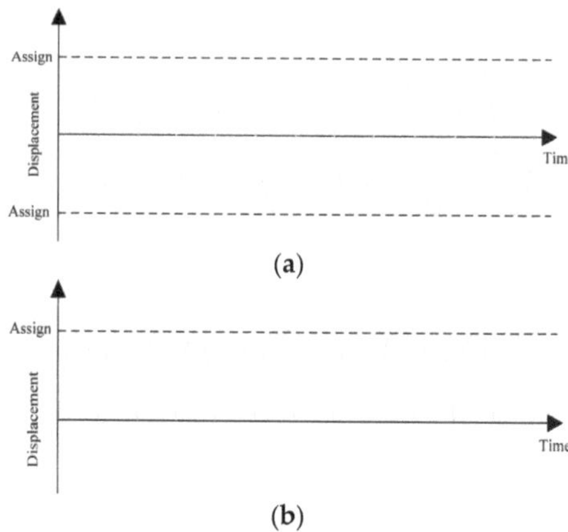

Figure 12. Two types of waveforms. (**a**) Sine wave, (**b**) half sine wave.

Based on 8.8-grade M8 hexagonal bolts carried out under sinusoidal wave control in the form of tangential cyclic displacement, when the machine reaches a set displacement, a certain tensile force is generated. However, this force presents a non-linear variation and cannot ensure a single variable for the experiment. Therefore, the use of a displacement signal ensures that the flat plate pulls the same displacement. In this experiment, the center position is 0 mm, the amplitude is 1 mm, the vibration frequency is 1 Hz, the target number of cycles is 500, and the force/displacement cycle is shown in Figure 13. In the initial vibration stage, the lower peak value of displacement was stable at about 0.2 mm, and the upper peak value of displacement showed fluctuations between 0.584 and 1 mm. Neither reached the set value, which led to changes in the tension of the universal testing machine. The reason for this is that the hole diameter of the nickel steel plate is 10 mm, the bolt diameter is 8 mm, and there is an unavoidable gap between the two. The buckling phenomenon occurs when reciprocating displacement is applied to the universal testing machine. The change in displacement will have repeated impacts on the center gap, which in turn causes the inaccuracy of the results. Therefore, the displacement signal of the sine wave control did not meet the set standard.

Adjusting the waveform to a half-sine eliminates the effects of simultaneous tensile and compressive loading of the nickel steel plate. Its displacement stability is shown in Figure 14. The torque of 22 Nm was applied in the three groups of experiments, and the initial pre-loading values were 12,398 N, 12,285 N, and 12,696 N, respectively. Due to the error of the experiment, the pre-loading fluctuates in a certain range. The total number of cycles at a tangential 1.0 mm displacement was 130. However, at 73 cycles, the bolt had loosened and had no tightening effect. The total number of cycles for a tangential displacement of 0.5 mm is 500, whereas for 400 cycles the pre-loading is less than 100. At 400 cycles, the value of pre-loading is less than 100. A total of 1000 cycles is required for 0.25 mm tangential displacement, and the pre-loading remains in the stable range. At the end of the experiment, 10,539 N of pre-loading remained, with a loss of only 15%. The experimental results show that the bolts remained robust under a small displacement and loosened faster under a large displacement.

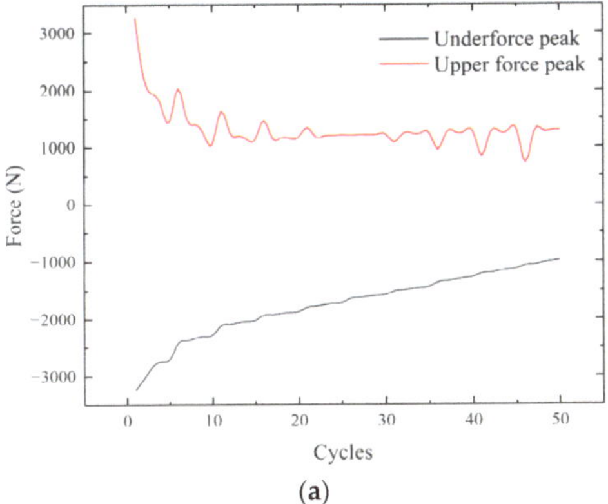
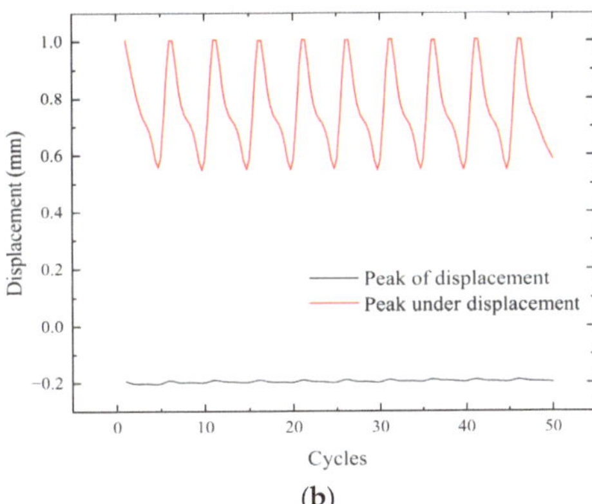

(**a**)

(**b**)

Figure 13. Waveform curve under sinusoidal waveform control. (**a**) Force change curve. (**b**) Displacement change curve.

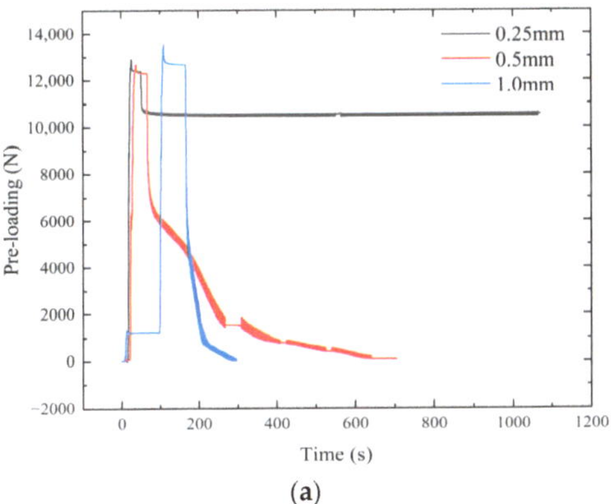
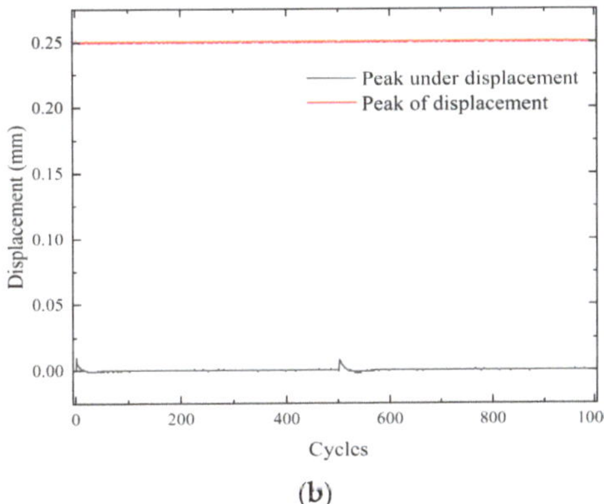

(**a**)

(**b**)

Figure 14. Waveform curve under half sine wave control. (**a**) Half sine wave displacement control. (**b**) Displacement change rule.

3.2. Study on Relaxation Law of Bolt Pre-Loading

3.2.1. Analysis of Relaxation Efficiency of Different Bolts

In order to more finely study the changes in pre-loading relaxation under bolt vibration conditions, this paper carries out a manual control of a universal testing machine in order to stretch and compress a nickel steel flat plate. This operation eliminates the effects of the half-sine wave. Based on nickel steel flat plates with a Ra1.6 roughness, the effects of different bolt strengths and different bolt head contact areas on pre-loading relaxation are investigated. The median of this experiment is 0 mm, the amplitude is 1 mm, the vibration frequency is 1 Hz, and the target number of cycles is 50 times. The relaxation law of pre-loading is shown in Figure 15, and the specific relaxation parameters are shown in Table 7. High-strength bolts have high anti-loosening properties. Under the same bolt strength grade, the outer hexagon bolt loosens faster than the inner hexagon bolt. The contact area of the bolt head has a certain influence on the relaxation of pre-loading.

The extent of damage to the threads of the different bolts is shown in Figure 16. The length of the bolt is 40 mm, and the severe wear is at the coupling of the upper and lower nickel steel plates. There is a gap at the coupling, and it is subjected to shear force, which results in a high degree of thread wear. The surface of the flat plate and the threads, the nut, and the screw are all in cross-scale phenomena. Under the control of a large displacement, after many cycles of vibration, it is easy to cause fractures here, which affect the safety performance of the equipment.

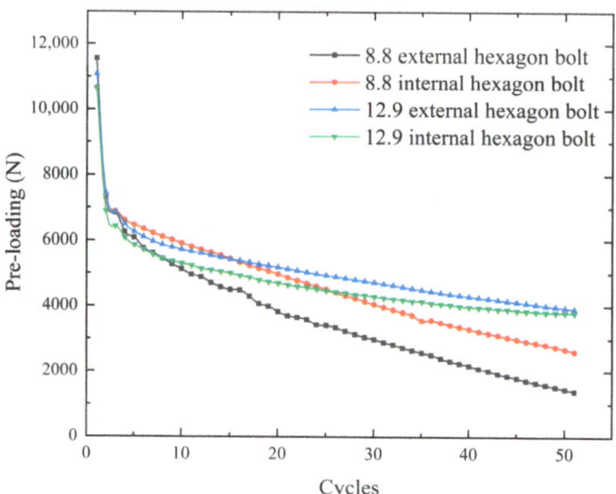

Figure 15. Pre-loading relaxation of different bolts.

Table 7. Pre-loading decay of different bolts.

Bolt	Torque (Nm)	Initial Pre-Loading (N)	Pre-Loading after Test (N)	Torque Coefficient	Pre-Loading Decay (%)
8.8 external hexagon bolt	22.17	11,567	1402	0.2396	87.9%
8.8 internal hexagon bolt	22.15	10,662	2622	0.2597	75.4%
12.9 external hexagon bolt	22.05	11,083	3803	0.2487	65.7%
12.9 internal hexagon bolt	22.0	10,661	3913	0.2579	63.3%

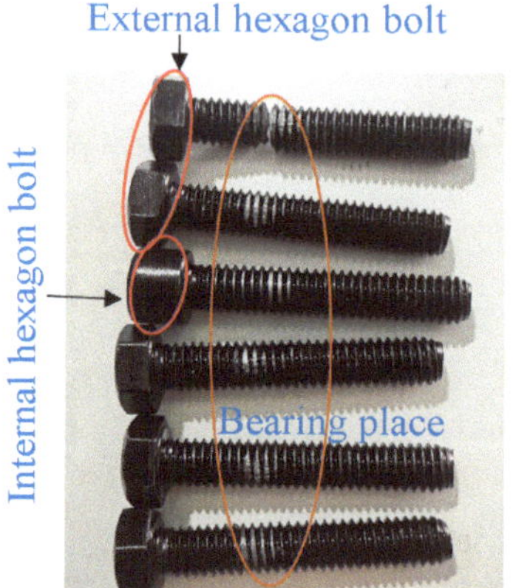

Figure 16. Degree of bolt damage.

During the tangential load loading process, the tangential tension of the universal testing machine versus time is shown in Figure 17. As the number of loading times increases, the force required to reach the set 1 mm tangential displacement gradually decreases. In the last cycle, the positive tension reaches 1085 N, which is 76.8% lower than the initial tension value. The negative pressure reaches 276 N, which is 90.6% lower than the initial pressure value, and the pressure attenuation is more. At this time, the bolt loosening degree is large, affecting the safety and reliability of the equipment.

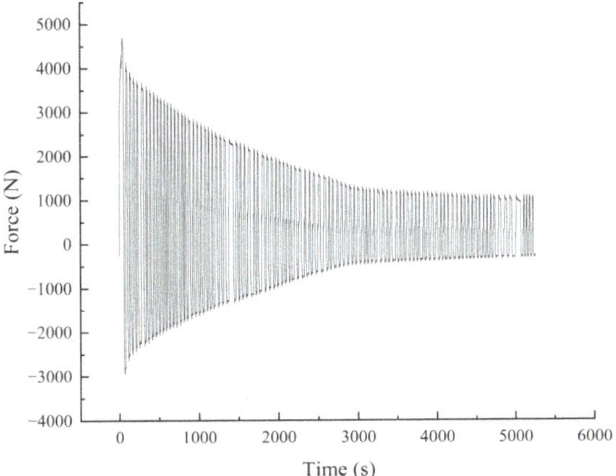

Figure 17. Force time variation relationship.

3.2.2. Effects of Different Tangential Displacements

A phase 22 Nm torque was applied to the same kinds of blots to study the bolt pre-loading relaxation law under 0.25 mm, 0.5 mm, 1.0 mm, 1.5 mm, and 2.0 mm tangential cyclic displacement. The tangential load vibration experiments were carried out 100 times to record the value of pre-loading when the tangential displacement was 0. The experimental results are shown in Figure 18a, and the whole process of bolt pre-loading changes is shown in Figure 18b. Under the action of a small load of 0.25 mm, the bolt can maintain a relatively good tightening effect. The pre-loading loss is only 22.0% after 100 cycles. With the increase in tangential displacement, the attenuation of bolt pre-loading increases. Under the action of a 2.0 mm tangential displacement, the bolt reached complete relaxation in only 50 cycles. In the first vibration cycle, the pre-loading was reduced by 50.2%, as shown in Table 8. It can be seen that under the action of a large tangential vibration load, the bolt pre-loading relaxation was faster. This affects the safety and reliability of equipment. In engineering practice, multiple bolts can be tightened in a nickel steel plate to avoid the impact of large load tangential displacement on complex equipment.

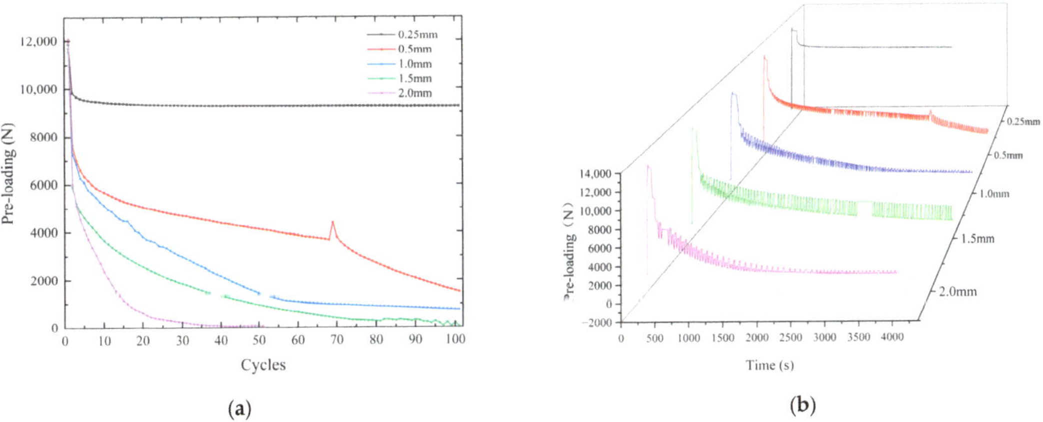

Figure 18. Pre-loading change rule under five kinds of tangential displacements. (**a**) Nodal variation curve of pre-loading cycle. (**b**) The whole process change curve.

Table 8. Specific attenuation of the bolt.

Displacement	1st	10th	50th	100th
0.25 mm	17.1%	20.5%	21.8%	22.0%
0.5 mm	36.2%	53.5%	65.6%	87.4%
1.0 mm	36.8%	57.1%	87.9%	93.6%
1.5 mm	49.6%	69.8%	92.1%	99.6%
2.0 mm	50.2%	82.7%	99.3%	

The hysteresis return line under a 1 mm tangential cyclic displacement is shown in Figure 19. Within a single cycle, the area enclosed by the curve is the energy dissipation value. As the cycle period increases, the curve gradually approaches the Y = 0 region, and its energy dissipation gradually decreases. The difference in energy dissipation between cycle 1 and cycle 2 is large, reflecting the large change in the value of bolt pre-loading in the next cycle. The maximum values of forces in different cycles in the figure are 4697 N, 3022 N, 1546 N, and 1030 N. The maximum force in 100 cycles is reduced by 78.1% compared to 1 cycle. The structure reaches the same tangential displacement, is subjected to a gradually decreasing maximum external tension, and has a lower degree of energy dissipation.

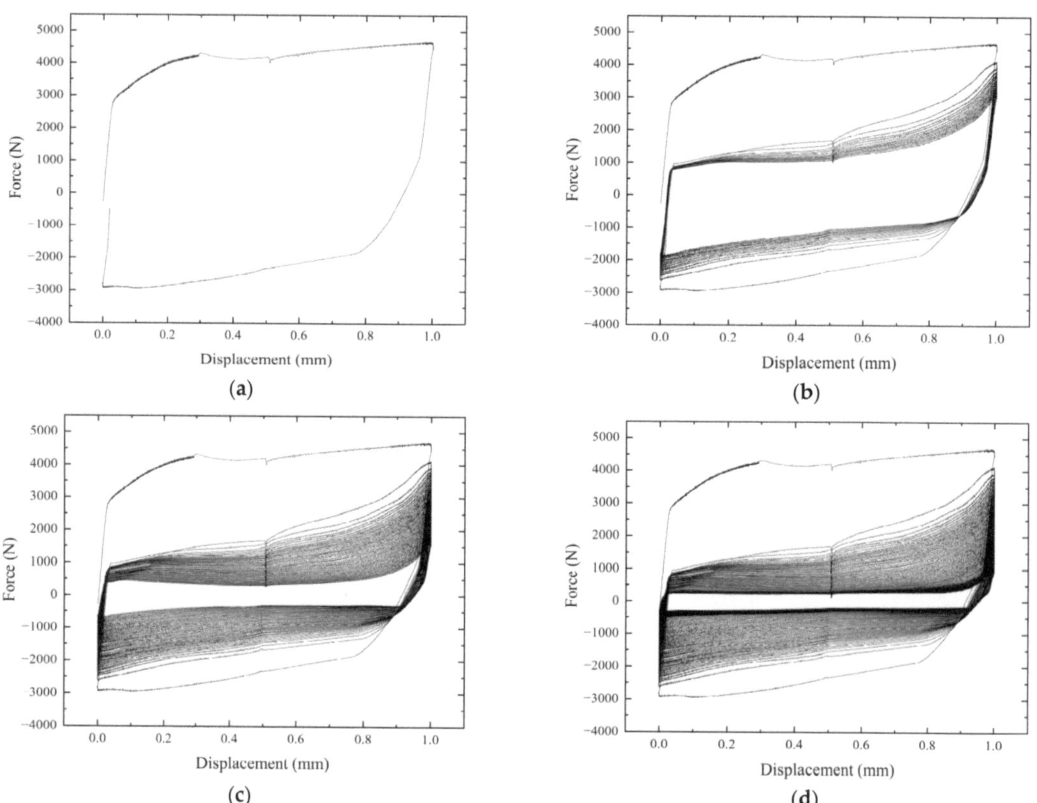

Figure 19. Hysteresis return line at a 1 mm displacement in tangential direction. (**a**) 1 cycle. (**b**) 15 cycles. (**c**) 50 cycles. (**d**) 100 cycles.

3.2.3. Effects of Different Initial Tightening Moments

Based on a Ra1.6 roughness nickel steel plate, bolt pre-loading relaxation experiments was carried out under different torques (4 Nm, 10 Nm, 16 Nm, 22 Nm, 28 Nm). The magnitude of the pre-loading at the tangential cyclic 0 mm displacement was recorded. The experimental results are shown in Figure 20a, and the curve of the whole process of pre-loading relaxation is shown in Figure 20b. In the process of stretching and compression displacement, the pre-loading has a small fluctuation. The initial pre-loading is 2105 N, 5963 N, 8402 N, 11,567 N, and 15,027 N, respectively. After 100 cycles of tangential 1 mm displacement, the remaining pre-loading is 78 N, 1845 N, 1568 N, 742 N, and 3523 N, respectively, and the specific decay percentages are shown in Table 9. Under the large displacement of 1 mm, the pre-loading generated by the 4 Nm torque is small and does not reach the tightening effect, resulting in its faster loosening. The 10 Nm torque has a good tightening effect, and the loss of pre-loading is only 69.1% after 100 vibration cycles. However, between 10 and 22 Nm, after the first cycle, there is little difference in the decay efficiency of pre-loading among the three. After several cycles, with the increase in torque, the attenuation speed of pre-loading increases, and the bolts can maintain a good pre-loading performance under the action of 28 Nm.

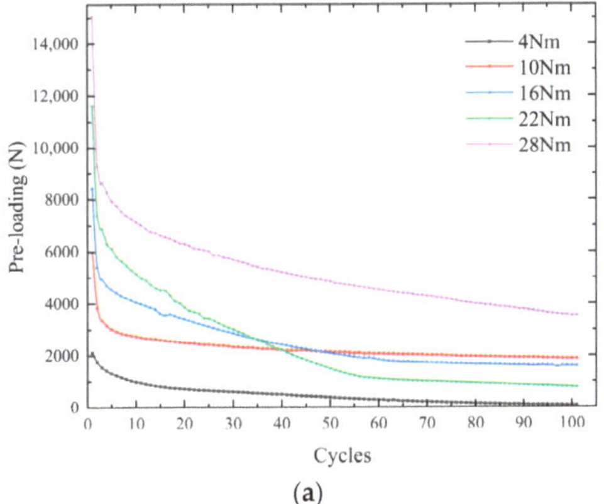

(a)

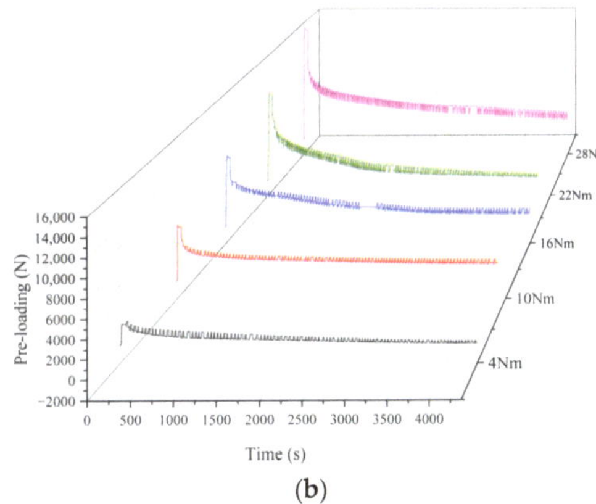

(b)

Figure 20. Pre-loading decay under different torques. (**a**) Nodal variation curve of pre-loading cycle. (**b**) Change curve of the whole process.

Table 9. Specific decay of bolts.

Torque	1st	10th	50th	100th
4 Nm	16.2%	54.7%	82.3%	96.3%
10 Nm	36.2%	55.2%	64.8%	69.1%
16 Nm	35.9%	52.8%	75.8%	81.3%
22 Nm	36.8%	57.1%	87.9%	93.6%
28 Nm	38.3%	53.5%	68.3%	76.6%

The backbone curve and stiffness degradation curve for a single stretch to 0.5 mm displacement are shown in Figure 21. The maximum force applied to the universal testing machine in the range of tangential displacement of 0–0.5 mm are 1852 N, 3285 N, 4770 N, 5486 N, and 7246 N. The increase in the initial torque of the bolts leads to an increase in the tensile force of the universal testing machine, but there is no linear multiplication.

The initial stiffness increases along with the increase in the tightening torque, respectively: 5.0438×10^4 N/mm, 8.5500×10^4 N/mm, 9.2083×10^4 N/mm, 9.8404×10^4 N/mm, 11.5200×10^4 N/mm. The bolt stiffness will show a brief upward trend in the initial stage. As the tangential displacement increases, the stiffness gradually degrades. At the 0.5 mm tangential cyclic displacement, the bolt is subjected to shear action, and the stiffness begins to change.

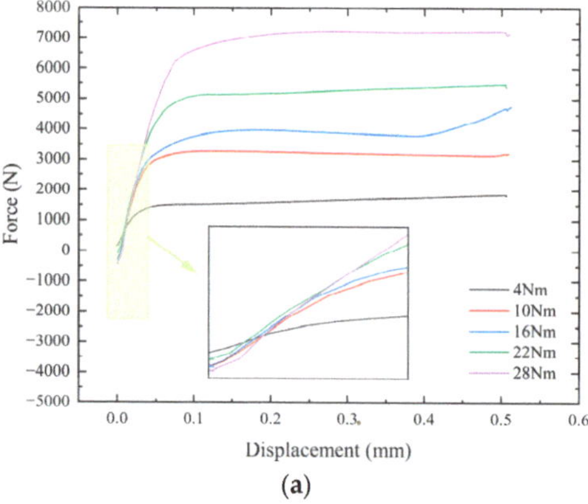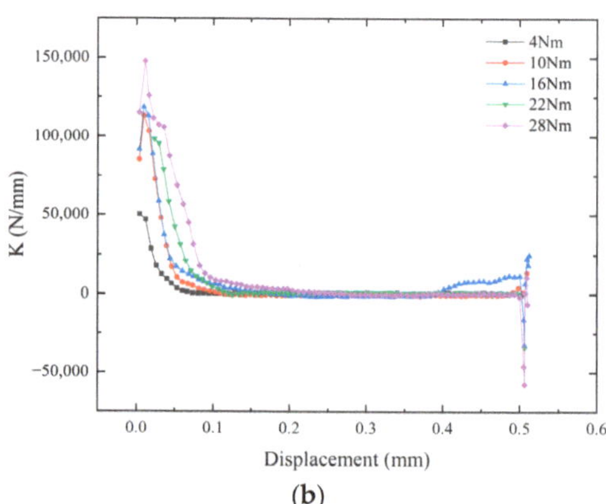

Figure 21. Energy change under different initial pre-loading conditions. (**a**) Backbone curve. (**b**) Stiffness degradation curve.

The bolt pre-loading degradation law presents two stages. The first stage is a rapid decline in bolt pre-loading, and the second stage is a slow decline. The first stage is caused by excessive local stress caused by the steel plate extrusion bolt and excessive pressure on the annular support surface of the fastener head. At the same time, the material yields and undergoes plastic deformation, which in turn leads to the relaxation of bolt pre-loading. It can be easily obtained from the data analysis that the phase transition process occurs at about 5–10 cycles. The second stage is when the bolts are subjected to large loads in the tangential cycle, and the bolted structure is most prone to rotational loosening. The next most prone to loosening is torque loading, and axial loading is the least prone to loosening. It is worth noting that the occurrence of relative rotation is not instantaneous and requires some accumulation. Local sliding is caused by a long accumulation, so the second stage is a slow descent process.

3.3. Predictive Study of Bolt Pre-Loading Performance

3.3.1. Performance Prediction under Multiple Displacements

The mathematical function of different tangential cyclic displacements is investigated for the case of a 22 Nm external tightening torque. The pre-loading relaxation law function is constructed using the allometric model function and the nine-stage polynomial function. The mathematical functions for five different tangential cyclic loads are shown in Figure 22.

The Allometricl model function is in good agreement with the experimental values after several cyclic loadings. The fitting accuracies are 0.71719, 0.72726, 0.88981, 0.85451, and 0.94252, respectively. There are large deviations in the data in 10–40 cycles, and the deviations are small in other cycles. With the increase in the tangential cyclic displacement, the parameter a in the power function increases gradually, and the parameter b decreases gradually. For a small displacement of 0.25 mm, the maximum prediction error is 1.2%. In the case of a 1.5 mm large displacement, the maximum prediction error is 38.2%. There are obvious differences in the curve equations, indicating that the relaxation change rules of the pre-loading are different. The nine-stage polynomial function improves the fitting accuracy by increasing the order of the polynomial. The fitting accuracy under a 0.25–2 mm displacement is 0.90364, 0.96814, 0.98343, 0.95656, 0.98215, respectively. Its fitting accuracy is higher than the alometric model function. The fitting accuracy of the two functions is poor when the displacement load is 0.25 mm, mainly because the relaxation of the pre-loading is relatively slow and the region is gradually stable.

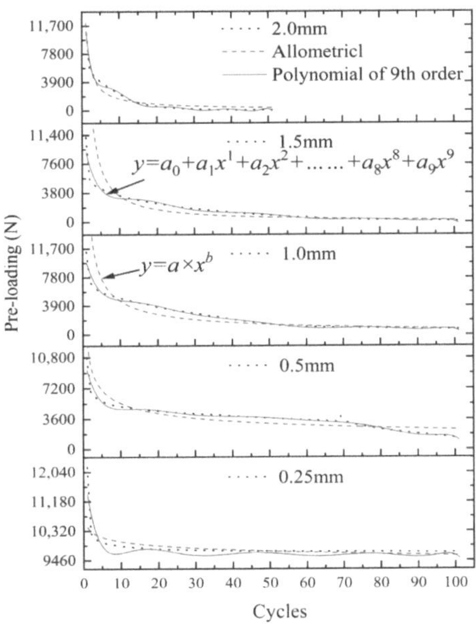

Figure 22. Mathematical function fitting for different tangential cyclic displacements.

3.3.2. Performance Prediction under Multiple Initial Torques

Based on the 1 mm tangential cyclic displacement, the mathematical function expression of the change in different initial pre-loadings with the number of cycles is constructed. The pre-loading relaxation law function and the parameters in the discrimination function can be constructed using two methods. The results of parameter identification under different initial pre-loadings are shown in Figure 23. The fitting accuracy of the fitting function in 4–28 Nm are 0.88923, 0.89613, 0.90541, 0.87723, and 0.94667. There are deviations in 10–40 cycles. The 35-cycle deviation in the 4 Nm torque is 31.3%, but the difference in pre-loading is small, which is in the normal range. Parameter a shows fluctuating changes under different initial pre-loads, and parameter b is more stable. It shows that the initial torque only affects parameter a. Overall, the curve parameters are well identified. Under the same tangential cyclic displacement, the curve patterns are approximately the same. The fitting accuracies of the nine-stage polynomial function for 4–28 Nm are 0.99818, 0.93917, 0.97441, 0.98343, and 0.95481, respectively. The fitting effects are all above 95%, which have a good accuracy. However, its polynomial contains 10 parameters to be identified, while the allometric model function has only 2 parameters to be identified, which leads to more complicated results.

In Origin 2022, pre-loading data plots are constructed and fitted, and the parameters are identified by setting the function. Based on the least squares method, numerical simulation techniques were used to determine the appropriate parameters of the function. The objective function describes the variation of the data through the optimal parameters. In this case, the nine-stage polynomial is used to select the optimal value by increasing the order from 1–9. The fitting effect is judged by the goodness of fit and root mean square error. Combined with the above studies, the allometric model function fits the worst, at 71.7% for the small displacement condition. Other than that, the allometric model function and the nine-stage polynomial function can reach more than 85.5% and 90.4%. This proves the accuracy of this paper for the prediction of bolt pre-loading relaxation.

Combined with the above studies, the bolt pre-loading relaxation under the action of tangential cyclic displacement can be divided into two stages. The first stage is the stage of rapid decrease of bolt pre-loading, and the second stage is the slow decrease process. The relaxation law in a specific case can be expressed by the allometric model function and nine-stage polynomial function. It can provide theoretical support for the safety and reliability of large and complex equipment.

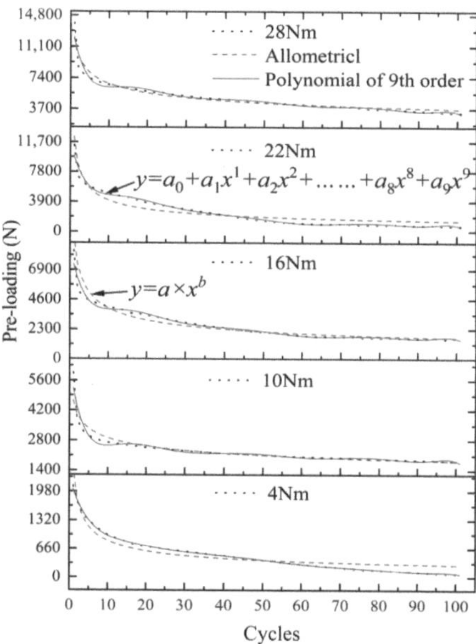

Figure 23. Mathematical function fitting for different initial pre-loadings.

4. Conclusions

Based on the problem that there are more bolt connection structures in complex equipment and the structure is prone to failure, this paper is based on the ring force transducer, universal experimental machine, and other equipment, with a simplified high-strength Z-shaped nickel steel plate used as the object of study. The relaxation law of bolt pre-loading for complex equipment is studied. The maximum tightening torque of the bolt was calculated, and the bolt pre-loading relaxation experiment was designed. The range of torque coefficients between bolt pre-loading and torque were calibrated, and they were all between 0.2312 to 0.2722, which was in line with the design range. The effects of different waveforms, nickel steel plate surface roughnesses, tangential displacement frequencies, bolts, torque magnitudes, and cyclic tangential displacements on pre-loading relaxation were investigated. A study of bolt performance prediction methods was carried out. The experimental results showed the following:

(1) The half sine wave has better stability. The rougher the nickel steel plate, the more pre-loading decay, but the decay rate is about the same. However, the rate of decay is approximately the same. The nickel steel plate of Ra6.3 is rougher than the plate of Ra0.8, with more more pre-loading attenuation, and the rate of decay is approximately the same. The contact time between the threaded sub and the screw hole of the nickel steel flat plate is longer under low-frequency loading. The connection interface is stressed for a longer time, resulting in more loosening of the bolt pre-loading. For the same bolt strength class, the outer hexagon bolts loosen faster than the inner hexagon bolts, and the high-strength bolts have a better stability.

(2) The pre-loading decay rate of the bolts increases as the tangential cyclic displacement increases. At a larger tangential cyclic displacement of 2 mm, the bolt is completely loosened and not tightened after 50 cycles of tangential vibration. At the smaller tangential cyclic displacement of 0.25 mm, the bolt loosened more slowly. After many cycles, it still maintained a good tightening effect. At a smaller initial torque, the bolt was not tightened. A 28 Nm torque had a better tightening effect, and between

10 and 22 Nm, after the first cycle, the three pre-loading decline efficiency was not significantly different. After several cycles, along with the increase in torque, the pre-loading decay rate increased.

(3)   The bolt pre-loading degradation law shows two stages. The first stage is a rapid decline in bolt pre-loading, and the second stage is a slow decline process. The first stage is due to the steel plate extrusion bolt resulting in excessive local stress, and fastener head ring support surface pressure caused by excessive yielding of the material makes the material undergo plastic deformation. This in turn leads to relaxation of the bolted joint pre-loading. The second stage is due to the slow decline phase resulting in micromotor wear. Wear cannot be instantaneous and needs to accumulate gradually.

(4)   Parameter identification of pre-loading relaxation curves was carried out. Mathematical functions that can express different tangential cyclic displacements and different initial pre-loadings were constructed. Under the small displacement condition, the allometric model function fit the worst, reaching 71.7%. Otherwise, the allometric model function was able to predict more than 85.5%, which required the use of least squares to identify the two unknown parameters. The nine-stage polynomial function fitting accuracy can reach more than 90.4%, which requires the use of least squares to identify 10 parameters. The complexity of the two is different, but both can better characterize the pre-loading relaxation law under specific conditions. The accuracy of this paper's prediction of bolt pre-loading relaxation is demonstrated.

(5)   In this paper, the pre-loading relaxation experiments of many kinds of bolts are carried out, and the study of relaxation based on the allometric model function and nine-stage polynomial function is carried out. In this paper, the bolt performance under the condition of a large tangential external load can be predicted. It is helpful to improve the safety and reliability of complex equipment, the aerospace industry, and other equipment, and to provide theoretical support for it.

Author Contributions: X.L. and M.Z. conceived and designed the project. C.L. performed the experiments and analyzed the data. S.L. wrote and edited the manuscript. S.W. and Z.L. reviewed and edited the manuscript. All authors have read and agreed to the published version of the manuscript.

Funding: This research received no external funding.

Institutional Review Board Statement: Not involving humans or animals.

Informed Consent Statement: Not applicable.

Data Availability Statement: Data sharing does not apply to this article.

Conflicts of Interest: The authors declare that they have no conflicts of interest.

References

1.   Guo, Z.; Lu, N.; Zhu, F.; Gao, R. Effect of preloading in high-strength bolts on bolted-connections exposed to fire. *Fire Saf. J.* **2017**, *90*, 112–122. [CrossRef]

2.   Jin, J.; Wu, X.; Zheng, J. Improvement of prestressing method for bolted joints. *J. Nav. Eng. Univ.* **2010**, *22*, 20–24.

3.   Gong, H.; Liu, J.; Ding, X. Study on the mechanism of preload decrease of bolted joints subjected to transversal vibration loading. *Proc. Inst. Mech. Eng. Part B J. Eng. Manuf.* **2019**, *233*, 2320–2329. [CrossRef]

4.   Zhang, Z.; Liu, M.; Su, Z.; Xiao, Y. Quantitative evaluation of residual torque of a loose bolt based on wave energy dissipation and vibro-acoustic modulation: A comparative study. *J. Sound Vib.* **2016**, *383*, 156–170. [CrossRef]

5.   Miao, R.; Shen, R.; Zhang, S.; Xue, S. A review of bolt tightening force measurement and loosening detection. *Sensors* **2020**, *20*, 3165. [CrossRef] [PubMed]

6.   Yang, J.; Hao, K.; Zhong, W.; Wang, W.; Miao, G.; Zhang, Q. Study on the loosening behavior of bolted and riveted connections under shear cyclic loading. *Prog. Archit. Steel Struct.* **2024**, *26*, 52–61.

7.   Hu, P.; Jiang, B.; Bao, Y.; Liu, J.; Yuan, P. Preloading of composite single-bolt joint structures. *Comput. Aided Eng.* **2023**, *32*, 42–47+54.

8.   Li, S.; Gao, P.; Wang, Y.; Wang, W.; Zhao, Z. Design and performance analysis of spindle bearing preload control assembly. *Comb. Mach. Tools Autom. Mach. Technol.* **2023**, *10*, 164–168.

9. Bibel, G.; Ezell, R. An improved flange bolt-up procedure using experimentally determined elastic interaction coefficients. *J. Press. Vessel. Technol.* **1992**, *114*, 439–443. [CrossRef]

10. Wang, W.; Xu, H.; Ma, Y.; Liu, H. Self-relaxation mechanism of bolted joints under vibration conditions. *Vib. Shock* **2014**, *33*, 198–202.

11. Kong, Q.; Li, Y.; Wang, S.; Yuan, C.; Sang, X. The influence of high-strength bolt preload loss on structural mechanical properties. *Eng. Struct.* **2022**, *271*, 114955. [CrossRef]

12. Liu, Y.; Zhu, M.; Lu, X.; Wang, S.; Li, Z. Simulation of Preload Relaxation of Bolted Joint Structures under Transverse Loading. *Coatings* **2024**, *14*, 538. [CrossRef]

13. Yang, J.; DeWolf, J.T. Mathematical model for relaxation in high-strength bolted connections. *J. Struct. Eng.* **1999**, *125*, 803–809. [CrossRef]

14. Shen, Y.; Xiao, Y. Effect of micromotion damage on preload relaxation of composite bolted joints. *Acta Mater. Compos. Sin.* **2019**, *36*. [CrossRef]

15. Guo, K.; Zhao, Y.; Wang, W. A passive and wireless smart washer for monitoring bolt pretightening force. *Meas. Sci. Technol.* **2022**, *34*, 035903.

16. Yuan, B.; Sun, W.; Wang, Y.; Zhao, R.; Mu, X.; Sun, Q. Study on bolt preload measurement: An error compensation model for ultrasonic detection based on solid coupling. *Measurement* **2023**, *221*, 113484. [CrossRef]

17. Zhao, Q.; Wang, J.; Liao, M. Experimental study on influence of different tightening positions on bolt preload. *Adv. Eng. Technol. Res.* **2022**, *1*, 523. [CrossRef]

18. Li, P.; Chen, Q.; Zhou, Q.; Zhang, D. Design of Remote Monitoring System for Pre-Tightening Force of Anchor Bolts. *J. Phys. Conf. Ser.* **2021**, *2005*, 012234.

19. Yang, Z.; Huo, L. Bolt preload monitoring based on percussion sound signal and convolutional neural network (CNN). *Nondestruct. Test. Eval.* **2022**, *37*, 464–481. [CrossRef]

20. Baghalian, A.; Senyurek, V.Y.; Tashakori, S.; McDaniel, D.; Tansel, I.N. A novel nonlinear acoustic health monitoring approach for detecting loose bolts. *J. Nondestruct. Eval.* **2018**, *37*, 24. [CrossRef]

21. Wang, F.; Chen, Z.; Song, G. Smart crawfish: A concept of underwater multi-bolt looseness identification using entropy-enhanced active sensing and ensemble learning. *Mech. Syst. Signal Process.* **2021**, *149*, 107186. [CrossRef]

22. Huynh, T.-C.; Dang, N.-L.; Kim, J.-T. Preload monitoring in bolted connection using piezoelectric-based smart interface. *Sensors* **2018**, *18*, 2766. [CrossRef]

23. Wang, C.; Wang, N.; Ho, S.-C.; Chen, X.; Pan, M.; Song, G. Design of a novel wearable sensor device for real-time bolted joints health monitoring. *IEEE Internet Things J.* **2018**, *5*, 5307–5316. [CrossRef]

24. Pan, Q.; Pan, R.; Chang, M.; Xu, X. A shape factor based ultrasonic measurement method for determination of bolt preload—ScienceDirect. *NDT E Int.* **2020**, *111*, 102210. [CrossRef]

25. Sun, Q.; Yuan, B.; Mu, X.; Sun, W. Bolt preload measurement based on the acoustoelastic effect using smart piezoelectric bolt. *Smart Mater. Struct.* **2019**, *28*, 055005. [CrossRef]

26. Park, J.-H.; Kim, T.; Lee, K.-S.; Nguyen, T.C.; Kim, J.-T. Novel bolt-loosening detection technique using image processing for bolt joints in steel bridges. In Proceedings of the 2015 World Congress on Advances in Structural Engineering and Mechanics (ASEM15), Incheon, Republic of Korea, 25–29 August 2015.

27. Akbar, M.A.; Qidwai, U.; Jahanshahi, M.R. An evaluation of image-based structural health monitoring using integrated unmanned aerial vehicle platform. *Struct. Control Health Monit.* **2019**, *26*, e2276.1-20. [CrossRef]

28. Song, D.; Xu, X.; Qiu, S.; Ou, Y.; Chen, W. Bolt fastener detection based on improved YOLOv5. *J. Nav. Eng. Univ.* **2024**, *4*, 1–5.

29. ISO 898-2:2022; Mechanical Properties of Fasteners Made of Carbon Steel and Alloy Steel—Part 2: Nuts with Specified Property Classes—Coarse Thread and Fine Pitch Thread. International Organization for Standardization: Geneva, Switzerland, 2022.

30. Qin, D.; Xie, L. *Modern Mechanical Design Manual*; Chemical Industry Press: Beijing, China, 2011.

31. GB/T 5783-2016; Standardization Administration of the State. Dimension and Shape Tolerance of Mechanical Products. Standards Press of China: Beijing, China, 2016.

32. Motosh, N. Development of Design Charts for Bolts Preloaded up to the Plastic Range. *J. Eng. Ind.* **1976**, *98*, 849. [CrossRef]

33. Nassar, S.A.; Sun, T.; Zou, Q. The Effect of Coating and Tightening Speed on the Torque-Tension Relationship in Threaded Fasteners. *Theriogenology* **2006**, *83*, 1304.

34. Persson, E.; Roloff, A. Ultrasonic tightening control of a screw joint: A comparison of the clamp force accuracy from different tightening methods. *Arch. Proc. Inst. Mech. Eng. Part C J. Mech. Eng. Sci.* **2016**, *230*, 2595–2602. [CrossRef]

35. Wang, S.-A.; Zhu, M.; Xie, X.; Li, B.; Liang, T.-X.; Shao, Z.-Q.; Liu, Y.-L. Finite Element Analysis of Elastoplastic Elements in the Iwan Model of Bolted Joints. *Materials* **2022**, *15*, 5817. [CrossRef] [PubMed]

36. Jia, X.; Li, H.; Yuan, W. Experimental study on the influencing factors of torque coefficient of high strength bolts. *Machinery* **2004**, *S1*, 8–9.

37. Xin, Y.J.; Cai, P.C.; Li, P.; Qun, Y.; Sun, Y.T.; Qian, D.; Cheng, S.L.; Zhao, Q.X. Comprehensive analysis of band gap of phononic crystal structure and objective optimization based on genetic algorithm. *Phys. B Condens. Matter* **2023**, *667*, 415157. [CrossRef]

38. Gelman, A.; Imbens, G. Why High-Order Polynomials Should Not Be Used in Regression Discontinuity Designs. *J. Bus. Econ. Stat.* **2019**, *37*, 447–456. [CrossRef]

39. Guo, H. A simple algorithm for fitting a Gaussian function [DSP tips and tricks]. *IEEE Signal Process. Mag.* **2011**, *28*, 134–137. [CrossRef]
40. Xie, Y.; Xiao, Y.; Lv, J. Relaxation behavior of bolted joints induced by material and contact creep: Elastic-viscoplastic analysis method. In Proceedings of the Chinese Mechanics Conference, Guangzhou, China, 27–30 July 2019; pp. 1255–1262.

sensors

MDPI

Article

A New Method for Bearing Fault Diagnosis across Machines Based on Envelope Spectrum and Conditional Metric Learning

Xu Yang [1], Junfeng Yang [2,*], Yupeng Jin [3] and Zhongchao Liu [1]

[1] School of Intelligent Manufacturing, Nanyang Institute of Technology, Nanyang 473004, China; yangxuican@bjtu.edu.cn (X.Y.); lzc@nyist.edu.cn (Z.L.)
[2] School of Electrical Engineering, Beijing Jiaotong University, Beijing 100044, China
[3] School of Aerospace, Mechanical and Mechatronic Engineering, The University of Sydney, Sydney, NSW 2050, Australia
* Correspondence: yangjunfeng@bjtu.edu.cn

Abstract: In recent years, most research on bearing fault diagnosis has assumed that the source domain and target domain data come from the same machine. The differences in equipment lead to a decrease in diagnostic accuracy. To address this issue, unsupervised domain adaptation techniques have been introduced. However, most cross-device fault diagnosis models overlook the discriminative information under the marginal distribution, which restricts the performance of the models. In this paper, we propose a bearing fault diagnosis method based on envelope spectrum and conditional metric learning. First, envelope spectral analysis is used to extract frequency domain features. Then, to fully utilize the discriminative information from the label distribution, we construct a deep Siamese convolutional neural network based on conditional metric learning to eliminate the data distribution differences and extract common features from the source and target domain data. Finally, dynamic weighting factors are employed to improve the convergence performance of the model and optimize the training process. Experimental analysis is conducted on 12 cross-device tasks and compared with other relevant methods. The results show that the proposed method achieves the best performance on all three evaluation metrics.

Keywords: fault diagnosis; conditional metric learning; envelope spectrum; convolutional neural network

Citation: Yang, X.; Yang, J.; Jin, Y.; Liu, Z. A New Method for Bearing Fault Diagnosis across Machines Based on Envelope Spectrum and Conditional Metric Learning. *Sensors* **2024**, *24*, 2674. https://doi.org/10.3390/s24092674

Academic Editor: Jongmyon Kim

Received: 27 March 2024
Revised: 13 April 2024
Accepted: 16 April 2024
Published: 23 April 2024

1. Introduction

Bearings are among the most often used elements in mechanical devices, serving the function of supporting and carrying rotating shafts [1–3]. Bearings play a crucial role in mechanical equipment as they reduce friction and wear, ensuring stable operation of the equipment. Therefore, bearing fault diagnosis is of vital importance in maintaining the normal functioning of mechanical equipment [4–6].

Currently, there are numerous approaches available for bearing fault diagnosis. Vibration analysis technology is one of the most widely used, with its core function being to detect bearing vibration signals and judge whether there are faults in the bearings based on signal characteristics. This technology has become the preferred choice due to its efficiency. Traditionally, fault diagnosis methods focus on the analysis of pulse impact intervals in vibration signals to distinguish different fault types. In the contemporary landscape of fault diagnosis, an array of algorithms have been developed based on the principles of mechanical fault theory. These methods encompass diverse techniques, such as resonance demodulation [7], envelope demodulation [8,9], generalized demodulation [10], and order ratio analysis [11]. The recent surge in the field of bearing fault diagnosis can be attributed to the continuous advancements in deep-learning technologies. This evolution has led to the validation and widespread adoption of innovative methods, including convolutional neural networks [12], autoencoders [13], recurrent neural networks [14], generative adversarial networks [15], and graph neural networks [16]. It is noteworthy that these models have stringent requirements for data, which require the data distribution of the training set and the test set to remain consistent. However, in practical engineering applications, due to changes in rotation speed, load, and sensor installation position, the data of the training set and test set may experience shifts. Therefore, unsupervised fault diagnosis approaches based on transfer learning have emerged. These methods can be divided into two major categories according to different application scenarios: unsupervised cross-domain learning on the same device and cross-domain learning on different devices. The emergence of this method provides a new approach to overcoming data drift issues, making fault diagnosis more feasible.

The scenario addressed by unsupervised domain adaptation within the same device is when the source domain and target domain data come from different vibration data under varying rotational speeds or loads. Many scholars have proposed numerous solutions to tackle the cross-domain fault diagnosis problem. Li et al. [17] constructed a deep convolutional neural network and used the maximum mean discrepancy based on multiple kernels (MK-MMD) to reduce the domain feature distance between multiple layers of the neural network, significantly improving the diagnostic performance. The method was validated using training bogie-bearing data. Chen et al. [18] employed domain adversarial training techniques to minimize the differences between the source domain and target domain data. They applied this approach in a deep transfer convolutional neural network and conducted extensive domain shift experiments on gearbox and bearing datasets. Li et al. [19] addressed the issue of low diagnostic accuracy due to insufficient training data by utilizing deep generative models to synthesize fault signals under the condition of known healthy data. The generated fault signals were then used in the domain adaptation process and validated for effectiveness using two different bearing fault datasets. Xiao et al. [20] utilized simulated fault mechanism data to construct a data- and physics-coupled fault diagnosis model, reducing the dependence on experimental setups. The proposed improved Joint Maximum Mean Discrepancy (JMMD) simultaneously aligned the conditional distribution and marginal distribution. The results showed that the proposed method achieved unsupervised domain adaptive fault diagnosis. In the scenario where the fault categories differ between the source and target domain datasets, Han et al. [21] proposed an intrinsic and extrinsic domain generalization network. This network combined label loss, triple loss, and adversarial loss functions to achieve gearbox fault diagnosis in unseen operating conditions within the target domain.

The scenario addressed by unsupervised domain adaptation across different devices is when the source domain and target domain data come from different devices' vibration data. Some scholars have proposed feasible solutions. Guo et al. [22], employing transfer learning techniques and adversarial training, introduced a deep convolutional transfer learning network (DCTLN) that adeptly diagnosed bearing faults across three disparate devices. Liu et al. [23] considered both rotational speed shifts and cross-device fault diagnosis tasks and proposed a deep adversarial subdomain adaptive network (DASAN). Experimental results demonstrated the effectiveness of DASAN. Wang et al. [24] proposed a subdomain adaptive transfer learning network (SATLN) by taking into account adaptive marginals and conditional distributional bias and incorporating dynamic weighting elements. This network was validated to achieve an average diagnostic accuracy of 90.19%. It is worth mentioning that in addition to cross device scenarios, Yu et al. provided excellent fault diagnosis methods from three aspects: incremental learning [25], model interpretability [26], and an online fault diagnosis system based on an integrated learning strategy [27].

The above research results indicate that current cross-domain fault diagnosis within the same device can achieve high diagnostic effectiveness. However, methods for cross-device fault diagnosis face significant deviations between the source domain and target domain data. Moreover, most models ignore discriminative information under marginal distribution, resulting in subpar diagnostic accuracy. In response to the multifaceted challenges at hand, this research introduces a pioneering methodology designed to diagnose bearing faults across a spectrum of devices. The essence of this methodology lies in the amalgamation of envelope spectrum analysis and the sophisticated principles of conditional metric learning. Primarily, the methodology undertakes the transformation of temporal vibration signals into their frequency-domain manifestations through the adept application of envelope spectrum analysis. Subsequently, a cutting-edge convolutional neural network model is meticulously crafted, incorporating a deep Siamese transfer approach, while being intricately grounded in the foundational principles of conditional metric learning. This innovative framework not only enhances the diagnostic accuracy of bearing faults but also showcases a nuanced understanding of the intricate interplay between envelope spectrum analysis and conditional metric learning. Finally, the proposed method is validated in six domain adaptation tasks across three different devices and compared with current advanced cross-device fault diagnosis methods. The main innovations are as follows:

(1) In order to address the issues of data discrepancies and domain biases in cross-device fault diagnosis, we innovatively introduced envelope spectrum analysis in our research. This method aims to reduce the differences in data generated by different devices at the frequency domain level, thereby optimizing data consistency and enhancing the expression of fault characteristics.

(2) We adopted a feature transfer strategy based on the conditional kernel Bures metric, which further reduces the biases between data from different domains and provides a solid foundation for precise training of diagnostic models.

(3) To enhance the optimization of the training process and the accuracy of diagnosis, we implemented dynamic weight learning technology. This technology adjusts the weight distribution in real-time during the learning process to respond to the importance of different categories and features, ensuring that the model achieves optimal performance in various fault diagnosis tasks.

(4) To comprehensively demonstrate the effectiveness of our proposed method, we conducted cross-device fault diagnosis research on two public bearing fault datasets and one private dataset. We carefully designed six different diagnostic tasks and compared five advanced fault diagnosis methods using three quantitative metrics. Through this rigorous experimental design and evaluation, our method demonstrated its effectiveness and superiority in various tasks.

The chapters are arranged as follows: Section 2 provides the definition of the fault diagnosis problem discussed in this article. Section 3 discusses the fault diagnosis methods employed in this study. Section 4 presents the results of fault diagnosis and compares the performance of the proposed method. Finally, Section 5 provides a comprehensive summary of the entire study.

2. Problem Formulation

The pivotal objective is to discern and classify the fault states of varied devices, distinguishing between normal operational states and those indicative of faults. Notwithstanding, practical implementation encounters formidable hurdles, notably the paucity of labeled samples specific to the target domain. To surmount this challenge, a strategic recourse involves harnessing the available labels from the source domain. This proactive approach serves as the bedrock for constructing a robust fault diagnosis model, proficient in extrapolating and predicting the labels associated with the target domain's data. This adaptive methodology enhances the applicability and efficacy of bearing fault diagnosis in real-world scenarios.

Data $D_S^{h\_s} = \left\{ (x_s^{(i)}, y_s^{(i)}) \right\}_{i=1}^{N}$ from device A constitute the source domain, while data $D_T^{h\_t} = \left\{ (x_t^{(i)}) \right\}_{i=1}^{M}$ from device B constitute the target domain. N and M quantitatively represent the samples within the source and target domains. Health states are distinguished as h_s and h_t. Labels $\left\{ (y_s^{(i)}) \right\}_{i=1}^{N}$ are accessible for the $D_S^{h\_s} = \left\{ (x_s^{(i)}) \right\}_{i=1}^{N}$, while the data $D_T^{h\_t} = \left\{ (x_t^{(i)}) \right\}_{i=1}^{M}$ lack such annotations. There are significant differences in the distribution of P and Q data, leading to the occurrence of a domain shift $P \neq Q$. This paper undertakes the challenge of formulating an unsupervised fault diagnosis model. The focus is on addressing cross-device fault diagnosis tasks, specifically from device A to device B (target domain).

3. Methods

In this dedicated section, we meticulously expound upon the foundational tenets governing envelope spectrum analysis, a pivotal facet in the realm of fault diagnosis. The ensuing discourse delves into the intricacies of our meticulously crafted deep Siamese convolutional neural network model. Subsequently, we introduce the theoretical underpinnings of conditional kernel Bures (CKB), a sophisticated framework augmenting our analytical prowess. Following this, a comprehensive exploration of the dynamic weighting mechanism ensues, contributing to the nuanced understanding of our proposed methodology.

3.1. Envelope Spectrum

Envelope spectrum analysis is a commonly used signal analysis method in mechanical fault diagnosis. By performing envelope analysis on vibration signals, periodic components in mechanical systems can be effectively extracted.

Generally, the fault characteristic frequency in the envelope spectrum can preliminarily identify the fault type, and the formula is as follows [28]:

$$\begin{cases} f_0 = \frac{1}{2}Z(1 - \frac{d}{D}\cos\alpha)f_r \\ f_i = \frac{1}{2}Z(1 + \frac{d}{D}\cos\alpha)f_r \\ f_b = \frac{D}{2d}[1 - (\frac{d}{D})^2\cos^2\alpha]f_r \\ f_c = \frac{1}{2}[1 - \frac{d}{D}\cos\alpha]f_r \end{cases} \tag{1}$$

where D is the bearing pitch diameter; d is the rolling element diameter; f is the rotation frequency; f_0, f_i, f_b, f_c represent the outer ring, inner ring, rolling element, and cages fault characteristic frequencies, respectively; and α is the contact angle between the rolling element and the raceway.

Although traditional envelope spectrum analysis provides a feasible method for fault feature extraction, it is usually limited to linear, stationary signals. However, in practical applications, many bearing fault signals are non-stationary and contain noise, requiring more advanced analysis techniques. The method we propose is based on conditional metric learning, which exhibits better performance in analyzing non-stationary signals containing complex noise and interference because it considers the potential non-linear features related to faults in the signal. At the same time, our method also utilizes unsupervised domain adaptation techniques to optimize the model's generalization ability in new domains (such as data from different equipment or operating conditions), which may not be achievable with envelope spectrum analysis.

This method is based on the principle of Fourier transform, decomposing the signal into multiple frequency components. Then, the amplitude variations of these frequency components are analyzed using envelope detection techniques to obtain an envelope spectrum. The original time-domain signal is transformed into a one-dimensional frequency spectrum. The analytical signal $z(t)$ of the signal $x(t)$ is built below:

For a single classification signal, its phase function can be written as follows:

$$z(t) = x(t) + jH(x(t)) \tag{2}$$

$$a(t) = |z(t)| = \sqrt{x^2(t) + H^2(t)} \tag{3}$$

Subsequently, the fast Fourier transform is used to convert $a(t)$ into a frequency domain signal, resulting in an envelope spectrum signal:

$$x_k = \sum_{n=0}^{N-1} x_n e^{-i2\pi kn/N} \tag{4}$$

$e^{i2\pi/N}$ is a primitive $N-th$ root of 1.

3.2. Deep Siamese Convolutional Neural Network

Depicted in Table 1 is the underlying framework of the deep Siamese convolutional neural network model explored in this paper. A feature extractor G_f and classifier G_c constitute the proposed model. The implementation of batch normalization ensures stable training dynamics by normalizing the input of each layer. Meanwhile, the activation function, ReLU, introduces non-linearity to the model, enabling it to capture complex patterns in the data. In the classifier, the presence of two fully connected layers facilitates the hierarchical learning of abstract features, contributing to the model's discriminative capabilities. This carefully designed architecture aims to optimize the extraction of distinctive features and enhance the discriminatory power of the model.

Table 1. The structure of the proposed network.

Networks Structure	Layer	Parameters Setting
Feature extractor	Conv-1	Kernels 16-64×1, Stride 8, Padding 1; Batch Normalization; ReLU; Maxpooling 2×1, Stride 2
	Conv-2	Kernels 32-3×1, Stride 1, Padding 1; Batch Normalization; ReLU; Maxpooling 2×1, Stride 2
	Conv-3	Kernels 64-3×1, Stride 1, Padding 1; Batch Normalization; ReLU; Maxpooling 2×1, Stride 2
	Conv-4	Kernels 64-3×1, Stride 1, Padding 1; Batch Normalization; ReLU; Maxpooling 2×1, Stride 2
	Conv-5	Kernels 64-3×1, Stride 1, Padding 1; Batch Normalization; ReLU; Maxpooling 2×1, Stride 2
	Conv-6	Kernels 1024-3×1, Stride 1; Batch Normalization; ReLU; Maxpooling 2×1, Stride 2
Classifier	Linear-1	Node: 256
	Linear-2	Node: No.category; Softmax

3.3. Conditional Kernel Bures

The CKB, a new measure of gauging conditional distribution disparities [29,30], finds its niche within the realm of Optimal Transport (OT). Operating as a statistically grounded and interpretable tool, CKB facilitates an in-depth exploration of the intricate knowledge transfer mechanisms inherent in transfer learning models. Robust validation across the domains of computer vision and pattern recognition attests to the efficacy and interpretability of CKB. The strategic infusion of CKB into cross-device mechanical fault diagnosis stems from its distinctive interpretability and adeptness in domain adaptation. This strategic integration aims to elevate the clarity of interpretative aspects related to features within the transfer model while concurrently mitigating data disparities across a spectrum of devices.

$$d_{\mathrm{CKB}}^2\left(R_{XX|Y}^s, R_{XX|Y}^t\right) = \mathrm{tr}\left(R_{XX|Y}^s + R_{XX|Y}^t - 2R_{XX|Y}^{st}\right) \tag{5}$$

where

$$R_{XX|Y}^{st} = \sqrt{\sqrt{R_{XX|Y}^s} R_{XX|Y}^t \sqrt{R_{XX|Y}^s}} \tag{6}$$

$R_{XX|Y}$ is the conditional covariance operator, and its calculation formula is

$$R_{XX|Y} = R_{XX} - R_{XY} R_{YY}^{-1} R_{YX} \tag{7}$$

R_{XY} is the cross-covariance operator [31]

$$R_{XY} = \mathbb{E}_{XY}[(\phi(X) - \mu_X) \otimes (\psi(Y) - \mu_Y)] \tag{8}$$

μ_X and μ_Y are the means. The non-linear mappings $\phi(X)$ and $\psi(Y)$ for X and Y. Under the condition $X = Y$, the relationship holds $R_{XX} = R_{YY} = R_{XY}$. We opt for an equivalent transformation [30].

$$
\begin{aligned}
\hat{d}^2_{\text{CKB}}&\left(\hat{R}^s_{XX|Y}, \hat{R}^t_{XX|Y}\right) \\
&= \text{tr}\left[G^s_X(\varepsilon n I_n + G^s_Y)^{-1}\right] + \varepsilon \text{tr}\left[G^t_X(\varepsilon m I_m + G^t_Y)^{-1}\right] \\
&\quad - \frac{2}{\sqrt{nm}}\left\|(H_m C_t)^T K^{ts}_{XX}(H_n C_s)\right\|_*
\end{aligned}
\tag{9}
$$

In the domain of fault diagnosis, the variable n represents the sample size within the source domain, while m designates the corresponding quantity within the target domain. The regularization coefficient $H_n = I_n - \frac{1}{n}1_n 1_n^T$ is denoted by $\varepsilon > 0$. I_n and H_m both assume an n-dimensional form, with the latter manifesting as a diagonal matrix replete with one 1_n and the former representing an explicit kernel matrix $(K^{ts}_{XX})_{ij} = k_\mathcal{X}\left(x^t_i, x^s_j\right)$. The kernel norm is aptly denoted as $\|\cdot\|_*$, and the interrelation between C_s and G^s_Y is succinctly expressed through the ensuing equation. This formulation establishes a foundational understanding within the realm of fault diagnosis, elucidating the crucial parameters governing the relationship between source and target domains.

$$
\begin{aligned}
\varepsilon n\left(G^S_Y + \varepsilon n I_n\right)^{-1} &= \\
U_s D_s U_s^T = U_s \sqrt{D_s}\left(U_s \sqrt{D_s}\right)^T &= C_s C_s^T
\end{aligned}
\tag{10}
$$

U_s, D_s are the eigenvector and eigenvalue matrices. Employing the Formula (9) facilitates the quantification of the conditional distribution distance pertaining to the features of data originating from both the source and target domains. This calculated distance serves as a crucial metric for assessing the alignment between domains. Figure 1 visually presents the demonstration of the efficacy of the conditional kernel Bayes (CKB) methodology in navigating the intricacies of this conditional distribution alignment. This visualization not only reinforces the empirical findings but also provides a tangible representation of the method's utility in the diagnostic context.

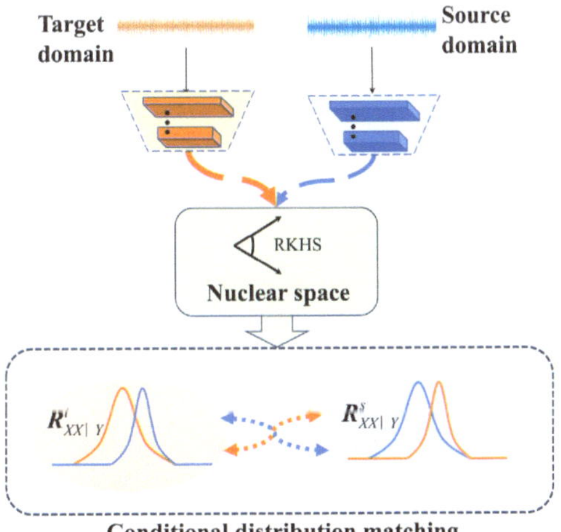

Figure 1. The schematic diagram of CKB.

3.4. Dynamic Weight Mechanism

Within the domain of profound learning, the loss metric is utilized to gauge the incongruity between the prognostications of the model and the factual outcomes. Generally, the loss function's coefficients remain unaltered, signifying uniform loss consideration for all instances. Nevertheless, instances arise where uniformity must be forsaken, compelling the incorporation of a mechanism that imbues dynamism into the weights.

The dynamic weight mechanism refers to the personalized allocation of loss weights for different samples. There are several common dynamic weight mechanisms:

(1) Category-based dynamic weight mechanism: Different weights are set for samples of different categories to adjust their contributions to the loss function. For example, larger weights can be assigned to samples of minority categories to make the model pay more attention to these samples.

(2) Difficulty-based dynamic weight mechanism: Higher weights can be assigned to samples that are more difficult, forcing the model to focus more on these challenging samples. Typically, the difficulty of a sample can be measured by the difference between its loss value and the average loss value of the training set.

(3) Adaptive learning rate-based dynamic weight mechanism: Usually, as the model continues to train, the learning rate gradually increases. Therefore, an adaptive learning rate mechanism is needed to adjust the weights of each sample.

There are two commonly used dynamic weight updating strategies, and their computation formulas are as follows:

$$\lambda = \frac{2}{\exp\left(\frac{-10 \, \text{epoch}}{\text{max\_epoch}}\right)} - 1 \tag{11}$$

$$\lambda^* = \frac{-4}{\sqrt{\frac{\text{epoch}}{\text{max\_epoch} + 1}} + 1} + 4 \tag{12}$$

where, epoch is the iteration period, max_epoch is the maximum iteration period. The experimental results below indicate that the weight strategy of Formula (11) is more effective.

3.5. Proposed Cross-Machine Fault Diagnosis Method

3.5.1. Overall Loss Function

The overall loss function during model training is a combination of three components, labeling loss, entropy loss, and domain loss. In the initial stage, we conduct supervised training utilizing data labeled within the source domain. This procedure is effectuated through the application of the label loss function.

$$\begin{aligned}
L_C &= \mathbb{E}_{(x_s^{(i)}, y_s^{(i)}) \in D_S^{h_s}}[-\log(\hat{y}_c^{(n)})] \\
&= -\frac{1}{N} \sum_{n=1}^{N} \sum_{c=1}^{C} y_c^{(n)} \log \frac{\exp\left(\hat{y}_c^{(n)}\right)}{\sum_{\tilde{c}=1}^{C} \exp\left(\hat{y}_{\tilde{c}}^{(n)}\right)}
\end{aligned} \tag{13}$$

N stands for the total count of samples, while C denotes the number of categories. $y_c^{(n)}$ serves as the sign function, producing either 0 or 1. Moreover, $\hat{y}_c^{(n)}$ signifies the output value at the c-th node of the fully connected layer 2 for the n-th sample.

Next, the entropy loss function is used to constrain the output uncertainty of the target domain data. This process is unsupervised. The computation formula is as follows:

$$\begin{aligned}
L_E &= \mathbb{E}_{(x_t^{(i)}, y_t^{(i)}) \in D_T^{h_t}}[-\log(\hat{y}_c^{(m)})] \\
&= -\frac{1}{M} \sum_{m=1}^{M} \sum_{c=1}^{C} \hat{y}_c^{(m)} \log \frac{\exp\left(\hat{y}_c^{(m)}\right)}{\sum_{\tilde{c}=1}^{C} \exp\left(\hat{y}_{\tilde{c}}^{(m)}\right)}
\end{aligned} \tag{14}$$

Within this framework, M is the overall sample count, with C indicating the number of categories. $y_c^{(m)}$ acts as the sign function, resulting in a binary outcome of 0 or 1.

Ultimately, $L_{\text{CKB}} = \hat{d}^2_{\text{CKB}}\left(\hat{R}^s_{XX|Y}, \hat{R}^t_{XX|Y}\right) = 0$. The approximation of these marginal distributions can be accomplished through the application of the maximum mean discrepancy loss, as formulated below.

$$
\begin{aligned}
L_{\text{MMD}} &= \hat{d}^2_{\text{MMD}}(X, Y) \\
&= \sup_{f \in \mathcal{F}} \left(\frac{1}{N} \sum_{i=1}^{N} f\left(x_s^{(i)}\right) - \frac{1}{M} \sum_{i=1}^{M} f\left(x_t^{(i)}\right) \right)
\end{aligned}
\tag{15}
$$

where $\mathcal{F}$ is a class function in RKHS, and sup (*) represents the supremum. The overall loss function of the model is as follows:

$$
L_{\text{all}} = L_{\text{C}} + \lambda^*(L_{\text{E}} + L_{\text{MMD}} + L_{\text{CKB}})
\tag{16}
$$

The holistic refinement of the model's parameters is systematically achieved through the minimization of L_{all}. This intricate optimization procedure is facilitated by the judicious application of the backpropagation algorithm in conjunction with the Ranger optimizer. Importantly, the learning rate is meticulously established at 0.002, and the comprehensive training regimen spans a predetermined 200 iterations.

3.5.2. Training and Testing Procedure of the Proposed Fault Diagnosis Framework

Step 1: The proposed process diagram for interpretable mechanical fault diagnosis is shown in Figure 2, and the specific steps are summarized as follows.

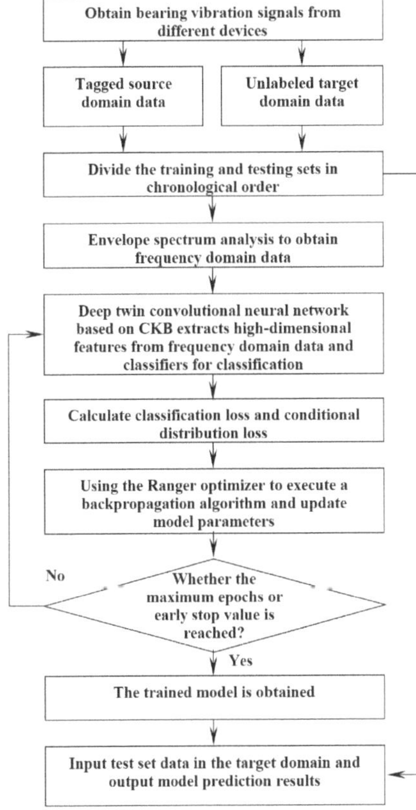

Figure 2. The flow chart of training and testing for the proposed framework.

Step 2: Obtain vibration signals from different mechanical equipment and divide the source domain and target domain data into training and testing sets in chronological order.

Step 3: Apply Hilbert envelope spectrum analysis to derive frequency domain details from both the sets designated for training and testing.

Step 4: Assemble the deep Siamese convolutional neural network, initializing the model parameters accordingly.

Step 5: Compute the overall loss function as shown in Formula (16), utilize the Ranger optimizer to perform the backpropagation algorithm, and update the model parameters.

Step 6: Execute the training process iteratively, culminating in the production of the trained model as the final outcome.

Step 7: Input the testing data from the target domain into the trained model and obtain the model's predicted results for the health condition of the bearings.

4. Results

Our primary aim is to substantiate the validity of the proposed methodology for cross-device fault diagnosis. The initial step involves a meticulous introduction to three distinct datasets integral to the cross-device fault diagnosis task, each containing vibration signals from motor bearings, as shown in Figure 3. Following the dataset introduction, a detailed exposition of the outcomes derived from both the proposed methodology and comparative approaches is presented, emphasizing the evaluation across three key metrics. To enhance comprehension, visualizations are employed to articulate diagnostic results derived from different methodologies. To delve deeper into the method's intricacies, ablation experiments are conducted, providing a nuanced understanding of its operational effectiveness. This approach not only contributes to the validation of the proposed methodology but also aligns with the conventions of the scholarly literature.

Figure 3. Data collection platform for (**a**) Case Western Reserve University, (**b**) NASA center's comprehensive bearing dataset, and (**c**) high-speed traction motor bearing failure data.

4.1. Dataset Introduction

(1) Dataset A: This set, procured from Case Western Reserve University, comprises vibration signals obtained from an acceleration sensor with a 12 kHz sampling frequency. The signals correspond to a motor with drive-end bearings and are categorized under normal, outer race fault, inner race fault, and rolling element fault conditions. Operating at 1750 r/min and sustaining a 2 HP load, the motor provides a rich dataset for analysis.

(2) Dataset B: Sourced from the NASA center's comprehensive bearing dataset, this dataset includes original vibration signals captured using a 20 kHz sampling frequency sensor. It likewise contains four health states. The motor's operational parameters are set at 2000 r/min rotation speed and a 26.6 kN load, with the unique characteristic that all data points were collected under conditions of severe failure.

(3) Dataset C: the high-speed traction motor bearing failure data, again, have four health states and the data details are shown in Table 2.

Table 2. The details of the bearing dataset.

Datasets	Bearing Type	Health Condition	Sampling Frequency	Speed, Load	Label
A	Motor bearing	Normal	12 kHz	1750 r/min 2HP	0
		Outer ring fault		1750 r/min 2 HP	1
		Inner ring fault		1750 r/min 2 HP	2
		Roller fault		1750 r/min 2 HP	3
B	Shaft support bearing	Normal	20 kHz	2000 r/min 26.6 kN	0
		Outer ring fault		2000 r/min 26.6 kN	1
		Inner ring fault		2000 r/min 26.6 kN	2
		Roller fault		2000 r/min 26.6 kN	3
C	High-speed traction motors bearing	Normal	10 kHz	2873 r/min 2.87 kN	0
		Outer ring fault		2873 r/min 3.09 kN	1
		Inner ring fault		2766 r/min 2.60 kN	2
		Roller fault		2765 r/min 2.57 kN	3

4.2. Comparative Methods and Experimental Settings

In the validation of our proposed methodology, a comparison is undertaken with five pre-existing techniques.

(1) Initially, the deep convolutional transfer learning network (DCTLN) distinguishes itself as a pioneering strategy for cross-device domain-adaptive fault diagnosis. It amalgamates domain adversarial training with the notion of maximum mean discrepancy (MMD).

(2) The Deep Adversarial Subdomain Adaptation Network (DASAN) achieves enhanced diagnostic accuracy through the utilization of a specialized loss function designed for subdomain adaptation.

(3) Based on the fault diagnosis model proposed in this paper, we modify the domain adaptation loss function to MMD, JMMD, and LMMD, forming comparative methods (3)–(5).

In the realm of experimental design, our cross-validation experiments span three meticulously chosen datasets, yielding six distinct cross-device transfer tasks: A→C, B→A, B→C, C→A, and C→B. On average across the five experiments, each health condition encompasses 1000 samples, each with a sequence length of 1200 points. The chronological division of training and testing sets adheres to a split ratio of 0.5. Of paramount importance is the meticulous preservation of equity in our comparative study. To this end, we faithfully adhere to the original parameter settings of DCTLN and DASAN as delineated in their respective papers. Concurrently, methods (3)–(5) align with the parameter configurations meticulously detailed in our paper. Specifically, the parameters and hyperparameters of our proposed method are meticulously specified: employing the Ranger optimizer to iteratively compute optimal model parameter values corresponding to the minimum loss function, with a learning rate of 2×10^{-3}, an L2 weight decay coefficient of 5×10^{-3}, a batch size of 128, and a total of 200 iterations.

4.3. Evaluation Metrics

In this paper, three evaluation metrics are used to quantify the diagnostic performance of different methods: the mean accuracy for fault identification (Acc), the F1-score (F1), and the average area under the receiver operating characteristic curve (AUC).

$$\text{Acc} = \frac{TP + TN}{TP + TN + FP + FN} \tag{17}$$

$$\text{F1} = \frac{2TP}{2TP + FP + FN} \tag{18}$$

TP (true positive), *TN* (true negative), *FP* (false positive), and *FN* (false negative) signify the relevant parameters.

4.4. Cross-Machine Diagnostic Results

Figure 4 displays the time-domain plots of vibration signals under different health conditions in three datasets. Through careful observation, it can be noticed that for each condition, the raw time-domain data exhibit unique vibration waveform patterns. These patterns are important indicators for understanding and monitoring the health status of the equipment, as they are directly related to the operational status of mechanical components and can provide early indications of possible equipment performance degradation. However, solely relying on visual inspection of these plots is insufficient to clearly define the corresponding vibration signals among different devices and their respective fault states. The observed differences in signal waveforms provide us with a preliminary basis for fault detection, but to improve diagnostic accuracy, further analysis and more sophisticated algorithms are needed to interpret these differences.

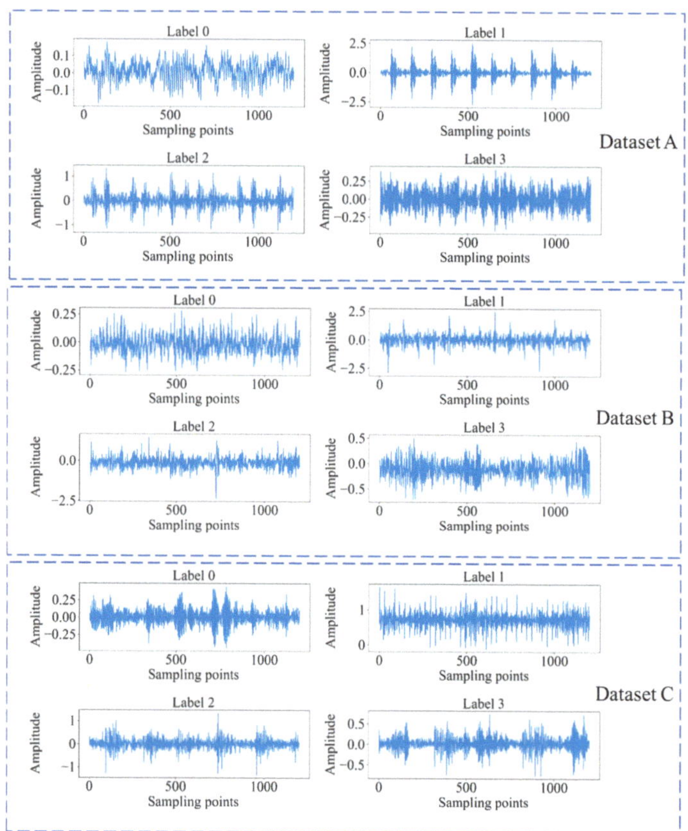

Figure 4. The time domain waveforms of bearing datasets.

In Figure 5, we present frequency-domain plots obtained through envelope spectrum analysis. Compared to time-domain data, frequency-domain plots provide a different perspective for observing the characteristics of vibration signals. Frequency domain analysis transforms time-series signals into displays of frequency components, allowing for clearer visualization of subtle cyclic variations, thereby revealing deeper mechanical fault indications. Figure 5 reveals the characteristic frequencies and energy distributions of vibration signals under different conditions in the frequency domain, which are crucial for identifying the types of faults. However, similar to the situation in Figure 4, although Figure 5 provides clearer signal differentiation, there is still insufficient direct correspondence between data with the same label from different devices. Therefore, using either time-domain or frequency-domain analysis alone is not sufficient; it is necessary to combine advanced domain adaptation techniques such as conditional metric learning to achieve high-accuracy cross-device fault diagnosis.

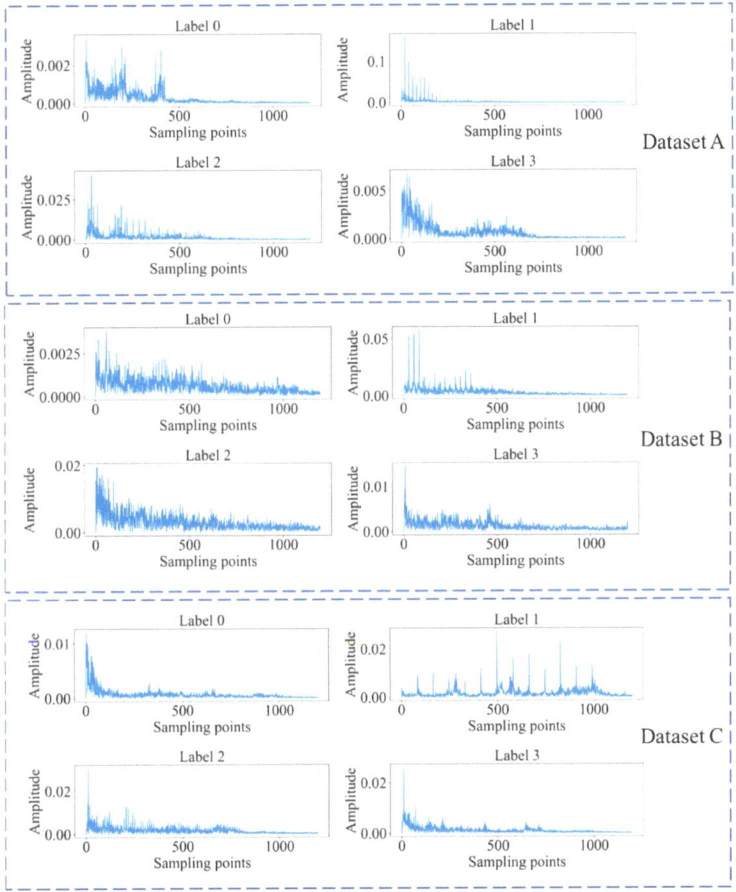

Figure 5. The frequency domain waveforms of bearing datasets.

Comparative studies were conducted using the proposed method and five other comparison methods. Each method was tested five times for each cross-device fault diagnosis task, and the mean and variance of the three-evaluation metrics were calculated. The tabulated data in Table 3 delineate the empirical outcomes derived from our rigorous experiments. The discerning feature of paramount significance is the exemplary diagnostic proficiency demonstrated by the proposed methodology across the comprehensive spectrum of the six designated tasks. It merits explicit acknowledgment that the proposed approach attains a pristine diagnostic accuracy of 100% in tasks entailing the transition from B to A and C to A. This accomplishment underscores the robustness and efficacy of the introduced diagnostic framework, thus substantiating its applicability in diverse scenarios within the field.

Table 3. The results of different methods.

Datasets	Bearing Type	DCTLN	DASAN	MMD	JMMD	LMMD	Proposed
A→B	Acc	85.69 ± 10.29	77.40 ± 22.73	95.67 ± 3.66	89.06 ± 9.02	86.58 ± 12.86	98.84 ± 1.10
	F1	0.85 ± 0.11	0.74 ± 0.26	0.96 ± 0.04	0.88 ± 0.10	0.85 ± 0.14	0.99 ± 0.01
	AUC	0.90 ± 0.07	0.85 ± 0.15	0.97 ± 0.02	0.93 ± 0.06	0.91 ± 0.09	0.99 ± 0.01
A→C	Acc	62.33 ± 6.22	71.82 ± 17.59	82.90 ± 8.98	71.99 ± 11.90	71.56 ± 16.03	98.34 ± 1.77
	F1	0.53 ± 0.07	0.68 ± 0.23	0.79 ± 0.13	0.66 ± 0.15	0.67 ± 0.19	0.98 ± 0.02
	AUC	0.75 ± 0.04	0.81 ± 0.12	0.89 ± 0.06	0.81 ± 0.08	0.81 ± 0.11	0.99 ± 0.01
B→A	Acc	76.65 ± 18.31	59.58 ± 20.76	96.63 ± 5.12	96.18 ± 4.19	99.09 ± 1.82	100.00 ± 0.00
	F1	0.71 ± 0.25	0.58 ± 0.21	0.97 ± 0.05	0.96 ± 0.05	0.99 ± 0.02	1.00 ± 0.00
	AUC	0.84 ± 0.12	0.73 ± 0.14	0.98 ± 0.03	0.97 ± 0.03	0.99 ± 0.01	1.00 ± 0.00
B→C	Acc	89.73 ± 10.82	75.25 ± 17.17	92.74 ± 10.82	68.76 ± 17.40	81.17 ± 15.03	98.51 ± 1.37
	F1	0.90 ± 0.11	0.69 ± 0.19	0.91 ± 0.15	0.65 ± 0.19	0.78 ± 0.18	0.99 ± 0.01
	AUC	0.93 ± 0.07	0.84 ± 0.11	0.95 ± 0.07	0.79 ± 0.12	0.87 ± 0.10	0.99 ± 0.01
C→A	Acc	93.13 ± 10.91	66.22 ± 11.77	94.98 ± 11.23	84.96 ± 13.73	85.43 ± 13.33	100.00 ± 0.00
	F1	0.91 ± 0.14	0.56 ± 0.15	0.93 ± 0.15	0.80 ± 0.18	0.81 ± 0.17	1.00 ± 0.00
	AUC	0.95 ± 0.07	0.77 ± 0.08	0.97 ± 0.07	0.90 ± 0.09	0.90 ± 0.09	1.00 ± 0.00
C→B	Acc	65.34 ± 14.66	95.33 ± 4.16	83.00 ± 16.23	78.27 ± 14.70	91.52 ± 12.77	99.13 ± 0.60
	F1	0.55 ± 0.20	0.95 ± 0.04	0.80 ± 0.20	0.75 ± 0.17	0.90 ± 0.16	0.99 ± 0.01
	AUC	0.77 ± 0.10	0.97 ± 0.03	0.89 ± 0.11	0.86 ± 0.10	0.94 ± 0.09	0.99 ± 0.00

Although DASAN considers fine-grained and discriminative features, it does not take into account discriminative information under marginal distribution. However, DASAN's overall performance is close to that of DCTLN. This may be due to the fact that different methods perform differently on different datasets, which reflects the matching between data and models. Similarly, the stability and generalizability of methodologies (3)–(5) based on MMD, JMMD, and LMMD exhibit notable variabilities. These empirical findings underscore the pronounced efficacy of the advanced cross-device fault diagnosis methodology proposed in this study. A comprehensive overview of the methodologies' performances across the three designated evaluation metrics is shown in Figure 6. The graphical representation unmistakably elucidates the markedly superior diagnostic outcomes attained by the innovative approach delineated in this research. The proposed methodology not only excels in diagnostic precision but also showcases a heightened robustness, offering a substantial contribution to the field of fault diagnosis in diverse device environments.

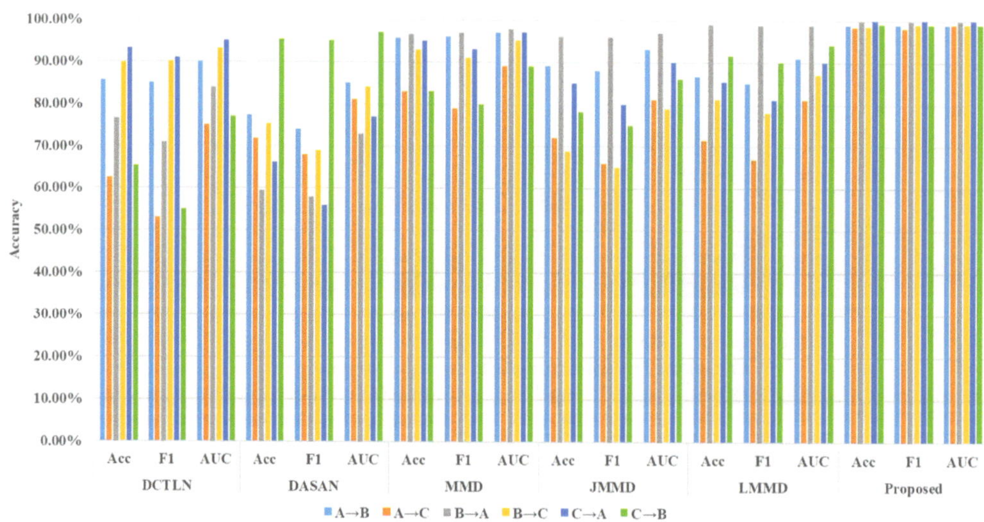

Figure 6. The histogram of diagnostic results.

4.5. Visualization Analysis

In the pursuit of visually elucidating the merits embedded in our proposed methodology, this section employs two crucial tools: confusion matrices and the t-SNE visualization algorithm. Centered on the B→C task, the nuanced insights provided by Figures 7 and 8 unravel the intricacies of the confusion matrices and t-SNE outcomes for the six methods, respectively. The confusion matrix results bring to the fore the capability of our proposed method to accurately predict and identify diverse health conditions. A deeper dive into the t-SNE results unveils the adeptness of our proposed method in harmoniously clustering data originating from both source and target domains, distinguished by shared labels. Of paramount significance are the conspicuously narrower intra-class distances and expansively broader inter-class separations, emblematic of the profound efficacy inherent in our approach. This notable achievement can be attributed to the meticulous integration of discriminative information within the marginal distribution, a unique hallmark of the CKB. Consequently, a palpable elevation in diagnostic performance is discerned. In synthesis, our proposed method unequivocally surmounts extant domain adaptation methodologies.

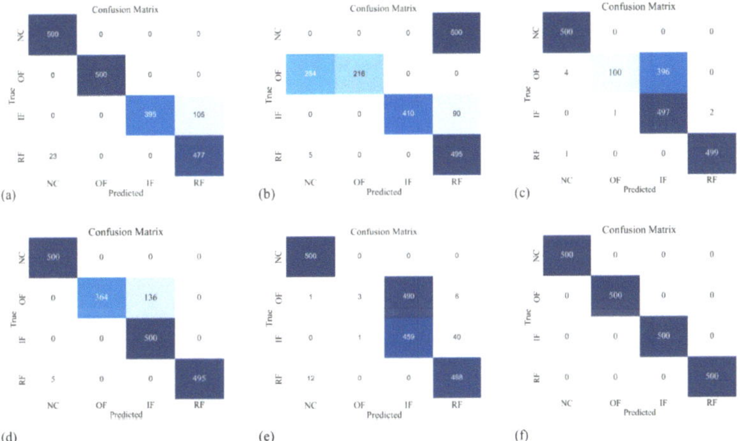

Figure 7. The confusion matrix results of: (**a**) DCTLN; (**b**) DASAN; (**c**) MMD; (**d**) JMMD; (**e**) LMMD; and (**f**) proposed method.

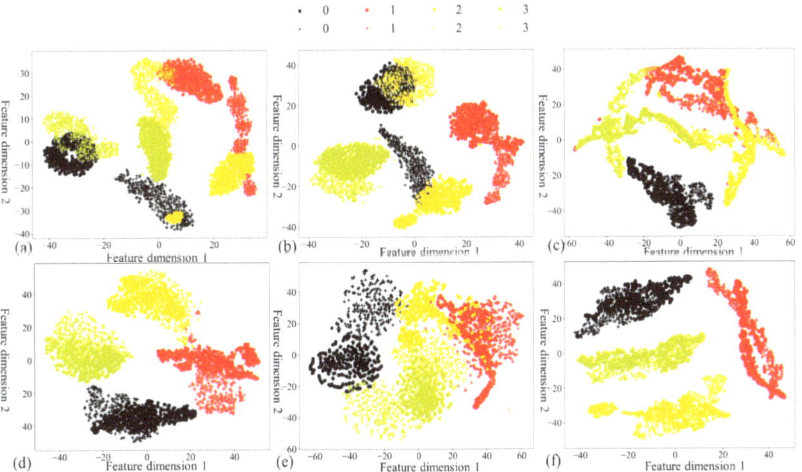

Figure 8. The t-SNE results of: (**a**) DCTLN; (**b**) DASAN; (**c**) MMD; (**d**) JMMD; (**e**) LMMD; and (**f**) proposed method.

4.6. Ablation Experiment

A thorough assessment of the rationality and effectiveness of our proposed methodology is conducted through a series of ablation experiments. Following the methodological framework, systematic ablation studies are executed on three critical elements: envelope spectrum analysis, various loss functions, and weight terms. The experimental design encompasses five distinct scenarios: an exclusion of the envelope spectrum, a neglect of the CKB + MMD loss function, an omission of the entropy loss function, a fixation of weight 1 for gamma_2 in Equation (10), and a dynamic adjustment of weight for gamma_1. Iterative repetitions of each experiment (five times) provide a robust basis for result analysis, with the means consolidated and graphically depicted in Figure 9. The results reveal that the model exhibits its least favorable performance when lacking the CKB + MMD loss function. Furthermore, it becomes apparent that entropy, despite its marginal impact, substantiates its essential role in augmenting the diagnostic model's performance. From this analytical exploration, it is concluded that the structural arrangement of our proposed methodology stands as a rational choice, the selection of pertinent loss functions is judicious, and the proposed method achieves the best diagnostic performance.

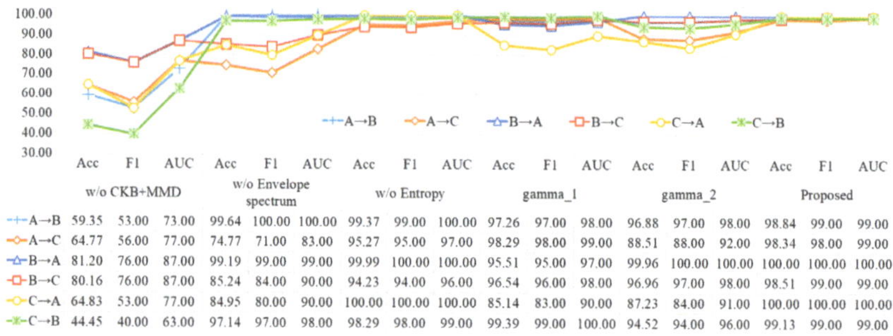

	Acc	F1	AUC	Acc	F1	AUC	Acc	F1	AUC	Acc	F1	AUC	Acc	F1	AUC	Acc	F1	AUC
	w/o CKB+MMD			w/o Envelope spectrum			w/o Entropy			gamma_1			gamma_2			Proposed		
A→B	59.35	53.00	73.00	99.64	100.00	100.00	99.37	99.00	100.00	97.26	97.00	98.00	96.88	97.00	98.00	98.84	99.00	99.00
A→C	64.77	56.00	77.00	74.77	71.00	83.00	95.27	95.00	97.00	98.29	98.00	99.00	88.51	88.00	92.00	98.34	98.00	99.00
B→A	81.20	76.00	87.00	99.19	99.00	99.00	99.99	100.00	100.00	95.51	95.00	97.00	99.96	100.00	100.00	100.00	100.00	100.00
B→C	80.16	76.00	87.00	85.24	84.00	90.00	94.23	94.00	96.00	96.54	96.00	98.00	96.96	97.00	98.00	98.51	99.00	99.00
C→A	64.83	53.00	77.00	84.95	80.00	90.00	100.00	100.00	100.00	85.14	83.00	90.00	87.23	84.00	91.00	100.00	100.00	100.00
C→B	44.45	40.00	63.00	97.14	97.00	98.00	98.29	98.00	99.00	99.39	99.00	100.00	94.52	94.00	96.00	99.13	99.00	99.00

Figure 9. Diagnosis results with different diagnosis models.

4.7. Parameter Sensitivity Analysis

To assess the impact of different parameter values on the model, this section conducted parameter sensitivity analysis experiments. Specifically, we analyzed the influence of different learning rates and batch sizes on the performance of the proposed method. The ranges of learning rates and batch sizes were $[1 \times 10^{-4}, 2 \times 10^{-3}, 2 \times 10^{-2}, 1 \times 10^{-1}, 5 \times 10^{-1}]$ and $[16, 32, 64, 128, 256]$, respectively, resulting in a total of 25 parameter combinations. The fault diagnosis results under three different performance indicators are shown in Figure 10a,b. It can be observed that different parameter combinations yield different diagnostic results. Based on the experimental results, the optimal parameter combination can be selected. In this study, the learning rate and batch size were chosen as 2×10^{-3} and 128, respectively.

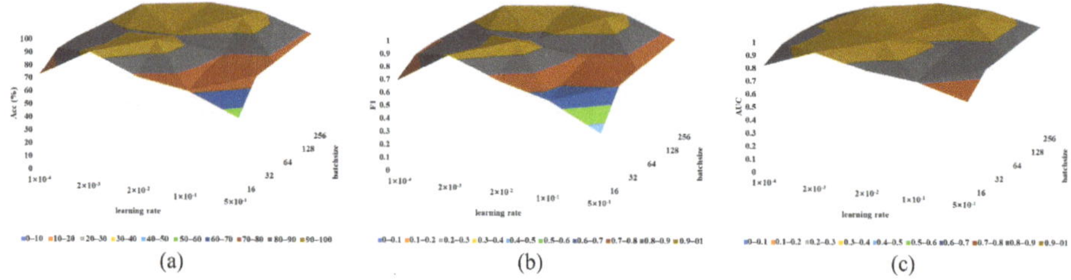

(a)   (b)   (c)

Figure 10. Parameter sensitivity analysis results mapfor (**a**) Case Western Reserve University, (**b**) NASA center's comprehensive bearing dataset, and (**c**) high-speed traction motor bearing failure data.

5. Conclusions

To solve the domain shift in bearing fault diagnosis, this paper proposes a novel cross-device bearing fault diagnosis method based on envelope spectrum analysis and conditional metric learning. Domain shift trials are performed on three different sets of equipment, and the following conclusions can be drawn from the analysis of the experimental results:

(1)   Envelope spectrum analysis, as a preprocessing step for the original vibration signals, can highlight fault information and improve the diagnostic accuracy of the model.

(2)   The application of the conditional kernel Bures metric aims to boost the efficacy of the model by minimizing the distributional gap. The dynamic weighting mechanism accelerates the model's optimization process and positively contributes to the improvement of diagnostic accuracy.

(3)   In six cross-device transfer tasks, the proposed method outperforms other domain adaptation methods in terms of three performance evaluation metrics, demonstrating stronger diagnostic capability.

Author Contributions: Conceptualization, J.Y. and X.Y.; methodology, Y.J.; software, Z.L.; validation, J.Y. and X.Y.; formal analysis, X.Y.; investigation, X.Y.; resources, Y.J.; data curation, Y.J.; writing—original draft preparation, Z.L.; writing—review and editing, J.Y.; visualization, J.Y.; supervision, Y.J.; funding acquisition, J.Y. All authors have read and agreed to the published version of the manuscript.

Funding: This research was partly funded by the Fundamental Research Funds for the Central Universities (No. 2023JBZX006 and 2022YJS155), Scientific and Technological Research Projects, Henan Province (No. 242102241053), Doctoral Research Startup Fund Project, Nanyang Institute of Technology.

Institutional Review Board Statement: Not applicable.

Informed Consent Statement: Not applicable.

Data Availability Statement: Data are contained within the article.

Conflicts of Interest: The authors declare no conflicts of interest.

References

1.   Heras, I.; Aguirrebeitia, J.; Abasolo, M.; Coria, I.; Escanciano, I. Load distribution and friction torque in four-point contact slewing bearings considering manufacturing errors and ring flexibility. *Mech. Mach. Theory* **2019**, *137*, 23–36. [CrossRef]

2.   Ambrożkiewicz, B.; Syta, A.; Gassner, A.; Georgiadis, A.; Litak, G.; Meier, N. The influence of the radial internal clearance on the dynamic response of self-aligning ball bearings. *Mech. Syst. Signal Process.* **2022**, *171*, 108954. [CrossRef]

3.   Gao, S.; Chatterton, S.; Pennacchi, P.; Han, Q.; Chu, F. Skidding and cage whirling of angular contact ball bearings: Kinematic-hertzian contact-thermal-elasto-hydrodynamic model with thermal expansion and experimental validation. *Mech. Syst. Signal Process.* **2022**, *166*, 108427. [CrossRef]

4.   Lee, J.; Wu, F.; Zhao, W.; Ghaffari, M.; Liao, L.; Siegel, D. Prognostics and health management design for rotary machinery systems—Reviews, methodology and applications. *Mech. Syst. Signal Process.* **2014**, *42*, 314–334. [CrossRef]

5.   Li, W.; Huang, R.; Li, J.; Liao, Y.; Chen, Z.; He, G.; Yan, R.; Gryllias, K. A perspective survey on deep transfer learning for fault diagnosis in industrial scenarios: Theories, applications and challenges. *Mech. Syst. Signal Process.* **2022**, *167*, 108487. [CrossRef]

6.   Zhang, T.; Chen, J.; Li, F.; Zhang, K.; Lv, H.; He, S.; Xu, E. Intelligent fault diagnosis of machines with small & imbalanced data: A state-of-the-art review and possible extensions. *ISA Trans.* **2022**, *119*, 152–171. [PubMed]

7.   Wang, W. Early detection of gear tooth cracking using the resonance demodulation technique. *Mech. Syst. Signal Process.* **2001**, *15*, 887–903. [CrossRef]

8.   Bauer, M.; Balaratnam, N.; Weidenauer, J.; Wagner, F.; Kley, M. Comparison of envelope demodulation methods in the analysis of rolling bearing damage. *J. Vib. Control* **2022**, *29*, 5009–5020. [CrossRef]

9.   Alessandro Paolo, D.; Luigi, G.; Alessandro, F.; Stefano, M. Performance of Envelope Demodulation for Bearing Damage Detection on CWRU Accelerometric Data: Kurtogram and Traditional Indicators vs. Targeted a Posteriori Band Indicators. *Appl. Sci.* **2021**, *11*, 6262. [CrossRef]

10.   Liu, D.; Cui, L.; Cheng, W. Flexible Generalized Demodulation for Intelligent Bearing Fault Diagnosis under Nonstationary Conditions. *IEEE Trans. Ind. Inf.* **2022**, *19*, 2717–2728. [CrossRef]

11.   Cheng, W.; Gao, R.X.; Wang, J.; Wang, T.; Wen, W.; Li, J. Envelope deformation in computed order tracking and error in order analysis. *Mech. Syst. Signal Process.* **2014**, *48*, 92–102. [CrossRef]

12. Jia, F.; Lei, Y.; Lu, N.; Xing, S. Deep normalized convolutional neural network for imbalanced fault classification of machinery and its understanding via visualization. *Mech. Syst. Signal Process.* **2018**, *110*, 349–367. [CrossRef]
13. Lu, N.; Yin, T. Transferable common feature space mining for fault diagnosis with imbalanced data. *Mech. Syst. Signal Process.* **2021**, *156*, 107645. [CrossRef]
14. Zhao, Z.; Li, T.; Wu, J.; Sun, C.; Wang, S.; Yan, R.; Chen, X. Deep learning algorithms for rotating machinery intelligent diagnosis: An open source benchmark study. *ISA Trans.* **2020**, *107*, 224–255. [CrossRef] [PubMed]
15. Zhang, T.; Chen, J.; Li, F.; Pan, T.; He, S. A Small Sample Focused Intelligent Fault Diagnosis Scheme of Machines via Multimodules Learning with Gradient Penalized Generative Adversarial Networks. *IEEE Trans. Ind. Electron.* **2021**, *68*, 10130–10141. [CrossRef]
16. Yang, C.; Zhou, K.; Liu, J. SuperGraph: Spatial-Temporal Graph-Based Feature Extraction for Rotating Machinery Diagnosis. *IEEE Trans. Ind. Electron.* **2022**, *69*, 4167–4176. [CrossRef]
17. Li, X.; Zhang, W.; Ding, Q.; Sun, J.-Q. Multi-Layer domain adaptation method for rolling bearing fault diagnosis. *Signal Process.* **2019**, *157*, 180–197. [CrossRef]
18. Chen, Z.; He, G.; Li, J.; Liao, Y.; Gryllias, K.; Li, W. Domain Adversarial Transfer Network for Cross-Domain Fault Diagnosis of Rotary Machinery. *IEEE Trans. Instrum. Meas.* **2020**, *69*, 8702–8712. [CrossRef]
19. Li, X.; Zhang, W.; Ding, Q. Cross-Domain Fault Diagnosis of Rolling Element Bearings Using Deep Generative Neural Networks. *IEEE Trans. Ind. Electron.* **2019**, *66*, 5525–5534. [CrossRef]
20. Xiao, Y.; Shao, H.; Han, S.; Huo, Z.; Wan, J. Novel Joint Transfer Network for Unsupervised Bearing Fault Diagnosis from Simulation Domain to Experimental Domain. *IEEE ASME Trans. Mechatron.* **2022**, *27*, 5254–5263. [CrossRef]
21. Han, T.; Li, Y.-F.; Qian, M. A Hybrid Generalization Network for Intelligent Fault Diagnosis of Rotating Machinery under Unseen Working Conditions. *IEEE Trans. Instrum. Meas.* **2021**, *70*, 3520011. [CrossRef]
22. Guo, L.; Lei, Y.; Xing, S.; Yan, T.; Li, N. Deep Convolutional Transfer Learning Network: A New Method for Intelligent Fault Diagnosis of Machines with Unlabeled Data. *IEEE Trans. Ind. Electron.* **2019**, *66*, 7316–7325. [CrossRef]
23. Liu, Y.; Wang, Y.; Chow, T.W.S.; Li, B. Deep Adversarial Subdomain Adaptation Network for Intelligent Fault Diagnosis. *IEEE Trans. Ind. Inf.* **2022**, *18*, 6038–6046. [CrossRef]
24. Wang, Z.; He, X.; Yang, B.; Li, N. Subdomain Adaptation Transfer Learning Network for Fault Diagnosis of Roller Bearings. *IEEE Trans. Ind. Electron.* **2022**, *69*, 8430–8439. [CrossRef]
25. Yu, W.; Zhao, C. Broad Convolutional Neural Network Based Industrial Process Fault Diagnosis with Incremental Learning Capability. *IEEE Trans. Ind. Electron.* **2020**, *67*, 5081–5091. [CrossRef]
26. Yu, W.; Zhao, C. Sparse Exponential Discriminant Analysis and Its Application to Fault Diagnosis. *IEEE Trans. Ind. Electron.* **2018**, *65*, 5931–5940. [CrossRef]
27. Yu, W.; Zhao, C. Online Fault Diagnosis for Industrial Processes with Bayesian Network-Based Probabilistic Ensemble Learning Strategy. *IEEE Trans. Autom. Sci. Eng.* **2019**, *16*, 1922–1932. [CrossRef]
28. Lucas, G.B.; de Castro, B.A.; Serni, P.J.; Riehl, R.R.; Andreoli, A.L. Sensors Applied to Bearing Fault Detection in Three-Phase Induction Motors. *Eng. Proc.* **2021**, *10*, 40. [CrossRef]
29. Ren, C.-X.; Luo, Y.-W.; Dai, D.-Q. BuresNet: Conditional Bures Metric for Transferable Representation Learning. *IEEE Trans. Pattern Anal. Mach. Intell.* **2022**, *45*, 4198–4213. [CrossRef]
30. Luo, Y.-W.; Ren, C.-X. Conditional Bures Metric for Domain Adaptation. In Proceedings of the 2021 IEEE/CVF Conference on Computer Vision and Pattern Recognition (CVPR), Nashville, TN, USA, 20–25 June 2021.
31. Song, L.; Huang, J.; Smola, A.; Fukumizu, K. Hilbert space embeddings of conditional distributions with applications to dynamical systems. In Proceedings of the 26th Annual International Conference on Machine Learning—ICML'09, Montreal, QC, Canada, 14–18 June 2009; ACM Press: New York, NY, USA, 2009.

Article

Preventing Forklift Front-End Failures: Predicting the Weight Centers of Heavy Objects, Remaining Useful Life Prediction under Abnormal Conditions, and Failure Diagnosis Based on Alarm Rules

Jeong-Geun Lee [1,2], Yun-Sang Kim [2] and Jang Hyun Lee [3,*]

1   Department of Smart Digital Engineering, INHA University, Incheon 22212, Republic of Korea; jeonggeun2.lee@doosan.com
2   Doosan Industrial Vehicle, Incheon 22503, Republic of Korea; yunsang.kim@doosan.com
3   Department of Naval Architecture and Ocean Engineering, INHA University, Incheon 22212, Republic of Korea
*   Correspondence: jh_lee@inha.ac.kr

Abstract: This paper addresses the critical challenge of preventing front-end failures in forklifts by addressing the center of gravity, accurate prediction of the remaining useful life (RUL), and efficient fault diagnosis through alarm rules. The study's significance lies in offering a comprehensive approach to enhancing forklift operational reliability. To achieve this goal, acceleration signals from the forklift's front-end were collected and processed. Time-domain statistical features were extracted from one-second windows, subsequently refined through an exponentially weighted moving average to mitigate noise. Data augmentation techniques, including AWGN and LSTM autoencoders, were employed. Based on the augmented data, random forest and lightGBM models were used to develop classification models for the weight centers of heavy objects carried by a forklift. Additionally, contextual diagnosis was performed by applying exponentially weighted moving averages to the classification probabilities of the machine learning models. The results indicated that the random forest achieved an accuracy of 0.9563, while lightGBM achieved an accuracy of 0.9566. The acceleration data were collected through experiments to predict forklift failure and RUL, particularly due to repeated forklift use when the centers of heavy objects carried by the forklift were skewed to the right. Time-domain statistical features of the acceleration signals were extracted and used as variables by applying a 20 s window. Subsequently, logistic regression and random forest models were employed to classify the failure stages of the forklifts. The F1 scores (macro) obtained were 0.9790 and 0.9220 for logistic regression and random forest, respectively. Moreover, random forest probabilities for each stage were combined and averaged to generate a degradation curve and determine the failure threshold. The coefficient of the exponential function was calculated using the least squares method on the degradation curve, and an RUL prediction model was developed to predict the failure point. Furthermore, the SHAP algorithm was utilized to identify significant features for classifying the stages. Fault diagnosis using alarm rules was conducted by establishing a threshold derived from the significant features within the normal stage.

Keywords: PHM; CBM; diagnosis; lightGBM; random forest; contextual diagnosis; RUL; forklift

Citation: Lee, J.-G.; Kim, Y.-S.; Lee, J.H. Preventing Forklift Front-End Failures: Predicting the Weight Centers of Heavy Objects, Remaining Useful Life Prediction under Abnormal Conditions, and Failure Diagnosis Based on Alarm Rules. *Sensors* **2023**, *23*, 7706. https://doi.org/10.3390/s23187706

Academic Editors: Dong Wang, Shilong Sun and Changqing Shen

Received: 15 August 2023
Revised: 1 September 2023
Accepted: 3 September 2023
Published: 6 September 2023

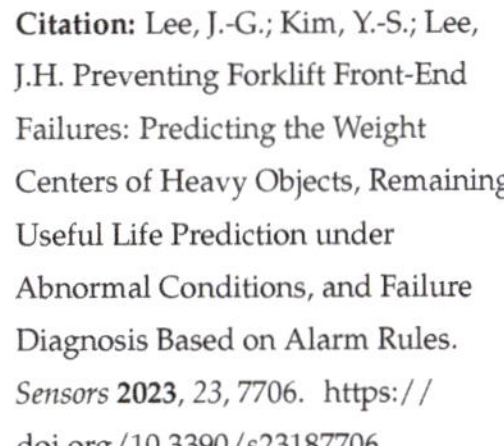

1. Introduction

This paper focuses on the challenge of mitigating durability degradation and accurately predicting failures in forklifts operating under demanding conditions. As classified by the American Society of Mechanical Engineers (ASME), forklifts, powered vehicles used for various material handling tasks, often encounter durability and performance issues while maneuvering heavy loads across logistics warehouses, construction sites, factories, and similar environments. These issues escalate service and maintenance costs, introduce safety hazards, and potentially endanger human lives. To address these challenges, the integration of prognostics and health management (PHM) technologies become essential. PHM, a comprehensive framework leveraging sensors, aims to assess system health, diagnose anomalies, and predict remaining useful life [1]. By encompassing condition monitoring, assessment, fault diagnosis, and prediction, PHM optimizes decision making for condition-based maintenance [2]. It necessitates the application of a systematic methodology to select appropriate feature engineering techniques and algorithms tailored to the specific context [3]. Notably, the study by Meng et al. [4] presented a comprehensive overview of research trends in PHM, particularly focusing on lithium-ion batteries. Categorizing prediction approaches into physics-based, data-driven, and hybrid categories, their work provides valuable insights into the diverse avenues within the field.

Recognizing the limitations of traditional reliability analysis, which relies on mean time to failure data and failure probability distributions, contemporary efforts are directed towards harnessing sensor-based PHM technologies to overcome these challenges. Previous studies in reliability and fault diagnosis have predominantly focused on statistical analyses under average load conditions or known failure scenarios [5–9]. However, the advent of the internet of things (IoT) is driving the transition to sensor-based PHM in a variety of domains as it evolves into a new phase of asset management [10]. Physics-based PHMs formalize the complex behavior of equipment and systems to help diagnose and predict failures when mechanical systems exceed predefined physical thresholds. In contrast, sensor-based methodologies have attracted considerable attention due to their ability to proficiently cope with complex troubleshooting scenarios due to their powerful representation and automated feature learning capabilities [11]. However, sensor-based PHMs face several challenges. Data collected from multiple sensors are prone to noise during acquisition and transmission. Since each type of failure produces different failure signals, signals and failures may not correspond exactly one-to-one in general [12]. The limitations of directly monitoring raw signals in deep-learning-based PHM are recognized, necessitating a transformation process termed feature engineering [12,13]. This process encompasses noise filtration, statistical feature extraction, frequency conversion, and context-dependent conditional reduction, thereby enhancing the data's meaningfulness and applicability.

Within the data-driven and deep-learning-based PHM framework, considerable research is dedicated to fault diagnosis and the prediction of remaining useful life. While PHM technologies aim for high fault detection accuracy and predictive capabilities, the practicality of models is also contingent on achieving a lightweight design. Ding et al. [14] introduced a lightweight multiscale convolutional network tailored for bearing fault diagnosis in edge computing scenarios, specifically targeting train bogie bearings. To complement model-based lightweighting efforts, feature-based lightweighting is pursued through dimension reduction. Lee et al. [3] emphasized the importance of effective feature engineering and data dimensionality reduction in PHM.

Addressing the challenge of imbalanced data situations, Zhang et al. [15] proposed an integrated multi-task intelligent bearing fault diagnosis method, leveraging representation learning under unbalanced sample conditions. This approach is particularly relevant given the common scenario of skewed sample ratios between fault and normal data. Furthermore, the scarcity of fault data arising in distinct environments necessitates innovative solutions. To overcome this data scarcity challenge, techniques such as data augmentation via generative adversarial networks (GANs) [16] and sensor signal transformation [17] have been advanced. Numerous studies have contributed comprehensive insights into remaining useful life (RUL) prediction within the PHM process [18–28]. RUL prediction encompasses four pivotal technical stages: data collection, construction of health indices, health state segmentation, and RUL prediction [18]. These stages synergistically form a comprehensive framework for accurate prediction.

The Industrial Truck Association (ITA) classifies forklifts into eight distinct product families, ranging from Class I to Class VIII. These classifications are based on specific usage characteristics and structural differentiations. For the purpose of this investigation, a PHM study was undertaken, focusing on the front-end structure responsible for lifting heavy loads. This structural component serves as the core element within ITA Class I-type electric counterbalance forklifts. Durability evaluation of forklifts typically involves testing under average loads and standardized conditions, conforming to established standards such as ISO 3691-1, ANSI/ITSDF B56.1, EN 1726-1, and ASME B56.1. However, the applicability of these conventional statistical approaches can be limited in non-standard operational settings where anomalies can arise. In real-world work environments, forklifts often encounter abnormal loading conditions that can result in rollovers or structural failures. Moreover, when latent strength deficiencies accumulate during assembly, the risk of structural failure becomes significantly elevated.

Among the various safety concerns linked with forklifts, the most significant hazard is the occurrence of vehicle rollovers. These incidents commonly occur when maneuvering heavy objects that exceed the forklift's designated capacity or when handling unbalanced or improperly centered loads. Furthermore, the repetitive use of forklifts while carrying unbalanced heavy loads can lead to performance degradation over time. This cumulative deterioration can eventually culminate in safety accidents. This particular failure mode operates outside the established standard and stands as an outlier load condition. It resides beyond the boundaries of average reliability ranges and is, therefore, not explicitly accounted for in typical design considerations. Paradoxically, these abnormal failures frequently occur within real-world operational environments. This inherent discrepancy necessitates the development of abnormal failure prevention technologies. These technologies should extend beyond the confines of existing reliability-based lifetime management approaches. Specifically, there is a need for a PHM technique that leverages machine learning and deep learning methodologies, utilizing sensor data to detect changes in forklift operational states. This enables the accurate classification and prediction of impending failures. Consequently, this study introduces an effective PHM approach designed to mitigate safety accidents stemming from abnormal forklift usage conditions. By addressing these anomalies, the aim is to enhance forklift durability, thereby curbing maintenance costs and fostering a value chain that prioritizes accident prevention.

This paper provides a systematic presentation of forklift failure diagnosis and prediction, aiming to ensure the reliability of the forklift front-end, as illustrated in Figure 1. The following overview outlines the content of each section.

- In Section 2, vibration data were collected under weight-imbalance conditions, and a subsequent feature engineering process was executed to facilitate a comprehensive classification of the centers of heavy objects. The primary objective of this classification study was to diagnose weight imbalance, a critical factor influencing forklift durability. By addressing this imbalance, the study aimed to proactively prevent factors leading to durability degradation in forklifts, thereby minimizing the risk of vehicle rollover.
- Continuing in Section 3, vibration and sound data were collected from forklifts operating under abnormal weight-imbalance conditions. Employing another feature engineering process alongside health stage classification, this section further delves into the realm of anomaly detection and classification.
- Section 4 is devoted to conducting remaining useful life (RUL) analysis and an alarm-rule-based fault diagnosis, serving the overarching goal of enhancing operational maintenance. The RUL prediction was based on the probability model established in Section 3, enabling accurate predictions. Additionally, significant features were extracted from both classifiers and features previously generated in Section 3. This information was then employed for an early front-end failure diagnosis through an alarm-rule-based approach.
- Lastly, Section 5 encapsulates the findings and conclusions, summarizing the contributions derived from the investigation.

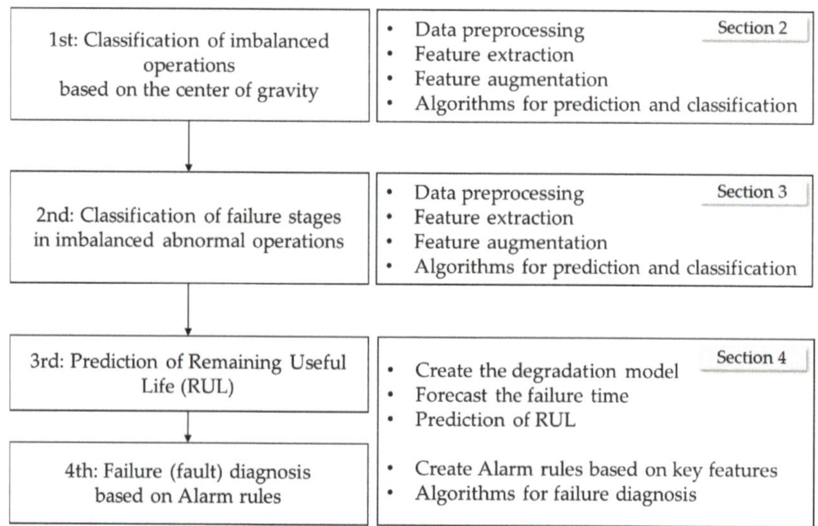

Figure 1. Outline of procedure used to predict the failure of forklift front-end.

2. Diagnosing and Classifying the Weight Center of Heavy Objects Carried by Forklifts

This section addresses the diagnosis of weight imbalances, which have a direct impact on the overall durability of forklifts. To achieve this, a structured approach was followed that included several essential steps: data preprocessing, feature engineering, data augmentation, and careful selection and evaluation of appropriate machine learning models. The overall goal was to develop a comprehensive method to detect the center of gravity, as shown in Figure 2.

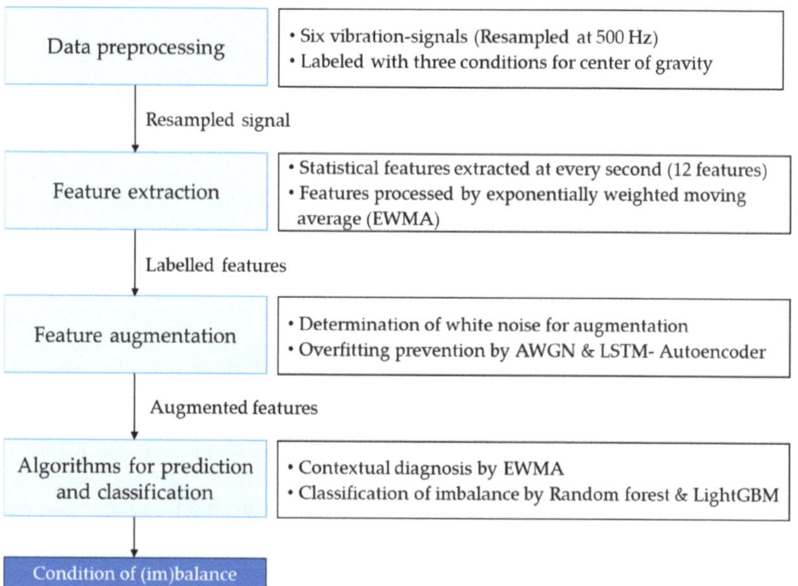

Figure 2. Classification procedure of imbalanced operations based on the center of gravity.

2.1. Experimental Data Acquisition and Feature Engineering

Vibration (acceleration) data were acquired from the forklift's front-end structure and presented a process to diagnose and classify the weight center of heavy objects carried by forklifts. The front-end structure of the forklift consists of a mast, backrest, carriage, and forks, as shown in Figure 3, and the acceleration signals were measured from the outer beam of the mast.

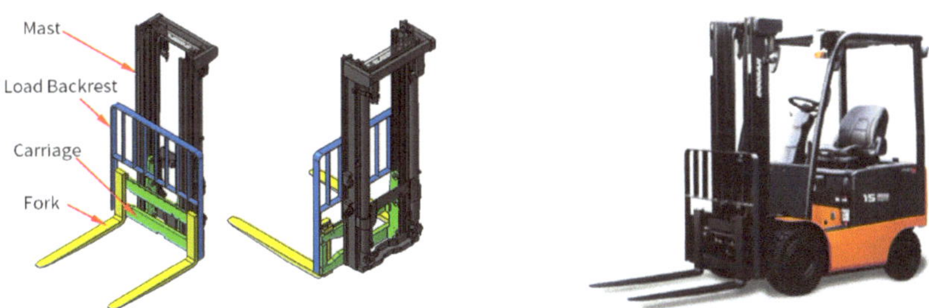

Figure 3. Electric-powered counterbalance forklift of ITA Class I type.

In the measurement experiment for data acquisition, the weight center of heavy objects carried by the forklift was measured in three configurations: center, left, and right. The condition segments of the dataset were classified and organized into center, left, and right according to each center of gravity condition. Two embedded devices (one on the left and one on the right) were attached to the front-end structure of the forklift truck to measure the vibration acceleration in three axes (x, y, z) (sampling rate 500 Hz), as shown in Figure 4. Considering the load conditions under which the forklift operates, the operating environments of the two datasets (datasets 1 and 2) were simulated in the experiments while maintaining a state that included ground noise. In dataset 1 (only driving mode), the vehicle was loaded with 3200 kg of weight and traveled 80 m at maximum speed, as shown on the left in Figure 5. All measurements were taken for approximately 20 min for center, left, and right condition segments to eliminate data imbalances within condition segments. In dataset 2 (complex mode), the forklift made a round trip of 80 m, as shown on the right in Figure 5, and added lifting, lowering, and back-and-forth tilting tasks at the end of the trip. In dataset 2, approximately 32 min of data were acquired, which is 12 min longer than in dataset 1. Similarly, approximately 32 min of data were acquired for the center, left, and right condition segments to eliminate the data imbalance.

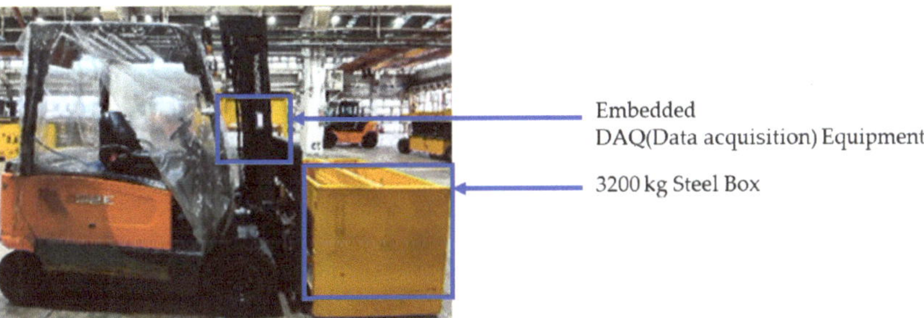

Figure 4. The location of DAQ and the apparatus of a forklift carrying object.

Six acceleration signals (2 sensors × (x, y, z accelerations)) were included in Data sets 1 and 2. In Dataset 2, for one acceleration signal, 960,000 feature vectors were collected at 500 (Hz) × 32 (min) × 60 (s/min). Given the large data size, which could lead to inefficient analysis, this study referred to previous studies [27,28] to handle the data. Eight features were extracted from each window at one-second intervals: min, max, peak to peak, mean (abs), rms (root mean square), variance, kurtosis, and skewness. In addition, four features were added by combining the max, rms, and mean (abs) features: crest factor, shape factor, impulse factor, and margin factor, as listed in Table 1.

Figure 5. Two operational scenarios for test datasets: driving only (**left**), complex mode (**right**).

Table 1. Description of features (crest factor, shape factor, impulse factor, and margin factor).

Feature	Dimension
Crest factor	max/RMS
Shape factor	RMS/mean (abs)
Impulse factor	max/mean (abs)
Margin factor	max/mean ((abs)^2)

Through the above process, 12 features were extracted, and the measured data were compressed for efficient analysis. The data compression process transformed the dimensionality of the data from 960,000 × 6 × 1 to 1920 × 6 × 12. In this way, the number of feature vectors in the data was reduced, but the number of features was increased twelve-fold, resulting in 72 features. The aim was to enable effective data processing and facilitate the diagnosis of the weight center of heavy objects at a one-second interval. Furthermore, the data range of the feature vectors was scaled from 0 to 1 using min–max normalization. Furthermore, an exponentially weighted moving average (EWMA) with a window size of 2 to 3 s was used to smooth the noise signals of the generated features. This approach helped to minimize the noise and outliers in the feature data. Moving averages average out the effects of past data, and exponentially weighted moving averages have the advantage of exponentially attenuating these effects. Therefore, they are used for tasks such as time series forecasting and noise reduction [29,30]. The exponentially weighted moving average is used in a way that adjusts the value of alpha (α) based on the window size, as shown in Equation (1). As described in Equation (3), this method smooths out noise in the feature vector to minimize outliers while also reducing the influence of past vectors.

$$\alpha = \frac{2}{window\ size + 1}, \quad window\ size \geq 1 \tag{1}$$

$$y_0 = x_0 \tag{2}$$

$$y_t = \frac{x_t + (1 - \alpha)x_{t-1} + (1 - \alpha)^2 x_{t-2} + \ldots + (1 - \alpha)^t x_0}{1 + (1 - \alpha) + (1 - \alpha)^2 + \ldots + (1 - \alpha)^t} \tag{3}$$

The parameters employed within Equations (1) through (3) are listed below:

- α: the weight coefficient of exponentially moving average.
- *window size*: the number of data points used by EWMA.
- y_t: the exponentially weighted moving average value at the current time t.
- x_t: the input data value at the current time t.
- $x_{t-1}, x_{t-2}, \ldots, x_0$: the input data values at past time points.
- $(1 - \alpha)^t, (1 - \alpha)^{t-1}, \ldots, 1$: the weights of past time points.
- $1 + (1 - \alpha) + (1 - \alpha)^2 + \ldots + (1 - \alpha)^t$: the sum of weights.

The exponentially weighted moving average was applied to take advantage of its ability to minimize outliers in the feature vector. Figure 6 shows the 'min' feature extracted from the x-acceleration signal using the EWMA technique. Tables 2 and 3 contrast features extracted from the dataset before and after the application of EWMA. They showcase the initial data state alongside the effects of EWMA, including smoothing and value adjustments.

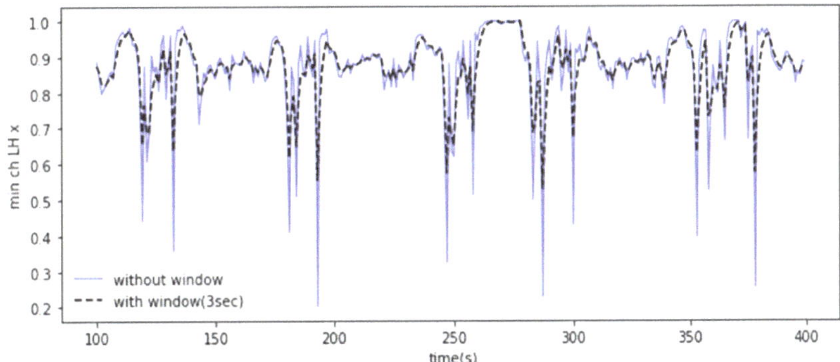

Figure 6. EWMA-processed feature (min) extracted from the x-acceleration signal.

Table 2. Table of features before applying the exponentially weighted moving average.

Features	Feature Vectors without Exponentially Weighted Window								
	1	...	101	102	103	104	105	...	m
Min	.	...	0.885433	0.843096	0.797650	0.811161	0.827289	...	.
Max	.	...	0.078264	0.080768	0.106396	0.108621	0.116061	...	.
Peak to Peak	.	...	0.093902	0.112324	0.146313	0.142308	0.140491	...	.
Mean (abs)	.	...	0.351697	0.401478	0.407600	0.368202	0.429075	...	.
RMS	.	...	0.267110	0.309412	0.318085	0.291022	0.342365	...	.
Variance	.	...	0.073500	0.098043	0.103476	0.086930	0.119690	...	.
Kurtosis	.	...	0.004885	0.008043	0.011320	0.016485	0.014369	...	.
Skewness	.	...	0.477752	0.469798	0.491220	0.486768	0.492277	...	.
Crest Factor	.	...	0.064416	0.039363	0.096850	0.127109	0.100406	...	.
Shape Factor	.	...	0.049605	0.060854	0.070667	0.080790	0.088413	...	.
Impulse Factor	.	...	0.033882	0.023190	0.053079	0.070003	0.057815	...	.
Margin Factor	.	...	0.003583	0.002052	0.003671	0.005534	0.003556	...	.

Table 3. Table of features after applying the exponentially weighted moving average.

Features	Feature Vectors with Exponentially Weighted Window (3 s)								
	1	...	101	102	103	104	105	...	m
Min	.	...	0.875394	0.859245	0.828448	0.819804	0.823547	...	.
Max	.	...	0.076663	0.078716	0.092556	0.100588	0.108325	...	.
Peak to Peak	.	...	0.096910	0.104617	0.125465	0.133807	0.137189	...	.
Mean (abs)	.	...	0.365030	0.383254	0.395427	0.381814	0.405444	...	.
RMS	.	...	0.276765	0.293089	0.305587	0.298304	0.320335	...	.
Variance	.	...	0.079095	0.088569	0.096022	0.091476	0.105583	...	.
Kurtosis	.	...	0.005647	0.006845	0.009082	0.012784	0.013576	...	.
Skewness	.	...	0.479047	0.474423	0.482821	0.484795	0.488536	...	.
Crest Factor	.	...	0.052390	0.045876	0.071363	0.099236	0.099821	...	.
Shape Factor	.	...	0.048446	0.054650	0.062658	0.071724	0.080069	...	.
Impulse Factor	.	...	0.027911	0.025551	0.039315	0.054659	0.056237	...	.
Margin Factor	.	...	0.002978	0.002515	0.003093	0.004314	0.003935	...	.

Through the previous feature engineering process, as shown in Table 4, 3699 feature vectors were generated from dataset 1 (only driving) in approximately 20 min, and 5755 feature vectors were derived from dataset 2 (complex mode), measured for approximately 32 min. In total, 9454 datasets were obtained, which were further divided into training and test datasets at a 7:3 ratio. The training and test datasets comprised 6617 and 2837 samples, respectively, with each feature vector containing 72 features. An attempt was made to use the training dataset to develop machine learning classifier models and check the performance of the machine learning models on the test dataset.

Table 4. Configuration of feature vectors in the dataset.

	Label: Center	Label: Left	Label: Right	Sum
Dataset 1	1244	1209	1246	3699
Dataset 2	1910	1928	1917	5755
Sum	3154	3137	3163	9454

The number of feature vectors in the dataset was also augmented to minimize overfitting during the training process. Data augmentation was performed using additive white Gaussian noise (AWGN) and long short-term memory autoencoder (LSTM AE), which expanded the training dataset to a maximum of 19,851 samples (Table 5).

Table 5. Dataset size based on feature augmentation.

Feature Combination	No. of Training Data	No. of Test Data
1. Original	6617	2837
2. Original + LSTM AE	13,234	2837
3. Original + AWGN	13,234	2837
4. Original + LSTM AE + AWGN	19,851	2837

AWGN was applied by referring to prior studies [31], and the target signal-to-noise ratio (SNR) was set to 20 dB. An additional dataset could be generated by mixing noise with the original data, as shown in Figure 7. AWGN is a method of adding noise with a Gaussian distribution to the input or output signal of a system. SNR serves as a scale that quantifies the ratio between the signal's strength and the noise level. A higher SNR value corresponds to a more robust signal, reducing the relative impact of noise. AWGN based on SNR can be expressed as follows [31]:

$$N(t) = A\sqrt{\frac{10^{-SNR/10}}{2}} \cdot w(t) \tag{4}$$

- $N(t)$: the noise of the signal.
- A: the magnitude of the input or output signal.
- $SNR/10$: the standard deviation σ of the noise, calculated from the SNR value.
- $w(t)$: the white noise, which follows a Gaussian distribution with a mean of 0 and a variance of 1.

The autoencoder is a neural network that can use unlabeled training data to learn a code that efficiently represents the input data. This type of coding is useful for dimensionality reduction because it typically has much lower dimensionality than the input. In particular, it works as a powerful feature extractor that can be used for the unsupervised pre-training of deep neural networks. An autoencoder consists of an encoder that converts the input to an internal representation and a decoder that converts the internal representation back to output [32]. The output result is called reconstruction because the autoencoder reconstructs the input. This study used the mean square error (MSE) as the reconstruction loss in training. LSTM, an artificial recurrent neural network, was designed to address the vanishing gradients in traditional recurrent neural networks (RNNs) [33]. As the number of hidden layers in a neural network and the number of nodes in each layer increase, the last layer is trained while the initial layer is not trained.

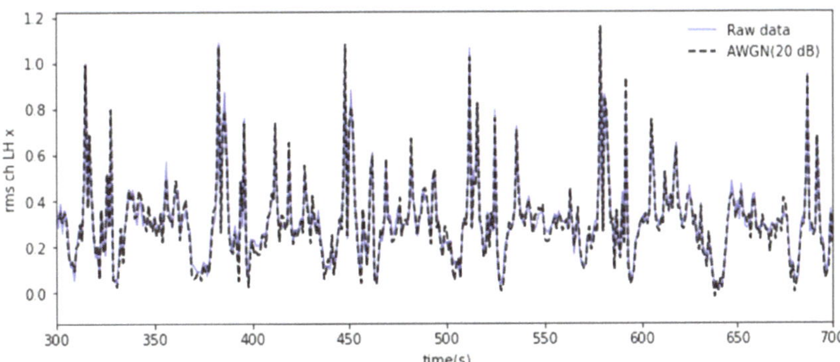

Figure 7. Results of noise mixed augmentation using AWGN.

This long-term dependency problem arises from the vanishing gradient problem, where the gradients tend to converge to zero during the gradient propagation process, particularly when the data length increases during the training phase of RNNs [34,35]. On the other hand, unlike traditional RNNs, LSTM can effectively overcome the vanishing gradient problem by incorporating long- and short-term state values in the learning process, enabling successful learning even with long training durations [36,37]. In this study, an LSTM autoencoder consisting of two LSTM layers was implemented because time series data were used, as shown in Figure 8. Each layer is used as an encoder and decoder [38]. Furthermore, the repeat vector was used in the decoder part to restore the compressed representation to the original input sequence. The repeat vector function repeats the compressed latent space representation to produce a representation that matches the sequence length. This allows the decoder to use the compressed representation multiple times to reconstruct the original input sequence. Using the LSTM autoencoder, the original feature vectors were trained as input data, and the output vectors were used as the augmentation dataset. The output dataset was generated and replicated to minimize the MSE, resulting in a dataset with similar characteristics and patterns to the input dataset, as shown in Figure 9. The equations and parameter descriptions for the LSTM autoencoder are as follows [32,33].

$$h_t = LSTM_{encoder}(x_t, h_{t-1}) \tag{5}$$

$$z_t = f(W_z h_t + b_z) \tag{6}$$

$$h'_t = LSTM_{decoder}(z_t, h'_{t-1}) \tag{7}$$

$$\hat{x}_t = f(W_x h'_t + b_x) \tag{8}$$

$$MSE = \frac{1}{n}\sum_{i=1}^{n}(x_i - \hat{x}_i)^2 \tag{9}$$

- x_t: the input time series data.
- z_t: the output (latent variable) of the encoder.
- $\hat{x}_t$: the output (reconstructed time series data) of the decoder.
- h_t and h'_t: the hidden states of the LSTM.
- W_z, b_z, W_x, and b_x: the learnable parameters (weights and biases) of the model.

- f: the activation function, typically sigmoid or tanh function.
- MSE: the mean squared error, loss function.
- x_i: the ith element of the input data.
- $\hat{x}_i$: the ith element of the model's prediction (reconstructed data).
- n: the number of elements in the input data.

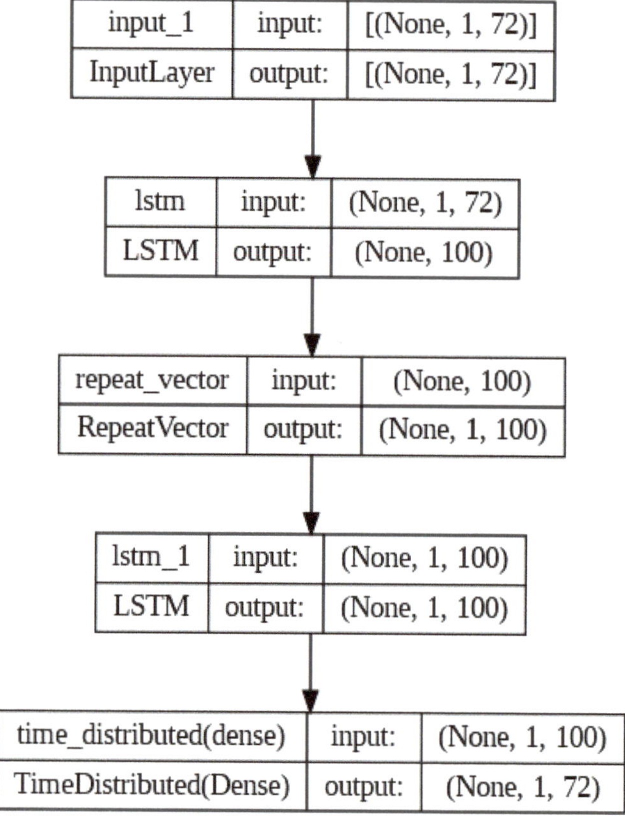

Figure 8. Structure of the LSTM autoencoder layers.

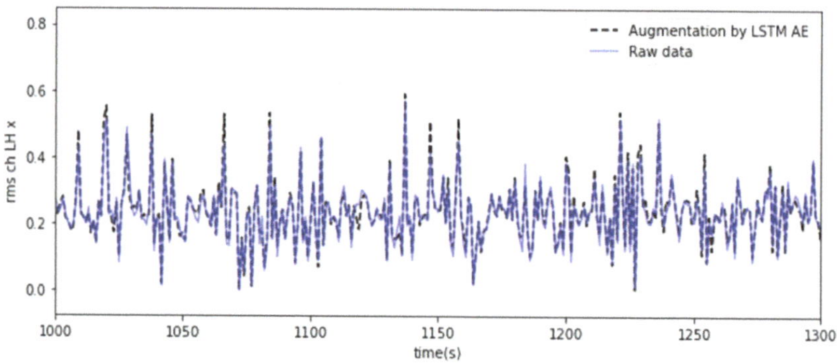

Figure 9. Result of feature augmentation using the LSTM autoencoder.

2.2. Result of Classification

To compare the accuracy of failure prediction, the selected classification algorithms were random forest [39] and LightGBM [40]. To enhance their performance, the 'Bayesian optimization' method was employed for hyperparameter tuning. Random forest, an ensemble technique, is rooted in decision trees and serves as a classifier. Decision trees build tree-like models based on input variables, efficiently growing the tree by identifying optimal splitting rules at each branch. However, the vulnerability of a single decision tree to overfitting can hinder its ability to generalize well. To address this concern, the random forest algorithm is applied to alleviate overfitting concerns. A decision tree, by itself, operates as a tree algorithm for data classification or prediction. It navigates the classification or prediction process by creating a tree structure grounded in the data, partitioning it into multiple child nodes through evaluations of specific conditions at each node. The criteria of these conditions are typically determined by metrics such as information gain (IG) or the Gini index ($Gini$). These metrics measure the impurity of class distribution at each node, selecting a splitting criterion that minimizes the difference in impurity before and after the split [39,40].

$$\text{Information gain}: \ IG(D_p, f) = I(D_p) - \sum_{j=1}^{m} \frac{N_j}{N_p} I(D_j) \tag{10}$$

$$\text{Gini index}: \ Gini(D_p) = 1 - \sum_{k=1}^{K} p_k^2 \tag{11}$$

- D_p: the data of the parent node.
- D_j: the data of the jth child node.
- f: the splitting criterion variable.
- m: the number of child nodes generated after splitting.
- N_p: the number of data points in the parent node.
- N_j: the number of data points in the jth child node.
- K: the number of classes.
- p_k: the ratio of the kth class.

Random forest stands out as a notable ensemble learning technique, leveraging the power of decision trees. The methodology of random forest is structured around the collaborative efforts of multiple decision trees, which are subsequently aggregated to yield prediction results. The workflow of random forest unfolds as follows:

1. Bootstrap sample creation: The process starts by randomly selecting a subset of the input data to create what is known as a bootstrap sample. This sample comprises a distinct dataset consisting of data instances randomly extracted from the original input dataset.
2. Multiple decision tree generation: Next, numerous decision trees are generated, each stemming from a bootstrap sample. These decision trees come into existence with a random element, ensuring their diversity and independence.
3. Data prediction by decision trees: Each of the generated decision trees is then utilized to predict the input data. This prediction process is carried out individually for all the decision trees in the ensemble.
4. Aggregation of predictions: The prediction results obtained from the individual decision trees are aggregated. This aggregation can take the form of averaging the predictions or adopting a majority voting approach, depending on the task. The aggregated outcome serves as the foundation for the final predictions made by the random forest model.

By following these steps, random forest harnesses the collective insights of multiple decision trees, effectively enhancing prediction accuracy and generalization capabilities. Random forest addresses overfitting by generating multiple models from different data subsets. This diversification improves robustness against noise and uncertainties. The adjustment of the optimal number of decision trees and the splitting criteria can be performed through hyperparameter tuning. Typically, hyperparameters such as the splitting criteria of decision trees and the tree depth are set. The prediction function of random forest is as follow, where T is the number of generated decision trees and $f_t(x)$ represents the prediction function of the tth decision tree.

$$f(x) = \frac{1}{T}\sum_{t=1}^{T} f_t(x) \tag{12}$$

LightGBM is a machine learning model based on the gradient boosting decision tree (GBDT) algorithm. Gradient boosting works by improving the prediction model as new models compensate for the errors of the previous model. Therefore, multiple decision trees can be combined to develop a more robust model that minimizes overfitting. The working mechanism of lightGBM is similar to the conventional GBDT algorithm, but it utilizes the leaf-wise approach during the splitting process (Figure 10). This approach allows lightGBM to produce more unbalanced trees than the traditional level-wise approach, resulting in improved predictive performance. Furthermore, lightGBM includes the feature to perform splitting using only a subset of the data, ensuring faster processing speed for large-scale datasets. However, lightGBM may result in deeper trees, depending on their leaf-wise characteristics and hyperparameter settings, which may lead to deeper trees [41]. While this can improve the prediction accuracy of the training data, it may result in lower accuracy when predicting new data because of the overfitting problem. In the case of lightGBM, the aim was to minimize the overfitting problem through the feature augmentation conducted previously. The objective function and parameter descriptions for LightGBM are as follows [42]:

$$Obj(\Theta) = \sum_i l\left(y_i, \widehat{y_i y_i^{(t-1)}} + f_t(x_i)\right) + \sum_t \Omega(f_t) \tag{13}$$

$$\Omega(f_t) = \gamma T + \frac{1}{2}\lambda|w|^2 \tag{14}$$

- $l\left(y_i, \widehat{y_i^{(t-1)}} + f_t(x_i)\right)$: the loss functions used in the objective function.
- y_i: the actual value for the ith data.
- $\widehat{y_i^{(t-1)}}$: the prediction from the previous time step $(t-1)$.
- $f_t(x_i)$: the prediction of the tth tree for the ith data point x_i.
- $\Omega(f_t)$: a term used to regulate the complexity of the tree.
- T: the number of leaf nodes in the tree.
- γ: the cost parameter associated with the number of leaf nodes.
- λ: a coefficient that regulates the weight of leaf nodes.
- w: the weights assigned to the tree nodes.

Bayesian optimization is a method for finding the optimal solution that maximizes an arbitrary objective function. This optimization technique can be applied to any function for which observations can be obtained and is particularly useful for optimizing black-box functions with high cost and unknown shapes [43]. Therefore, Bayesian optimization is used mainly as a hyperparameter optimization method for machine learning models, taking advantage of the characteristics of such optimization techniques [44]. The optimized hyperparameters were derived through Bayesian optimization, as shown in Table 6. In Bayesian optimization, the aim is to identify the hyperparameter combination x^* that minimizes or maximizes the objective function $f(x)$. The objective function typically takes a form similar to Equation (15). $\eta(x)$ represents the actual value of the objective function for the hyperparameter combination x. $f(x)$ is defined as the sum of the actual objective function value $\eta(x)$ and the noise $\epsilon(x)$. Bayesian optimization involves experimentation with various hyperparameter combinations while modeling both the genuine objective function value $\eta(x)$ and the accompanying noise $\epsilon(x)$. Through iterative processes, the next hyperparameter combination to explore is forecast, advancing the optimization process.

$$f(x) = \eta(x) + \epsilon(x) \tag{15}$$

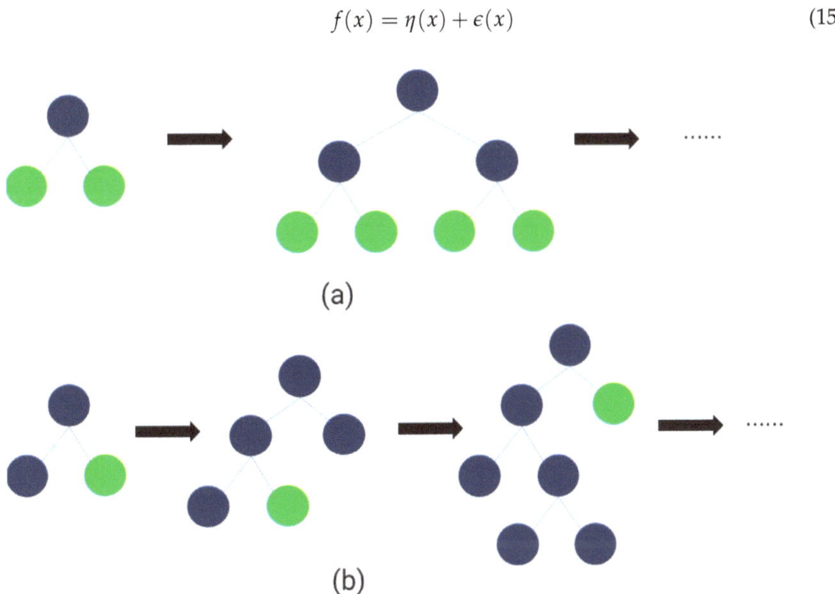

(a)

(b)

Figure 10. Two kinds of tree growth: (**a**) level-wise growth, (**b**) leaf-wise growth.

Table 6. Hyperparameter tuning results of the random forest and lightGBM.

Random Forest		LightGBM	
bootstrap	True	boosting_type	'gbdt'
ccp_alpha	0	class_weight	None
class_weight	None	colsample_bytree	1
criterion	'gini'	importance_type	'split'
max_depth	None	learning_rate	0.02
max_features	'auto'	max_depth	12
max_leaf_nodes	None	min_child_samples	20
max_samples	None	min_child_weight	0.001
min_impurity_decrease	0	min_split_gain	0
min_impurity_split	None	n_estimators	1000
min_samples_leaf	1	num_leaves	58
min_samples_split	2	random_state	None
min_weight_fraction_leaf	0	reg_alpha	0
n_estimators	130	reg_lambda	0
oob_score	False	silent	−1
random_state	42	subsample	0.8
warm_start	False	subsample_for_bin	200,000

Because forklifts move continuously, vibration data has the characteristics of time series data. The time series data and the state changes in the condition segment (center of heavy objects carried by forklift) do not depend on the state at a single point in time but on the past values. Therefore, the probability of classifying the conditioning segment was the EWMA to diagnose the conditioning segment contextually using the moving average instead of diagnosing the conditioning segment only by the probability at that time, as shown in Table 7. Contextual diagnosis in machine learning is a technique to diagnose by considering the context of the given data [45]. This provides a more profound understanding than simply analyzing and predicting data patterns. It simply considers the context of the data before and after, rather than individual data points, to help make an accurate diagnosis. In addition, it is used effectively for outlier detection in time series [46] and partial data [47]. This study attempted to minimize the effect of noise, such as outliers, using the exponentially weighted moving average for contextual diagnosis. In applying contextual diagnosis, this study examined the effects of the window size on the moving average. Figure 11 presents the learning and prediction process flow.

Table 7. Result of contextual diagnosis.

	Before Contextual Diagnosis				After Contextual Diagnosis		
Time	**Left (LLH)**	**Right (RRH)**	**Center**	**Time**	**Left (LLH)**	**Right (RRH)**	**Center**
1	0.00112	0.00014	0.99874	1	0.00112	0.00014	0.99874
2	0.00305	0.00044	0.99651	2	0.00241	0.00034	0.99725
3	0.00239	0.00070	0.99691	3	0.00240	0.00055	0.99706
4	0.00103	0.00012	0.99885	4	0.00167	0.00032	0.99801
5	0.00117	0.00022	0.99862	5	0.00141	0.00027	0.99832
6	0.02492	0.00198	0.97311	6	0.01335	0.00113	0.98552
7	0.06715	0.01467	0.91818	7	0.04046	0.00795	0.95158
8	0.06500	0.00303	0.93197	8	0.05278	0.00548	0.94174
9	0.13749	0.01739	0.84512	9	0.09522	0.01145	0.89333
10	0.07009	0.01556	0.91436	10	0.08264	0.01351	0.90385
.	...	...	...	.	...	...	...
.	...	...	...	.	...	...	...
.	...	...	...	.	...	...	...
m	...	...	...	m	...	...	...

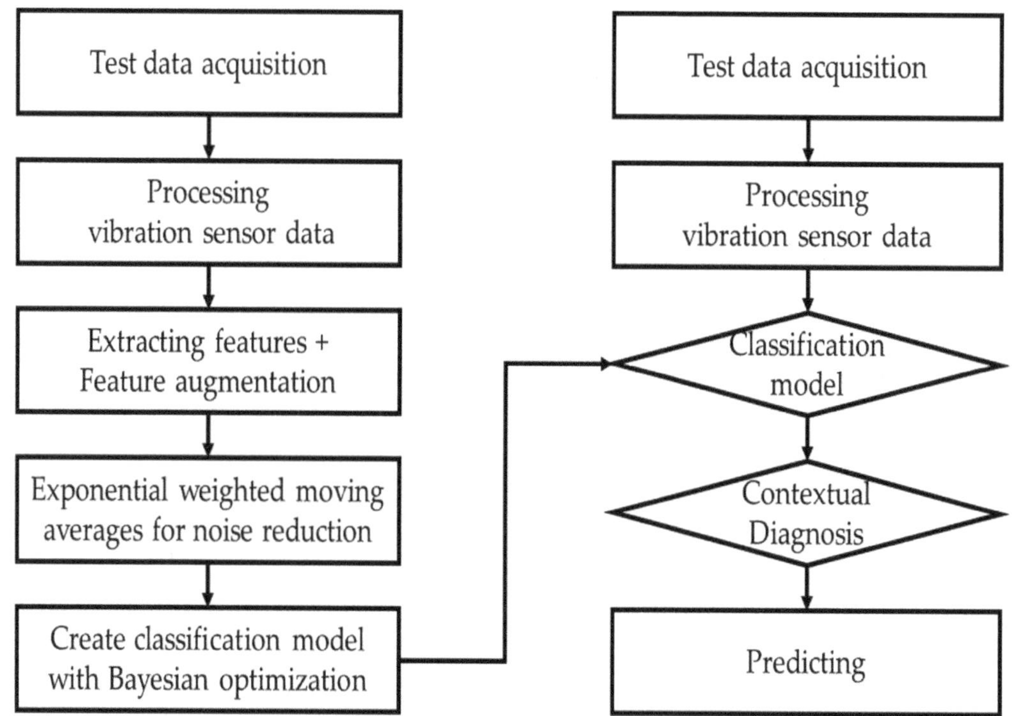

Figure 11. Schematic procedure of training and prediction.

As a result of contextual diagnosis through the exponentially weighted moving average, the classification probability of each condition segment predicted by machine learning changes, as listed in Table 7. Forklifts carry and transport unbalanced heavy objects during continuous movement or operation, and the centers of heavy objects do not fluctuate on a one-second basis. Therefore, a two to three-second window was used to calculate the moving average probability. As a result, when the condition segment was diagnosed as "center", applying a moving average to the probabilities in certain outlier segments diagnosed as "left" or "right" would result in lower values influenced by past data. Figures 12 and 13 present these probabilities as graphs as a function of time. From the observed results, the centers of heavy objects carried by the forklift were diagnosed more accurately by the generated classifier.

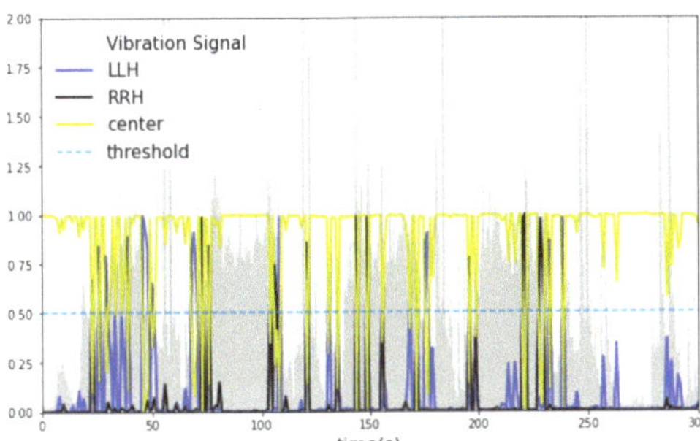
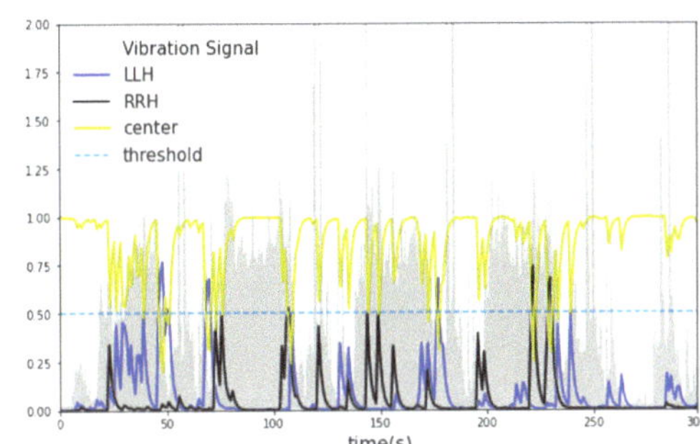

Figure 12. Contextual diagnosis graph of lightGBM (**left**: original, **right**: after moving average).

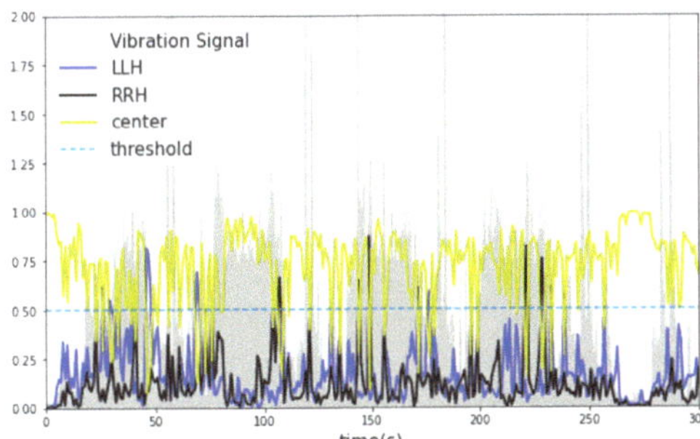
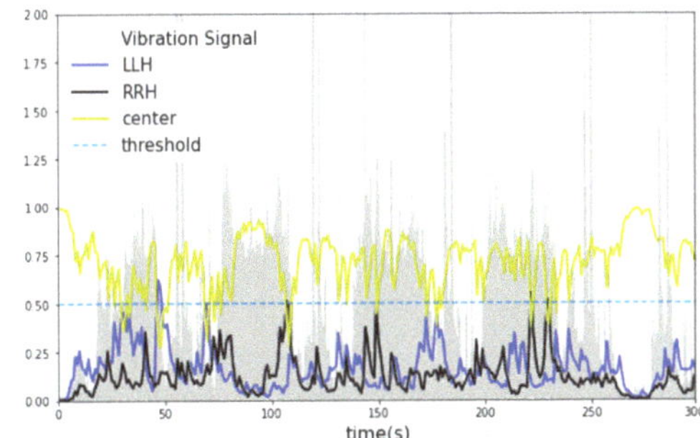

Figure 13. Contextual diagnosis graph of random forest (**left**: original, **right**: after moving average).

Because the data imbalance was minimized, the performance was compared using the accuracy score of 24 classifiers, including random forest and lightGBM (Table 8). First, the accuracy increased in all cases as the window size of the exponentially weighted moving average for smoothing the feature vector time series data increased and the alpha value decreased. The classification probabilities of the condition segment predicted by machine learning were subjected to the moving average to achieve a contextual diagnosis. In all cases, the classification accuracy of the condition segment increased gradually as the size of the moving average window increased, and the alpha decreased. On the other hand, although the data augmentation minimized the overfitting of the model, it resulted in the same or slightly lower accuracy.

Table 8. Result of case studies.

Case No.	Dataset	Machine Learning Model	Feature Moving Average Window Size	Smoothing Factor (α)	Contextual Diagnosis Accuracy Score (Probability Moving Average Window Size)		
					1 s	2 s	3 s
Case 1	Raw features	Random forest	1 s	1.00	0.7522	0.8135	0.8950
Case 2	Raw features	Random forest	2 s	0.67	0.8019	0.8622	0.9186
Case 3	Raw features	Random forest	3 s	0.50	0.8347	0.8752	0.9274
Case 4	With LSTM AE features	Random forest	1 s	1.00	0.7392	0.8047	0.8904
Case 5	With LSTM AE features	Random forest	2 s	0.67	0.7846	0.8470	0.9094
Case 6	With LSTM AE features	Random forest	3 s	0.50	0.8216	0.8713	0.9263
Case 7	With AWGN features	Random forest	1 s	1.00	0.7487	0.8238	0.9020
Case 8	With AWGN features	Random forest	2 s	0.67	0.7994	0.8601	0.9203
Case 9	With AWGN features	Random forest	3 s	0.50	0.8294	0.8819	0.9366
Case 10	With all features	Random forest	1 s	1.00	0.7487	0.8587	0.9362
Case 11	With all features	Random forest	2 s	0.67	0.8033	0.8897	0.9450
Case 12	With all features	Random forest	3 s	0.50	0.8305	0.9048	0.9563
Case 13	Raw features	lightGBM	1 s	1.00	0.7659	0.8453	0.9295
Case 14	Raw features	lightGBM	2 s	0.67	0.8223	0.8858	0.9496
Case 15	Raw features	lightGBM	3 s	0.50	0.8646	0.9129	0.9637
Case 16	With LSTM AE features	lightGBM	1 s	1.00	0.7621	0.8347	0.9098
Case 17	With LSTM AE features	lightGBM	2 s	0.67	0.8160	0.8773	0.9369
Case 18	With LSTM AE features	lightGBM	3 s	0.50	0.8488	0.9041	0.9521
Case 19	With AWGN features	lightGBM	1 s	1.00	0.7642	0.8421	0.9161
Case 20	With AWGN features	lightGBM	2 s	0.67	0.8379	0.8925	0.9485
Case 21	With AWGN features	lightGBM	3 s	0.50	0.8643	0.9122	0.9591
Case 22	With all features	lightGBM	1 s	1.00	0.7638	0.8389	0.9221
Case 23	With all features	lightGBM	2 s	0.67	0.8206	0.8883	0.9454
Case 24	With all features	lightGBM	3 s	0.50	0.8569	0.9115	0.9566

Applying moving average to the probabilities of the lightGBM model resulted in an overall increase in the scores (cases 13–24). This trend can be seen in the probability graphs diagnosing each condition segment in Figures 12 and 13. However, lightGBM, with its leaf-wise growth strategy, tended to overfit with increasing tree depth and often has probabilities highly skewed towards 0 or 1. Therefore, it was difficult to observe performance improvement in the combination of the augmented dataset and contextual diagnosis for the model (cases 13, 16, 19, and 22).

Compared to lightGBM, random forest exhibited relatively less overfitting and showed gradual fluctuations in probabilities corresponding to changes in the time series. This trend can be observed in Figure 13. The diagnosis of the condition segments was not sigmoidal but rather smooth and gradual because the probability was calculated by averaging the voting values of multiple randomly generated decision trees. These characteristics were expressed in classifiers utilizing the augmented dataset with AWGN and LSTM autoencoders. When applying AWGN and LSTM autoencoders to the dataset and contextual diagnosis in cases 10–12, the application of the probability moving average resulted in a higher score of 0.0331 (3.31%) compared to cases 1–3. On the other hand, the random forest score was lower than lightGBM in all cases when the moving average was not applied (without a contextual diagnosis).

By applying machine learning and contextual diagnosis, the diagnosis of the centers of heavy objects carried by forklifts was performed by the condition segment during the process of lifting, tilting, and moving heavy objects on uneven ground using a forklift. As a result, the random forest model (case 12) achieved a maximum accuracy of 0.9563, while the lightGBM model (case 24) achieved a maximum accuracy of 0.9566.

3. Abnormal Lifting Weight Stage Classification

In Section 3, vibration and sound data were acquired from forklifts operating under repeated abnormal weight lifting conditions. In addition, feature engineering and health stage classification of the data were performed following a systematic PHM procedure as depicted in Figure 14.

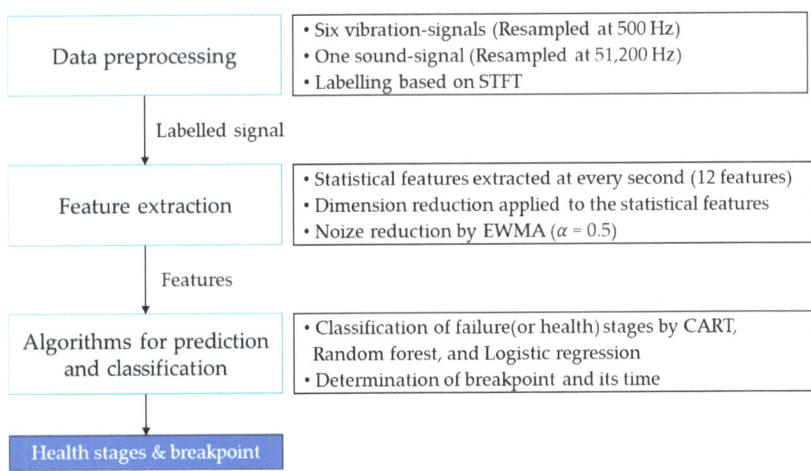

Figure 14. Classification of failure (health) stages in imbalanced abnormal operations.

3.1. Experimental Data Acquisition and Feature Engineering

Safety accidents and failure situations were induced by continuous lifting of unbalanced heavy objects in a laboratory environment. Data measurements and condition diagnoses were conducted for these situations. During the data measurement process, the forklift repeatedly lifted and lowered an unbalanced load of 1500 kg to the right at a consistent speed every 20 s for five hours. The forklift remained stationary throughout this process and did not perform any driving activity. In addition, four three-axis (x, y, z) acceleration sensors were mounted on the left and right sides of the front-end structure to collect vibration data (Figure 15). Repeated acceleration tests were performed until a failure condition occurred; the forklift swayed, and a loud noise was generated in the structure. In the actual operating environment, it is difficult to diagnose the condition and predict the lifespan by sound because of ambient noise. In this study, a microphone (sampling rate 51.2 kHz) was installed at the driver's position to eliminate ambient noise in an anechoic chamber environment, and labeling was performed using sound data. At the same time, the noise generated was measured by the forklift. These sound data were used to classify and label the failure stages into three stages: normal, failure entry, and failure.

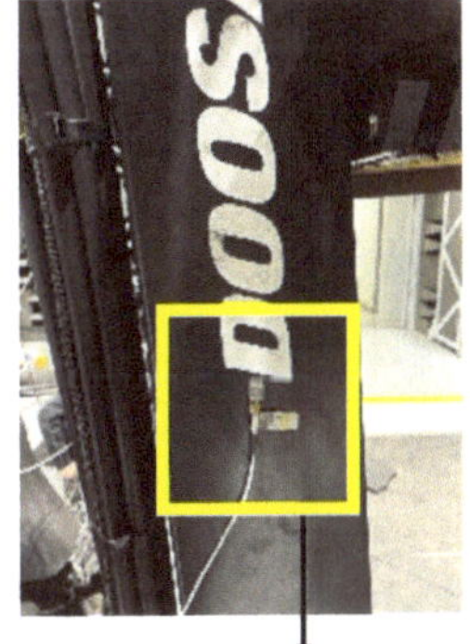

Figure 15. Images of the repeated abnormal lifting experiment in an anechoic chamber with microphone and vibration sensors.

The dataset obtained in the experiment contains 12 acceleration signals (four sensors × (x, y, z accelerations)). For each signal of one sensor, 512 (Hz) × 300 (min) × 60 (s/min) acceleration data were measured, resulting in a total of 9,216,000 data points. In addition, 921,600,000 (51,200 Hz × 300 min × 60 s/min) sound data were collected. As shown in Figure 16, the plotted sound signal obtained in the time domain made it difficult to track the state changes. By observing the frequency changes over time using the short-time Fourier transform (STFT), it was possible to determine if the forklift was operating over time. On the other hand, state changes and detailed differences were difficult to compare.

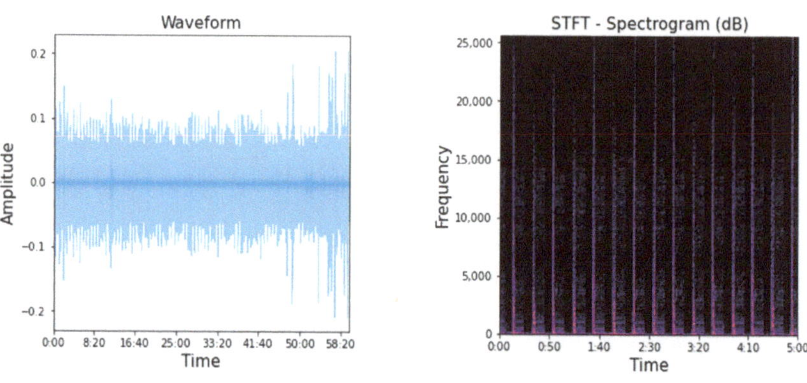

Figure 16. Waveform and short-time Fourier transform (STFT) of measured sound data.

The data size was large, and it was difficult to observe the distinctive state changes when analyzing the raw signals. Twelve features were extracted from the time-domain signal, including min, max, peak to peak, mean (abs), RMS, variance, kurtosis, skewness, crest factor, shape factor, impulse factor, and margin factor for every 20 s window to compress the data and solve these difficulties. Based on this, the dimensions of the acceleration signal and sound signal data were reduced to 900 × 12 × 12 and 900 × 1 × 12, respectively. As reported in Section 2, an exponentially weighted moving average (alpha 0.5) was applied to the reduced dataset to minimize the noise signal in the data. After feature engineering, the CART (classification and regression tree) algorithm [40] was used to classify the failure stages over time using sound features. The CART algorithm is a decision tree algorithm for classification and regression analysis that evaluates the importance of each variable based on the input data and prioritizes the important variables to produce a decision tree. The CART algorithm was used to derive 12 decision trees for each feature by pruning to prevent overfitting and classify the status into three levels (Figure 17). Significant branching points could be derived from eight of the 12 decision trees, and the failure stage was classified based on the average value of the branching points derived from the eight decision trees (Table 9). Based on the classification results, breakpoint 1 was determined to be at 1.18 h and breakpoint 2 at 1.77 h, which were used to distinguish the labeling of the data (Table 10). Furthermore, a comparison of the recorded forklift experimental video showed that beyond breakpoint 1, the forklift exhibited early failure symptoms (beginning to shake), demonstrating a complete failure state at breakpoint 2.

The 900 data generated were divided into 630 data for training and 270 data for testing. Given that the dataset is imbalanced, the performance was evaluated using the F1 score during the validation process. The F1 score is one of the metrics that evaluate the accuracy of a model and is calculated as the harmonic mean of precision and recall. When data are unbalanced, the prediction accuracy for a small number of data classifications is often high, but the prediction accuracy for a large number of data classifications is often low. In such situations, relying solely on accuracy may give the impression of high prediction accuracy for a small number of data classifications. On the other hand, the accuracy for most data classifications may be relatively lower, leading to inadequate overall model performance evaluation. The F1 score considers both precision and recall, which makes it more suitable for dealing with imbalanced classification problems [48–50]. Therefore, when the data are unbalanced, using the F1 score is a more accurate way to evaluate the model performance. Consequently, the F1 score was used to evaluate the performance in this unbalanced dataset.

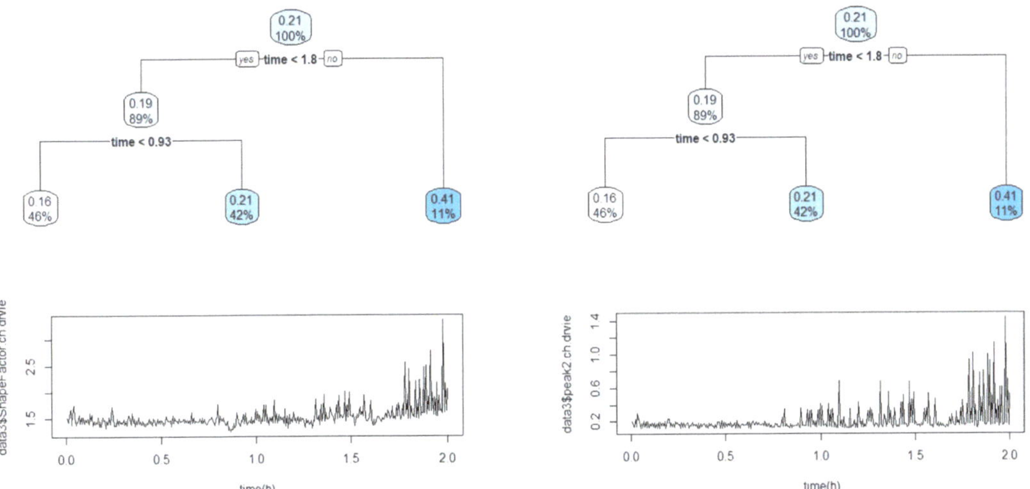

Figure 17. Visual examples of CART algorithm results.

Table 9. Stage-labeling results of the sound data features using the CART (classification and regression tree) algorithm.

Feature	Breakpoint 1	Breakpoint 2
Max	0.892	1.764
Min	0.925	1.775
Peak2	0.925	1.775
Skewness	1.308	1.831
Crest Factor	1.353	1.708
Shape Factor	1.308	1.775
Impulse Factor	1.353	1.764
Margin Factor	1.353	1.775
Average	1.177	1.771

Table 10. Labeling and structure of a dataset.

	Stage 1	Stage 2	Stage 3	Total
Phase	Normal	Semi-failure	Failure	-
Dataset Range (h)	0–1.18	1.18–1.77	1.77–5.0	0–5.0
Number of data	212	106	582	900

3.2. Stage Classification Result

In the previous steps, labels were assigned to instances using sound features. Leveraging these labels, logistic regression and random forest classifiers were trained. The dataset was divided into two distinct cases: case 1 utilized solely vibration features, encompassing 144 dimensions; in contrast, case 2 integrated both vibration and sound features, totaling 156 dimensions. While sound features were primarily introduced for labeling purposes, their inclusion in the training prompted additional cases to assess their impact on classifier performance. In both cases, logistic regression and random forest were selected as the machine learning models, with the details of random forest elaborated upon in Section 2.2. For multiclass classification, logistic regression employs the SoftMax function alongside the cross-entropy loss. In this context, the linear discriminant function $h_k(x)$ is defined as follows:

$$h_k(x) = w_k^T x_i + b_k \tag{16}$$

where w_k and b_k denotes the weights and bias for each class k, respectively, while x_i represents the ith input variable. $p(y = j|x_i)$, the probability of class j, predicted by the SoftMax function is as follows [32]:

$$p(y = j|x_i) = \frac{e^{h_j(x_i)}}{\sum_{k=1}^{K} e^{h_k(x_i)}} \tag{17}$$

The cross-entropy loss function, $J(\theta)$, is calculated as follows [32]:

$$J(\theta) = -\frac{1}{N}\sum_{i=1}^{N}\sum_{j=1}^{K} y_{ij}\log\widehat{y_{ij}} \tag{18}$$

- y_{ij}: the binary value indicating whether class j is the correct target for the ith data point.
- $\widehat{y_{ij}}$: the predicted probability of class j for the ith input data point.
- N: the total number of data points.
- K: the total number of classes.

LightGBM, used in the previous section, is prone to overfitting when the dataset size is small. LightGBM was deemed inappropriate because this dataset consists of 900 feature vectors and was not used in this study. No additional hyperparameter tuning was conducted, and only the default parameters provided by the scikit-learn module in Python were utilized. Based on the validation results of the test data shown in Table 11, the logistic regression classifier achieved an F1 score (macro) of 0.9599 for case 1 and 0.9790 for case 2. In the case of random forest, the F1 scores (macro) for cases 1 and 2 were 0.9116 and 0.9220, respectively. Additionally, Figure 18 depicts a confusion matrix, where the x-axis shows the actual (ground truth) class labels, and the y-axis shows the predicted class labels generated by a model. This confirms that it is possible to classify the forklifts as healthy or aging under repeated unbalanced load weight conditions based on acceleration signals.

Table 11. Validation result of stage classification.

Model	Dataset	Accuracy	F1 Score (Weighted)	F1 Score (Macro)
Logistic Regression	Case 1 (vibration feature)	0.9815	0.9814	0.9599
	Case 2 (vibration + sound feature)	0.9889	0.9891	0.9790
Random Forest	Case 1 (vibration feature)	0.9519	0.9523	0.9116
	Case 2 (vibration + sound feature)	0.9556	0.9556	0.9220

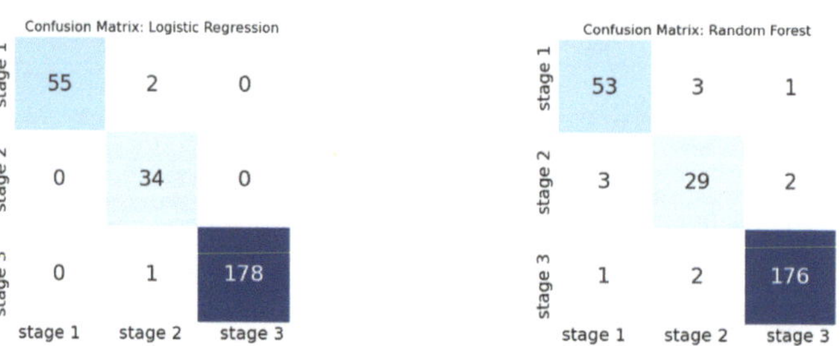

Figure 18. Confusion matrix result: (**left**) logistic regression, (**right**) random forest.

4. RUL Prediction and Fault Diagnosis with Alarm Rule for Abnormal Lifting

Numerous studies [18–28] have comprehensively explored RUL prediction aligned with fault diagnosis. Zhang et al. [19] introduced an inventive approach by parallelly integrating spatial and temporal features using a hybrid neural network. This model combined a 1D convolutional neural network (CNN) with a bidirectional gated recurrent unit (BiGRU). Furthermore, they assessed the limitations of RUL prediction using CNN, LSTM, and the transformer algorithm on aircraft turbofan engine data. To overcome these limitations, they introduced the integrated multi-head dual sparse self-attention network (IMDSSN), an architecture incorporating the ProbSparse self-attention network (MPSN) and LogSparse self-attention network (MLSN) components [20]. Other strategies in RUL prediction span a broad spectrum. Pham et al. [21] proposed RUL prediction for a methane compressor, employing flow system identification, proportional hazard modeling, and support vector machines. Loutas et al. [22] introduced an ε-support vector machine-based approach for estimating rolling bearing RUL. Gugulothu et al. [23] developed Embed-RUL, addressing differing patterns in embeddings between normal and degraded machines. This technique employs a sequence-to-sequence model based on RNNs to generate embeddings for multivariate time series subsequences. Hinchi et al. [24] introduced a method for bearing RUL prediction centered on a convolutional LSTM network. Niu et al. [25] proposed an RUL prediction technique utilizing a 1D-CNN LSTM network. Jiang et al. [26] presented a time series multiple channel convolutional neural network (TSMC-CNN) integrating CNN with LSTM and attention mechanisms. They further integrated TSMC-CNN with an attention-based long short-term memory (ALSTM) network for bearing RUL prediction. Saidi et al. [28] presented RUL prediction using support vector regression (SVR). These regression-based RUL predictions require well-structured feature vectors and substantial RUL data collection. To address this challenge, several studies have focused on performance enhancement, employing ensemble techniques and a range of model-based approaches, including CNN, LSTM, transformers, attention mechanisms, and SVR. While many works have explored convolutional LSTM networks and sequence-to-sequence models, Ley et al. [18] systematically outlined four technical processes including data collection, health index construction, health stage segmentation, and RUL prediction, with a dedicated focus on RUL.

Considering the algorithms proposed in the literature and the model of Ley et al. in [18], the present study combines appropriate algorithms in an integrated manner, adapted to the unique demands of abnormal lifting scenarios. This customized integration facilitates the model in capturing latent patterns and interactions arising from unbalanced load conditions, thus leading to enhanced RUL predictions. This section demonstrates real-time active diagnosis using the life model equation, even with limited data, thereby enhancing the practicality of the proposed approaches. As Section 3 discussed the classification of health stages, subsequently, in Section 4, the dataset and classifiers are applied to perform RUL prediction and fault diagnosis using alarm rules, thereby establishing a foundation for condition-based management [51]. This study also introduces real-time active forklift condition diagnosis using the life model equation, even with limited data. Moreover, alarm rules are devised utilizing key features extracted from the health stage classifier and designated as health indices. Figure 19 summarizes the RUL calculation process and the methods used.

4.1. Life Prediction Model and RUL Verification

In the case of logistic regression, as shown in Figure 20, the classification probabilities of the classifier were derived as either 0 or 1, making it difficult to track the progressive changes in the state. Therefore, it was not feasible to adopt degradation curves for logistic regression. On the other hand, as shown in Section 3, when the classification probabilities of the entire dataset belonging to each stage were visualized using a random forest classifier, a gradual change in state was observed in the graph (Figure 21). This represents the degradation state of the forklift using classification probabilities, and the probabilities of

Stages 1 and 3 were combined and averaged to generate the degradation curve (Figure 22). After analyzing the degradation curve, Stage 3 was reached after approximately 106 min, and the threshold was a probability of 0.7.

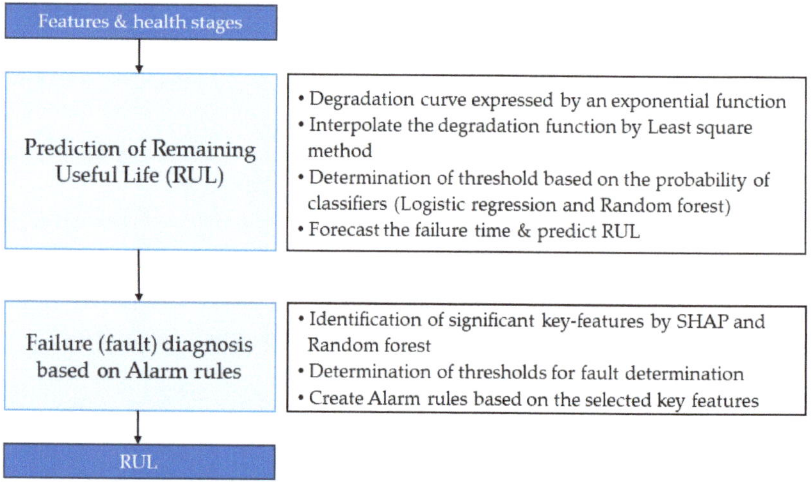

Figure 19. Prediction of RUL and fault diagnosis.

Figure 20. Probability of logistic regression classifier.

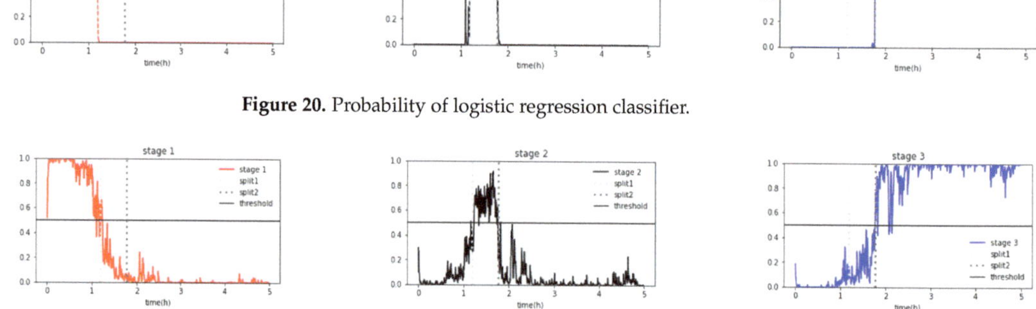

Figure 21. Probability of the random forest classifier.

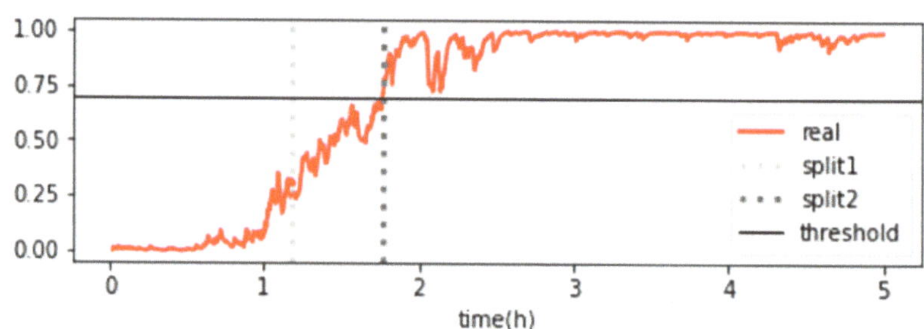

Figure 22. Degradation curve of forklift using the average probability of a random forest classifier.

The model for RUL was constructed using an exponential function equation, as follows.

$$y = a + b \times \exp(c \times x_t) \tag{19}$$

Exponential coefficients were determined using the method of least squares, focusing on data collected within the 30 min prior to the diagnostic time. The least squares method is a statistical tool used in regression analysis to determine model parameters that reduce the discrepancy (error) between observed data and the values expected by the model. The following equations and parameters illustrate the implementation of the least squares method with exponential functions, specifically in Equation (20), for a given dataset.

$$Dataset : ((x_1, y_1), (x_2, y_2), \ldots, (x_n, y_n)) \tag{20}$$

$$e_i = y_i - (a + b \times \exp(c \times x_{ti})) \tag{21}$$

$$E = \sum_{i=1}^{n} e_i^2 = \sum_{i=1}^{n} (y_i - (a + b \times \exp(c \times x_{ti})))^2 \tag{22}$$

- y_i: the observed value of the dependent variable at a point
- x_{ti}: the value of independent variable x_i at a time t.
- a, b, c: the parameters of the exponential function.
- e_i: the residuals associated with the exponential function's prediction.
- E: the objective function to compute the sum of squared residuals, which is minimized to determine the values of $a, b,$ and c.

Based on these characteristics, the exponential function continued to change with time, and the time when the y-value of the exponential function reached 0.7 was calculated to analyze the RUL. During the analysis of the RUL, the confidence interval was included by considering $Y(0.9x_t)$ to $Y(1.1x_t)$ in the pre- and post-prediction time points. The analysis showed that the early data predicted the failure rather early, as shown in Figure 23. The gap between the actual and the predicted RUL decreased gradually as the failure point approached (Figure 24).

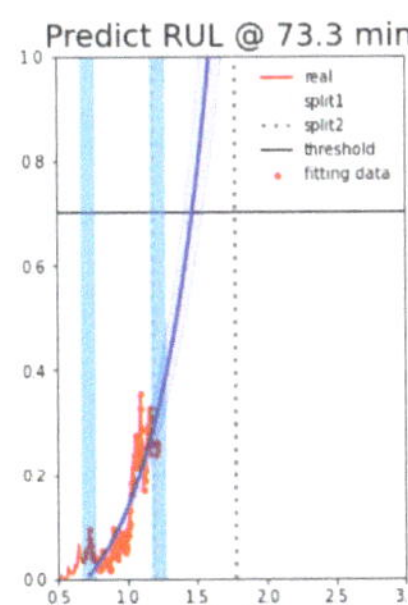

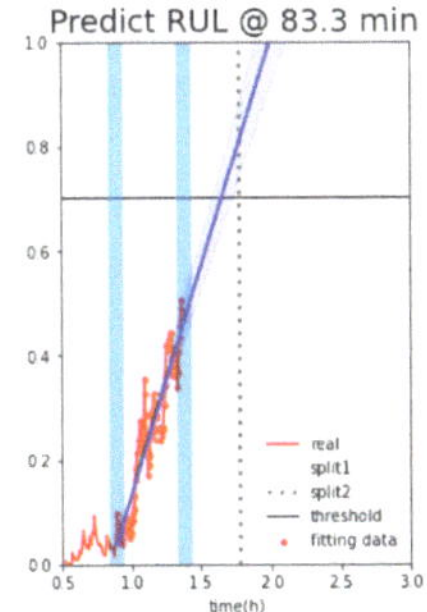

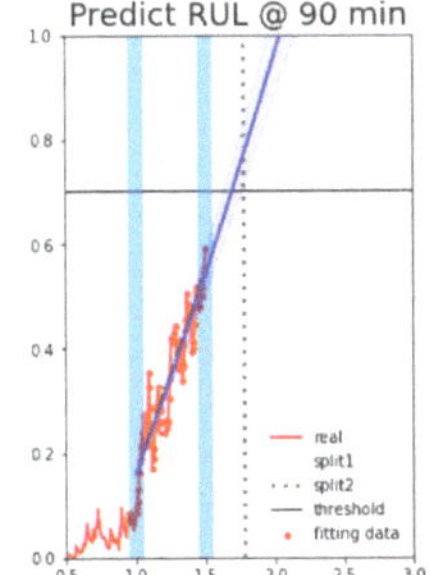

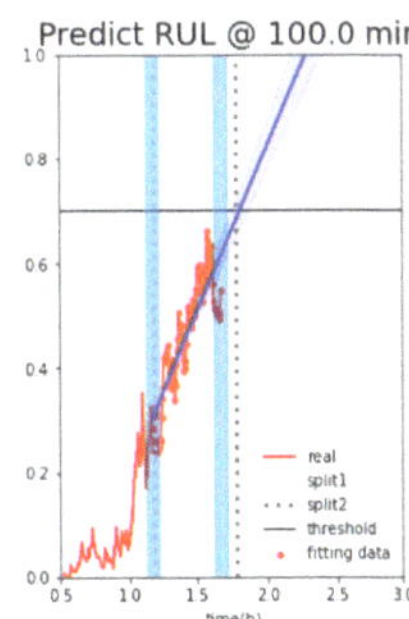

Figure 23. Degradation curve of forklift using the average probability of a random forest classifier.

4.2. Alarm-Rule-Based Fault Diagnosis

The SHAP (Shapley additive explanations) algorithm was used to diagnose faults based on the key features and alarm rules. The goal was to determine which features were critical for state classification using the SHAP algorithm with the previously trained random forest classifier. This analysis allowed the identification of the significant features used for state classification. SHAP is an algorithm used to analyze the contribution of features to the predictions made by machine learning models. It helps to determine the importance of each feature in the model's predictions. The SHAP algorithm considers all features necessary to explain the model predictions and calculates the impact of each feature on the prediction outcome [52]. Calculating the Shapley value for all feature combinations is computationally expensive. Therefore, the SHAP algorithm uses approximations that can be calculated relatively quickly for tree-based models, such as the random forests or XGBoost [29]. The SHAP algorithm provides various interpretation results, such as feature importance, feature contribution, and feature effect, which help to interpret the model and explain the prediction results reliably [52]. The SHAP algorithm improves the interpretability of machine learning models and plays a vital role in model development and helping users to understand the model prediction results [52].

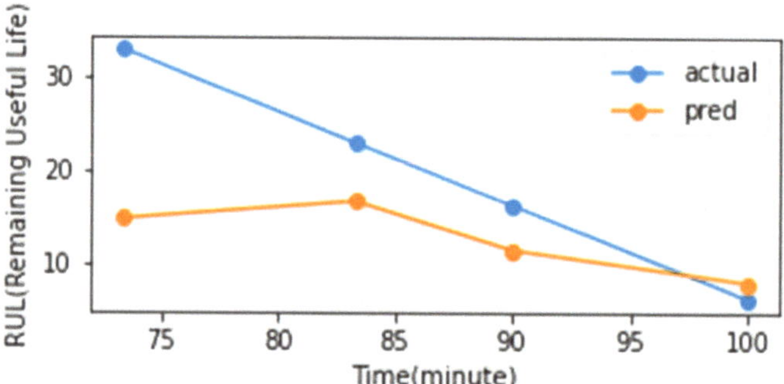

Figure 24. Remaining useful lifetime (RUL) result.

The features that contribute significantly to health diagnostics were identified using the SHAP algorithm. As shown on the left in Figure 25, the 'abs.mean ch In LH z' feature contributes the most to the health diagnosis. A progressive state change, similar to the degradation model graph, could be observed by observing the dispersion of the time series of the corresponding feature (Figure 25, right image). The 2-sigma and 3-sigma values of the 'abs.mean ch In LH z' feature were extracted from the distribution of the normal data range in stage 1. These values were then used as the threshold for fault diagnosis. Next, an alarm rule was set to diagnose a failure when the exponentially weighted moving average value of the feature (alpha 0.17) exceeds a threshold. As shown in Figure 25 (right), the 3-sigma threshold can diagnose failure just before failure, and the 2-sigma threshold can predict failure at the point of the precursor symptoms.

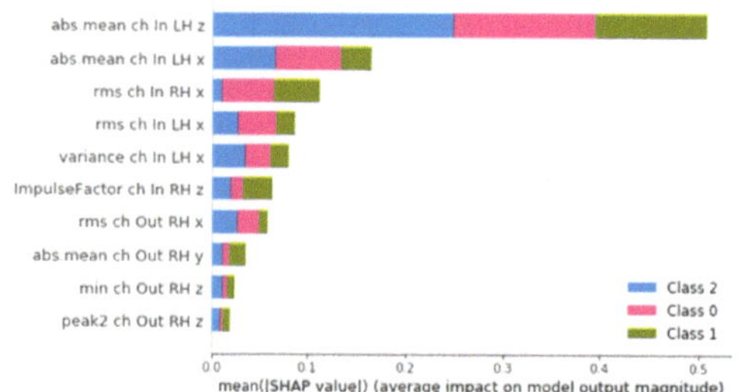
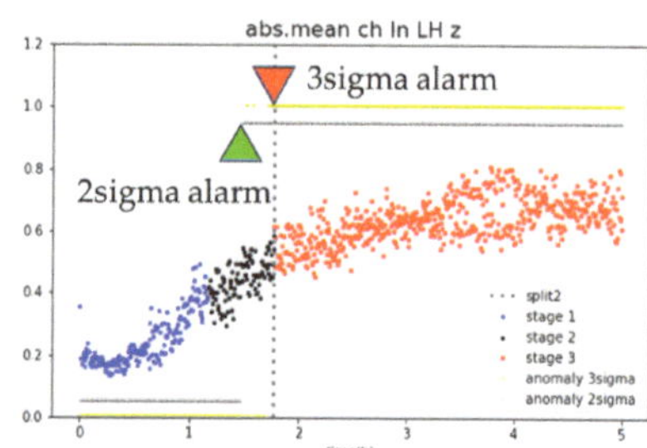

Figure 25. Results of SHAP and fault diagnosis based on the alarm rules.

5. Discussion

This study addresses the critical issue of preventing front-end failures in forklifts by predicting the center of gravity. The study begins by emphasizing the fault diagnosis of front-end failures in forklifts and the significance of accurately predicting the gravity center. In this pursuit, the study acquired acceleration signals during lifting, tilting, and moving heavy objects by the forklift on uneven ground and in various operating environments. These acceleration signals were captured from the outer beams on both the left and right sides of the forklift's front-end. To ensure effective processing and feature determination, time-domain statistical features were extracted and established as variables by applying a window at one-second intervals. To mitigate potential overfitting, the dataset was augmented with AWGN and LSTM autoencoders. Following these data enhancements, classifiers were used to accurately categorize the center of objects being transported by forklifts during their driving and working phases. This classification task was accomplished using the robust capabilities of the random forest and lightGBM models. The random forest model and lightGBM model were able to predict the center of gravity with an accuracy of 0.9563 and 0.9566, respectively. During the prediction of the gravity center, an exponentially weighted moving average was conducted to smooth out the noise of the features. As a result of applying this moving average, the dataset's outliers were reduced without necessitating complex noise filtering methods. Moreover, an observation was made: as the window size for the exponentially weighted moving average was increased, the accuracy of the machine learning models also improved. This suggests that a larger window captured more accurate data trends, enhancing predictive performance. By using data augmentation, overfitting was minimized, while maintaining similar or slightly lower accuracy scores. In the lightGBM model, the implementation of a moving average on the classification probabilities showed a tendency to improve the scores. In contrast, lightGBM tended to overfit with increased tree depth, resulting in biased classification probabilities. This made it difficult to improve performance when combining augmented datasets with contextual diagnosis. However, the random forest model showed less overfitting compared to lightGBM and showed gradual changes in classification probabilities over time. These trends were observed in the augmented dataset using the AWGN and LSTM autoencoders. Applying the augmented dataset and contextual diagnosis, along with the moving average to the classification probabilities, resulted in an average score improvement of 3.31%. Notably, the random forest score was lower than the lightGBM score without the moving average.

After predicting the center of gravity, the paper presented a procedure for forecasting the RUL during abnormal operational scenarios and diagnosing failures through the application of alarm rules. To forecast RUL during repetitive unbalanced load conditions, statistical features were extracted from acceleration data using a 20 s window. This data collection took place in an anechoic chamber, with simultaneous recording of sound signals via a microphone. The CART algorithm classified and labeled statistical features derived from the sound signals. Logistic regression and random forest models were then used for failure stage classification, achieving F1 scores of 0.9790 and 0.9220, respectively. Notably, logistic regression achieved the highest score among the classifiers. Conversely, when examining the classifier's probability change, the probabilities were often skewed toward 0 or 1. This skewed distribution made it difficult to accurately track the forklift's state changes. Consequently, monitoring the forklift's degradation status was accomplished through the random forest's generated classification probabilities. The results showed a gradual change in the condition of the forklift over time. Based on these results, the probabilities generated by the random forest for each stage were combined and averaged to create a degradation curve. The failure point was predicted by the failure threshold that was determined through degradation curve analysis.

In conclusion, predicting the gravity center of objects carried by the forklift yields insights into operations that impact forklift durability. This approach enhances equipment longevity and operational safety. Additionally, providing RUL data facilitates the development of operational plans and efficient maintenance by identifying critical features for setting failure thresholds.

Author Contributions: Conceptualization and methodology, J.-G.L. and Y.-S.K.; validation, formal analysis, investigation, and data curation, J.-G.L. and J.H.L.; research administration, J.H.L.; funding acquisition, J.H.L. All authors have read and agreed to the published version of the manuscript.

Funding: This research was supported by a grant from the National R&D Project "Development of fixed offshore green hydrogen production technology connected to marine renewable energy" funded by the Ministry of Oceans and Fisheries (1525013967).

Institutional Review Board Statement: Not applicable.

Informed Consent Statement: Not applicable.

Data Availability Statement: The data presented in this study are available upon request from the corresponding author. The data derived from the present study are only partially available for research purposes.

Conflicts of Interest: The authors declare no conflict of interest.

References

1. Tsui, K.L.; Chen, N.; Zhou, Q.; Hai, Y.; Wang, W. Prognostics and health management: A review on data driven approaches. *Math. Probl. Eng.* **2015**, *2015*, 793161. [CrossRef]
2. Sheppard, J.W.; Kaufman, M.A.; Wilmering, T.J. IEEE standards for prognostics and health management. In Proceedings of the 2008 IEEE AUTOTESTCON, Salt Lake City, UT, USA, 8–11 September 2008; pp. 97–103.
3. Lee, J.; Wu, F.; Zhao, W.; Ghaffari, M.; Liao, L.; Siegel, D. Prognostics and health management design for rotary machinery systems—Reviews, methodology and applications. *Mech. Syst. Signal Process.* **2014**, *42*, 314–334. [CrossRef]
4. Meng, H.; Li, Y.F. A review on prognostics and health management (PHM) methods of lithium-ion batteries. *Renew. Sustain. Energy Rev.* **2019**, *116*, 109405. [CrossRef]
5. Klyatis, L.M. *Accelerated Reliability and Durability Testing Technology*; John Wiley & Sons: Hoboken, NJ, USA, 2012; Volume 70.
6. Elsayed, E.A. Overview of reliability testing. *IEEE Trans. Reliab.* **2012**, *61*, 282–291. [CrossRef]
7. Kapur, K.C.; Pecht, M. *Reliability Engineering*; John Wiley & Sons: Hoboken, NJ, USA, 2014; Volume 86.
8. O'Connor, P.; Kleyner, A. *Practical Reliability Engineering*; John Wiley & Sons: Hoboken, NJ, USA, 2012.
9. Zio, E. Reliability engineering: Old problems and new challenges. *Reliab. Eng. Syst. Saf.* **2009**, *94*, 125–141. [CrossRef]
10. Kwon, D.; Hodkiewicz, M.R.; Fan, J.; Shibutani, T.; Pecht, M.G. IoT-based prognostics and systems health management for industrial applications. *IEEE Access* **2016**, *4*, 3659–3670. [CrossRef]
11. Zhang, L.; Lin, J.; Liu, B.; Zhang, Z.; Yan, X.; Wei, M. A review on deep learning applications in prognostics and health management. *IEEE Access* **2016**, *7*, 162415–162438. [CrossRef]
12. Fink, O.; Wang, Q.; Svensen, M.; Dersin, P.; Lee, W.J.; Ducoffe, M. Potential, challenges and future directions for deep learning in prognostics and health management applications. *Eng. Appl. Artif. Intell.* **2020**, *92*, 103678. [CrossRef]
13. Forman, G. An extensive empirical study of feature selection metrics for text classification. *J. Mach. Learn. Res.* **2003**, *3*, 1289–1305.
14. Ding, A.; Qin, Y.; Wang, B.; Jia, L.; Cheng, X. Lightweight multiscale convolutional networks with adaptive pruning for intelligent fault diagnosis of train bogie bearings in edge computing scenarios. *IEEE Trans. Instrum. Meas.* **2022**, *72*, 1–13. [CrossRef]
15. Zhang, J.; Zhang, K.; An, Y.; Luo, H.; Yin, S. An integrated multitasking intelligent bearing fault diagnosis scheme based on representation learning under imbalanced sample condition. *IEEE Trans. Neural Netw. Learn. Syst.* **2023**, 1–12. [CrossRef]
16. Bowles, C.; Chen, L.; Guerrero, R.; Bentley, P.; Gunn, R.; Hammers, A.; Dickie, D.A.; Hernández, M.V.; Wardlaw, J.; Rueckert, D. Gan augmentation: Augmenting training data using generative adversarial networks. *arXiv* **2018**, arXiv:1810.10863.
17. Ko, T.; Peddinti, V.; Povey, D.; Khudanpur, S. Audio augmentation for speech recognition. In Proceedings of the Sixteenth Annual Conference of the International Speech Communication Association, Dresden, Germany, 6–10 September 2015.
18. Lei, Y.; Li, N.; Guo, L.; Li, N.; Yan, T.; Lin, J. Machinery health prognostics: A systematic review from data acquisition to RUL prediction. *Mech. Syst. Signal Process.* **2018**, *104*, 799–834. [CrossRef]
19. Zhang, J.; Tian, J.; Li, M.; Leon, J.I.; Franquelo, L.G.; Luo, H.; Yin, S. A parallel hybrid neural network with integration of spatial and temporal features for remaining useful life prediction in prognostics. *IEEE Trans. Instrum. Meas.* **2022**, *72*, 1–12. [CrossRef]
20. Zhang, J.; Li, X.; Tian, J.; Luo, H.; Yin, S. An integrated multi-head dual sparse self-attention network for remaining useful life prediction. *Reliab. Eng. Syst. Saf.* **2023**, *233*, 109096. [CrossRef]
21. Pham, H.T.; Yang, B.S.; Nguyen, T.T. Machine performance degradation assessment and remaining useful life prediction using proportional hazard model and support vector machine. *Mech. Syst. Signal Process.* **2012**, *32*, 320–330.
22. Loutas, T.H.; Roulias, D.; Georgoulas, G. Remaining useful life estimation in rolling bearings utilizing data-driven probabilistic E-support vectors regression. *IEEE Trans. Reliab.* **2013**, *62*, 821–832. [CrossRef]
23. Gugulothu, N.; Tv, V.; Malhotra, P.; Vig, L.; Agarwal, P.; Shroff, G. Predicting remaining useful life using time series embeddings based on recurrent neural networks. *arXiv* **2017**, arXiv:1709.01073. [CrossRef]
24. Hinchi, A.Z.; Tkiouat, M. Rolling element bearing remaining useful life estimation based on a convolutional long-short-term memory network. *Procedia Comput. Sci.* **2018**, *127*, 123–132. [CrossRef]
25. Niu, J.; Liu, C.; Zhang, L.; Liao, Y. Remaining useful life prediction of machining tools by 1D-CNN LSTM network. In Proceedings of the 2019 IEEE Symposium Series on Computational Intelligence (SSCI), Xiamen, China, 6–9 December 2019; pp. 1056–1063.
26. Jiang, J.R.; Lee, J.E.; Zeng, Y.M. Time series multiple channel convolutional neural network with attention-based long short-term memory for predicting bearing remaining useful life. *Sensors* **2019**, *20*, 166. [CrossRef] [PubMed]

27. Ali, J.B.; Saidi, L.; Harrath, S.; Bechhoefer, E.; Benbouzid, M. Online automatic diagnosis of wind turbine bearings progressive degradations under real experimental conditions based on unsupervised machine learning. *Appl. Acoust.* **2018**, *132*, 167–181.

28. Saidi, L.; Ali, J.B.; Bechhoefer, E.; Benbouzid, M. Wind turbine high-speed shaft bearings health prognosis through a spectral Kurtosis-derived indices and SVR. *Appl. Acoust.* **2017**, *120*, 1–8. [CrossRef]

29. Holt, C.C. Forecasting seasonals and trends by exponentially weighted moving averages. *Int. J. Forecast.* **2004**, *20*, 5–10. [CrossRef]

30. Hunter, J.S. The exponentially weighted moving average. *J. Qual. Technol.* **1986**, *18*, 203–210.

31. Vaseghi, S.V. *Advanced Digital Signal Processing and Noise Reduction*; John Wiley & Sons: Hoboken, NJ, USA, 2008.

32. Aurélien, G. *Hands-on Machine Learning with Scikit-Learn, Keras, and TensorFlow*; O'Reilly Media, Inc.: Sebastopol, CA, USA, 2022.

33. Graves, A.; Graves, A. *Long Short-Term Memory. Supervised Sequence Labelling with Recurrent Neural Networks*; Springer: Berlin/Heidelberg, Germany, 2012; pp. 37–45.

34. Bengio, Y.; Simard, P.; Frasconi, P. Learning long-term dependencies with gradient descent is difficult. *IEEE Trans. Neural Netw.* **1994**, *5*, 157–166. [CrossRef]

35. Pascanu, R.; Mikolov, T.; Bengio, Y. On the difficulty of training recurrent neural networks. In Proceedings of the International Conference on Machine Learning, Atlanta, GA, USA, 16 June–21 June 2013; Volume 28, pp. 1310–1318.

36. Hochreiter, S. The vanishing gradient problem during learning recurrent neural nets and problem solutions. *Int. J. Uncertain. Fuzziness Knowl. Based Syst.* **1998**, *6*, 107–116. [CrossRef]

37. Gers, F.A.; Schmidhuber, J.; Cummins, F. Learning to forget: Continual prediction with LSTM. *Neural Comput.* **2000**, *12*, 2451–2471. [CrossRef]

38. Bao, W.; Yue, J.; Rao, Y. A deep learning framework for financial time series using stacked autoencoders and long-short term memory. *PLoS ONE* **2017**, *12*, e0180944. [CrossRef] [PubMed]

39. Breiman, L. Random forests. *Mach. Learn.* **2001**, *45*, 5–32. [CrossRef]

40. Breiman, L. *Classification and Regression Trees*; Routledge: London, UK, 2017.

41. Liang, W.; Luo, S.; Zhao, G.; Wu, H. Predicting hard rock pillar stability using GBDT, XGBoost, and LightGBM algorithms. *Mathematics* **2020**, *8*, 765. [CrossRef]

42. Ke, G.; Meng, Q.; Finley, T.; Wang, T.; Chen, W.; Ma, W.; Ye, Q.; Liu, T.Y. Lightgbm: A highly efficient gradient boosting decision tree. In Proceedings of the Advances in Neural Information Processing Systems 30 (NIPS 2017), Long Beach, CA, USA, 4–9 December 2017; Volume 30.

43. Greenhill, S.; Rana, S.; Gupta, S.; Vellanki, P.; Venkatesh, S. Bayesian optimization for adaptive experimental design: A review. *IEEE Access* **2020**, *8*, 13937–13948. [CrossRef]

44. Bergstra, J.; Bardenet, R.; Bengio, Y.; Kégl, B. Algorithms for hyper-parameter optimization. In Proceedings of the Advances in Neural Information Processing Systems 24 (NIPS 2011), Granada, Spain, 12–15 December 2011; Volume 24.

45. Singh, K.; Upadhyaya, S. Outlier detection: Applications and techniques. *Int. J. Comput. Sci. Issues IJCSI* **2012**, *9*, 307.

46. Weigend, A.S.; Mangeas, M.; Srivastava, A.N. Nonlinear gated experts for time series: Discovering regimes and avoiding overfitting. *Int. J. Neural Syst.* **1995**, *6*, 373–399. [CrossRef] [PubMed]

47. Kou, Y.; Lu, C.T.; Chen, D. Spatial weighted outlier detection. In Proceedings of the 2006 SIAM International Conference on Data Mining, Bethesda, MD, USA, 20–22 April 2006; Society for Industrial and Applied Mathematics: Philadelphia, PA, USA, 2006; pp. 614–618.

48. Sokolova, M.; Japkowicz, N.; Szpakowicz, S. Beyond accuracy, F-score and ROC: A family of discriminant measures for performance evaluation. In Proceedings of the AI 2006: Advances in Artificial Intelligence: 19th Australian Joint Conference on Artificial Intelligence, Proceedings 19, Hobart, Australia, 4–8 December 2006; Springer: Berlin/Heidelberg, Germany, 2006; pp. 1015–1021.

49. Jeni, L.A.; Cohn, J.F.; De La Torre, F. Facing imbalanced data—Recommendations for the use of performance metrics. In Proceedings of the 2013 Humaine Association Conference on Affective Computing and Intelligent Interaction, Geneva, Switzerland, 2–5 September 2013; pp. 245–251.

50. Sun, Y.; Wong, A.K.; Kamel, M.S. Classification of imbalanced data: A review. *Int. J. Pattern Recognit. Artif. Intell.* **2009**, *23*, 687–719. [CrossRef]

51. Kumar, S.; Mukherjee, D.; Guchhait, P.K.; Banerjee, R.; Srivastava, A.K.; Vishwakarma, D.N.; Saket, R.K. A comprehensive review of condition based prognostic maintenance (CBPM) for induction motor. *IEEE Access* **2019**, *7*, 90690–90704. [CrossRef]

52. Lundberg, S.M.; Lee, S.I. A unified approach to interpreting model predictions. In Proceedings of the Advances in Neural Information Processing Systems 30 (NIPS 2017), Long Beach, CA, USA, 4–9 December 2017; Volume 30.

Article

Research on Diesel Engine Fault Status Identification Method Based on Synchro Squeezing S-Transform and Vision Transformer

Siyu Li [†], Zichang Liu [†], Yunbin Yan, Rongcai Wang, Enzhi Dong and Zhonghua Cheng *

Shijiazhuang Campus of Army Engineering University of PLA, Shijiazhuang 050003, China; sy_li1988@163.com (S.L.); zc_liu1997@aeu.edu.cn (Z.L.); siyusi@126.com (Y.Y.); wrcpromising@aeu.edu.cn (R.W.); ez_dong@aeu.edu.cn (E.D.)
* Correspondence: a15032073178@sina.com; Tel.: +86-0311-87994538
[†] These authors contributed equally to this work.

Abstract: The reliability and safety of diesel engines gradually decrease with the increase in running time, leading to frequent failures. To address the problem that it is difficult for the traditional fault status identification methods to identify diesel engine faults accurately, a diesel engine fault status identification method based on synchro squeezing S-transform (SSST) and vision transformer (ViT) is proposed. This method can effectively combine the advantages of the SSST method in processing non-linear and non-smooth signals with the powerful image classification capability of ViT. The vibration signals reflecting the diesel engine status are collected by sensors. To solve the problems of low time-frequency resolution and weak energy aggregation in traditional signal time-frequency analysis methods, the SSST method is used to convert the vibration signals into two-dimensional time-frequency maps; the ViT model is used to extract time-frequency image features for training to achieve diesel engine status assessment. Pre-set fault experiments are carried out using the diesel engine condition monitoring experimental bench, and the proposed method is compared with three traditional methods, namely, ST-ViT, SSST-2DCNN and FFT spectrum-1DCNN. The experimental results show that the overall fault status identification accuracy in the public dataset and the actual laboratory data reaches 98.31% and 95.67%, respectively, providing a new idea for diesel engine fault status identification.

Keywords: synchro squeezing S-transform; vision transformer; diesel engine; fault status identification; reliability; time-frequency analysis

Citation: Li, S.; Liu, Z.; Yan, Y.; Wang, R.; Dong, E.; Cheng, Z. Research on Diesel Engine Fault Status Identification Method Based on Synchro Squeezing S-Transform and Vision Transformer. *Sensors* **2023**, *23*, 6447. https://doi.org/10.3390/s23146447

Academic Editors: Shilong Sun, Changqing Shen and Dong Wang

Received: 20 June 2023
Revised: 12 July 2023
Accepted: 13 July 2023
Published: 16 July 2023

1. Introduction

Diesel engines, as the main power source for heavy vehicles, construction machinery, ships, generator sets, tanks, etc., will greatly affect normal production and safe operation when they fail [1]. However, due to the high pressure and temperature generated during their operation, requiring high structural strength and stiffness of each relevant part, their reliability and safety gradually decrease with the increase in operation time [2]. The important components of a diesel engine will lose their original functions as their functions decay. Therefore, the safety and durability of diesel engines are receiving more and more attention, and the fault status identification of diesel engines has become a hot spot for research in related fields at home and abroad [3].

When identifying the status of a diesel engine, the original signal collected by the sensor is mainly a vibration signal. Because the collection of vibration signals is simple and fast, there is no need to disassemble the diesel engine body and change the diesel engine structure. Usually, the basic idea of fault status identification for diesel engines is firstly, collecting one-dimensional vibration data during the engine working condition, secondly, completing noise reduction and feature extraction of the original signal, and finally, completing fault status identification through a pattern recognition method [4].

Compared with traditional machine learning methods, deep learning has a powerful adaptive feature-learning capability to independently build the desired network model based on the sample data during the learning process, and has received much attention in the field of prognostics and health management [5]. Although deep learning is capable of training one-dimensional raw signals, two-dimensional images contain richer feature information and have received attention in this field as inputs for deep learning [6]. Compared with one-dimensional signals, two-dimensional images have two main advantages in fault status recognition: first, image data can consider multiple dimensions simultaneously when expressing information, whereas one-dimensional data can only consider one dimension of information, which is more one-sided [7]; secondly, images are easy to identify and classify, and the use of advanced algorithms to convert one-dimensional signals into two-dimensional images makes it easier to classify images from a visual perspective. Therefore, image data can provide more comprehensive and rich information for a wider range of application scenarios, such as image recognition, video analysis, and medical image processing [8]. Traditional time-frequency transformation methods, including short-time Fourier transform (STFT), continuous wavelet transform (CWT), S-transform (ST), etc., have achieved good results in studying time-frequency analysis [9]. After performing the time-frequency transform, the resulting feature maps are preprocessed and then input to a deep learning model for status recognition [10]. Liu et al. [11] converted diesel engine cylinder head vibration signals into time-frequency maps by STFT, which were input to an AlexNet network and ResNet-18 network for training, and achieved good fault classification by transfer learning algorithm. Xi et al. [12] used ST to convert diesel engine vibration signals into time-frequency maps and t-SNE to visualize fault features, which were input to an extreme learning machine classifier to intelligently classify diesel engine faults. Shen et al. [13] performed the Gabor transform on the vibration signal to obtain the time-frequency diagram of each operating status of the diesel engine. Mou et al. [14] converted the diesel engine vibration signal into a time-frequency map by smoothing the pseudo-Wigner distribution. However, the above method has problems such as fixed time-frequency resolution, little phase information, low resolution and poor energy aggregation [15]. The limited feature information contained in the conversion of the original vibration signal into an image by the above methods makes it difficult to effectively extract the time-dependent features of the vibration signals of the different statuses of the components to be monitored, which easily causes the loss of useful information and affects the recognition accuracy of the model [16]. Synchro squeezing wavelet transform (SSWT) is a combination of a synchro squeezing algorithm to squeeze the energy in a certain range around the center frequency of each frequency band to converge to the center frequency after wavelet transform to improve the resolution. Compared with the wavelet transform, the S-transform has better adaptivity and time-frequency resolution. The synchro squeezing S-transform (SSST), to a certain extent, solves the problems of poor adaptivity of SSWT and the low resolution of high-frequency low-amplitude signals, and has good effects in the time-frequency analysis of seismic signals and vibration signals.

In terms of pattern recognition technology, the convolutional neural network (CNN) is usually used to identify diesel engine fault status. Chen [17] established a diesel engine overall status identification model based on a support vector machine (SVM). However, SVM has the problem of being sensitive to parameter selection, which limits the application of the method in diesel engine condition identification. To solve the problems of index selection and the difficulty of weight determination in the traditional fault status identification method, Bai et al. [18] constructed a diesel engine fault status identification model based on CNN. Jiang et al. [19] addresses the problem that diesel engine faults are difficult to identify accurately under complex operating conditions, and the diesel engine vibration signals are fed into a one-dimensional CNN and a deep neural network of a long short-term network (LSTM) for training, which can be effective for status identification. Zhan et al. [20] proposed a fault identification method based on the combination of optimal variational mode decomposition (VMD) and an improved CNN. However, when classifying and recognizing images, the initial status parameters for the CNN can have a great impact on the network training, and a poor choice can cause the network to not work or potentially fall into local minima, underfitting, and overfitting. In 2019 researchers started to try to apply transformers to the CV domain, and finally in 2021, those involved proved that transformers have better scalability than CNNs, can handle sequential types of inputs, and are significantly better than CNNs when training larger models on larger datasets [21]. Alexey et al., proposed the vision transformer (ViT) model by directly applying the transformer architecture to image classification tasks, representing the input image as a feature vector that can be used for subsequent tasks; ViT significantly improves the performance of traditional image classification tasks [22].

Therefore, this paper takes diesel engines as the engineering research background, and proposes a diesel engine status recognition method that combines the advantages of SSST to represent time-varying nonlinear non-smooth signals with the excellent image classification ability of ViT to achieve the classification of diesel engine fault status in response to the current problem of inaccurate diesel engine status recognition. The main contributions and innovations of this paper are as follows:

(1)　Relying on the existing conditions in the laboratory, a pre-set fault experiment was carried out to realize the acquisition of diesel engine cylinder head vibration signals.

(2)　The original diesel engine vibration signal is represented as a time-frequency image by SSST, and the dependence of the vibration signal on time is mapped into the image feature space, so that the original feature information is retained in the time-frequency map as much as possible. Then, after applying the powerful learning ability of ViT to automatically extract the temporal and spatial features in the images, the fault status identification is completed.

(3)　The feasibility and effectiveness of the proposed diesel engine status recognition method is verified by means of public datasets and actual laboratory measurements.

The remaining sections are as follows: Section 2 introduces the relevant theories of SSST-ViT in detail; Section 3 provides the diesel engine pre-set fault experiments and the experimental data acquisition method, and the experimental results are analyzed and studied; In Section 4, the conclusions of this study and the outlook for future research work are presented.

2. Diesel Engine Fault Status Identification Method

In the diesel engine fault status identification method based on SSST-ViT, the original vibration signals collected are converted into time-frequency maps by the SSST, and the ViT network model is applied to identify each fault status. Therefore, in this section, SSST, the ViT network model and a diesel engine fault status identification method based on SSST-ViT are introduced.

2.1. Synchro Squeezing S-Transform

The ST of the acquired original diesel engine vibration signal is compressed synchronously and represented as a two-dimensional time-frequency map. Compared with CWT, ST has a better time-frequency discrimination effect. The result of the wavelet transform is a time-scale spectrum, and the result of ST is time-frequency spectrum, which is more intuitive and clearer, and is a reversible transform without signal loss [23]. In particular, it has an enhancement effect on the high-frequency weak amplitude components of the original signal, which is effective for weak signal testing and research analysis. The raw diesel engine signal collected by the vibration acceleration sensor is a typical one-dimensional time series, whose vertical coordinate is the amplitude corresponding to each sampling point and the horizontal coordinate is the time or sampling point. The original vibration signal cannot fully represent the fault status information of the diesel engine, and in order to effectively characterize the time-frequency characteristics of the original signal, the signal is converted into a time-frequency map, which can not only highlight the original characteristic information of the vibration signal, but can also further enhance the characteristic time series information [24].

Based on the principle of synchro squeezing transformation, the derivation process of SSST is as follows:

Define the ST equation of the signal $x(t)$ as

$$\mathrm{ST}_X(f,b) = \int_{-\infty}^{+\infty} x(t)\frac{|f|}{\sqrt{2\pi}}e^{-\frac{f^2(t-b)^2}{2}}e^{-i2\pi ft}\mathrm{d}t \tag{1}$$

where $S(f,b)$ is the time-frequency spectrum of $x(t)$, t is the time, f is the frequency, b is the displacement parameter, and i is the imaginary number.

$\varphi(t) = \frac{1}{\sqrt{2\pi}}e^{\frac{t^2}{2}}e^{i2\pi t}$, then Equation (1) yields

$$\mathrm{ST}_X(f,b) = |f|e^{-i2\pi fb}\int_{-\infty}^{+\infty} x(t)\overline{\varphi[f(t-b)]}\mathrm{d}t \tag{2}$$

$\overline{\varphi}$ and φ are complex conjugates, and by Parseval's theorem and Fourier transformations

$$\mathrm{ST}_X(f,b) = \frac{1}{2\pi}e^{-i2\pi fb}\int_{-\infty}^{+\infty} \hat{x}(\xi)\overline{\hat{\varphi}[(f^{-1}\xi)e^{ib\xi}]}\mathrm{d}\xi \tag{3}$$

where ξ is the angular frequency, $x(t)$ is obtained by Fourier transforming $\hat{x}(\xi)$, $\overline{\hat{\varphi}(\xi)}$ is the complex conjugate of $\varphi(t)$, and the single harmonic case: $x(t) = A\cos(2\pi f_0 t)$

$$\hat{x}(\xi) = A\pi[\delta(\xi - 2\pi f_0) + \delta(\xi + 2\pi f_0)] \tag{4}$$

S-transform is

$$\mathrm{ST}_X(f,b) = \frac{A}{2\pi}e^{-i2\pi(f-f_0)^b}\hat{\varphi}*(2\pi f^{-1}f_0) \tag{5}$$

The analysis shows that the energy of the time spectrum of $x(t)$ is distributed at $f = f_0$, but the actual time spectrum is around f_0 with a spurious spectral bandwidth. The goal of SSST is to obtain the real instantaneous frequency of $x(t)$ by converging the energy after compression.

Derivative of (5) with respect to time:

$$\frac{\partial}{\partial b}\mathrm{ST}_X(f,b) = -i\pi A(f - f_0)e^{-i2\pi(f-f_0)^b}\overline{\hat{\varphi}(2\pi f^{-1}f_0)} \tag{6}$$

The derivative of the time-frequency spectrum yields $\hat{f}$

$$\hat{f} = (f,b) = f + [i2\pi\mathrm{ST}_x(f,b)]^{-1}\frac{\partial}{\partial b}\mathrm{ST}_X(f,b) \tag{7}$$

According to $x(t) = A\cos(2\pi f_0 t)$, from (7) we can calculate

$$\hat{f} = (f,b) = f + [i2\pi\mathrm{ST}_x(f,b)]^{-1}\frac{\partial}{\partial b}\mathrm{ST}_X(f,b) = f + \frac{i\pi A(f - f_0)e^{-i2\pi(f-f_0)^b}\overline{\hat{\varphi}(2\pi f^{-1}f_0)}}{i2\pi\frac{A}{2}e^{-i2\pi(f-f_0)^b}\hat{\varphi}*(2\pi f^{-1}f_0)} = f_0 \tag{8}$$

The spectrum in the interval $\left[\hat{f}_c - \frac{\triangle}{2}\hat{f}_c, \hat{f}_c + \frac{\triangle}{2}\hat{f}_c\right]$ near the center frequency $\hat{f}_c$ is superimposed to obtain the simultaneous compression transform $\mathrm{SSST}(\hat{f}_c, b)$, which improves the resolution of the spectrum, and the expression is

$$\mathrm{SSST}_x(\hat{f}_c, b) = (\triangle\hat{f})^{-1}\sum_{f_j:|\hat{f}_c(f_j,b)-\hat{f}_c|\leq\frac{\triangle f_c}{2}} |\mathrm{ST}_X(f,b)|f_j\triangle f_j \tag{9}$$

where f_j is the discrete frequency of ST, $\triangle f_j = f_j - f_{j-1}$ interval, $\hat{f}_c$ and $\triangle \hat{f}_c$ are the center frequency and spectral bandwidth of the compressed interval. $\triangle \hat{f}_c = \hat{f}_c - \hat{f}_{c-1}$.

SSST is a loss-free invertible transform, $\text{SSST}_x(\hat{f}_c, b)$ can be expressed as $x(b)$, and its inverse transformation equation is

$$x(b) = \text{Re}\left[(C_\varphi C_\phi)^{-1} \sum_C \text{SSST}_X(\hat{f}_c, b) \triangle \hat{f}_c\right]$$
$$C_\varphi = 0.5 \int_0^\infty \overline{-\varphi(\xi)\xi^{-1}d\xi}, C_\phi = e^{-i[2\pi fb + \phi(f,b)]} f^2 \tag{10}$$

The time and frequency distributions from SSST are linear, and the original signal $x(t)$ can be calculated from the results of the synchronous compression transform by inverting the above equation.

2.2. Vision Transformer Network Model

The ViT network model was presented at ICLR2021, and the model consists of three modules including Embedding layer, Transformer Encoder, and MLP Head (which is eventually used for classification.) The Transformer model is based entirely on a self-attentive mechanism without any convolutional or recurrent neural network layers and is not subject to local interaction limitations [25]. ViT was the first Transformer model used to replace CNNs and applied to image classification [26]. Although Transformer was originally applied to sequence-to-sequence learning on text data, it has now been extended to various modern deep learning in areas such as vision, target detection and image segmentation [27].

Embedding layer: For the standard Transformer module, the input is required to be a sequence of tokens (vectors), i.e., a two-dimensional matrix; for image data, the data format is a three-dimensional matrix $[H, W, C]$, not what the Transformer wants. So, we need to carry out a transformation of the data using an Embedding layer, firstly dividing an image into a bunch of patches of a given size, and secondly mapping each patch into a one-dimensional vector by a linear mapping. This is achieved directly through a convolutional layer that flattens out the two dimensions, turning it into a two-dimensional matrix, which is what the Transformer wants. Before being input to the transformer encoder, a class token and Position Embedding need to be added, and a class token is inserted into the resulting pile of tokens specifically for classification.

Transformer Encoder: The main part is to normalize each token with layer norm, so as to simulate the whole sample data distribution.

Multi-head self-attention mechanism: This allows the model to focus on the information at different positions and complete the interaction information between sequences.

The structure of MLP consists mainly of a fully connected layer and an activation function. The output weights of the multi-headed self-attentive mechanism are received and compared to identify the fault types. The structure of the ViT model is shown in Figure 1.

According to the ViT model structure diagram, a ViT block runs in the following steps:

Step 1: Take the image size 224×224 as an example: as the input, the image is divided into fixed patches with size 16×16, so the number of patches generated is $224 \times 224 / 16 \times 16 = 196$, and the sequence of length 196 is obtained, the dimensions of patches are $16 \times 16 \times 3 = 768$, the dimensions of linear projection layer are $768 \times N$ ($N = 768$), so the dimension of the input after passing through the linear projection layer is still 196×768, that is, there are 196 tokens in total, and the dimension of each token is 768. There is also a class for classification, so the final dimension is 197×768. The image is converted into a sequence by the patch embedding layer.

Step 2: Positional encoding: ViT also needs to add positional encoding, which can be understood as a table with N rows; the size of N is the same as the length of the input sequence, each row represents a vector, and the dimension of the vector is the same as the dimension of the input sequence embedding (768). After adding the location encoding information, the dimension is still 197×768.

Step 3: LN/multi-head attention: The LN output dimension is still 197×768. With multi-head self-attention, if there is only one head, the dimension is 197×768. If there are 12 heads, $768/12 = 64$, the dimensions are 197×64. There are 12 groups in total, and finally the output of the 12 groups is stitched together again, and the output dimensions are 197×768. Then in a layer of LN, the dimensions are still 197×768.

Step 4: MLP: The dimension is enlarged and then reduced back (197×768 enlarged to 197×3072, then reduced to 197×768). After this, the block dimensions are still the same as the input (197×768), so it is possible to stack multiple blocks. Special characters (cls) corresponding to the output as the encoder output represent the final image presentation, followed by an MLP for image classification. The formula is as follows.

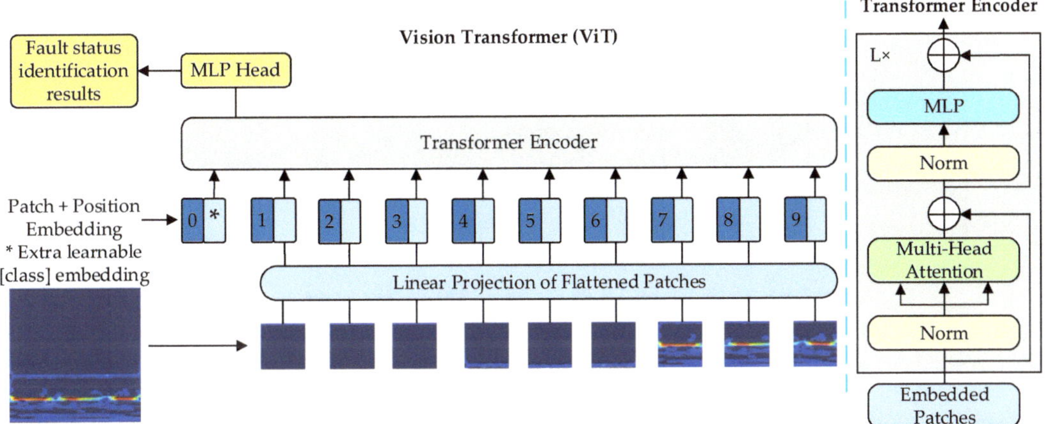

Figure 1. The architecture of the Vision Transformer.

Define the image x, $x \in R^{H \times W \times C}$, C is the number of channels, divided into N blocks of $P * P$ images, $N = \frac{H*W}{P*P}$.

$$Z_0 = \left[x_{\text{cls}}; x_p^1 E; x_p^2 E; \cdots ; x_p^N E \right] + E_{pos} E \in R^{p^2 \times c \times D}, E_{pos} \in R^{p(N+1) \times D} \tag{11}$$

At layer l ($l \in [1, N-1]$), the output is

$$Z_l' = MSA(LN(Z_{l-1})) + Z_{l-1} \tag{12}$$

$$Z_l = MLP(LN(Z'_l)) + Z_l' \tag{13}$$

$$y = LN(Z_L^0) \tag{14}$$

Among them, the multi-headed attention layer MSE completes the information interaction between image blocks. The classification of image y is completed by the MLP. After superimposing ViT Blocks several times, the result of fault status recognition is output.

2.3. Diesel Engine State Identification Based on SSST-ViT

The diesel engine status recognition method of SSST-ViT effectively integrates the advantages of SSST in characterizing time-varying nonlinear non-smooth signals with the excellent image recognition capability of ViT, which can achieve accurate and efficient state recognition. The model structure diagram is shown in Figure 2, and the specific steps are as follows:

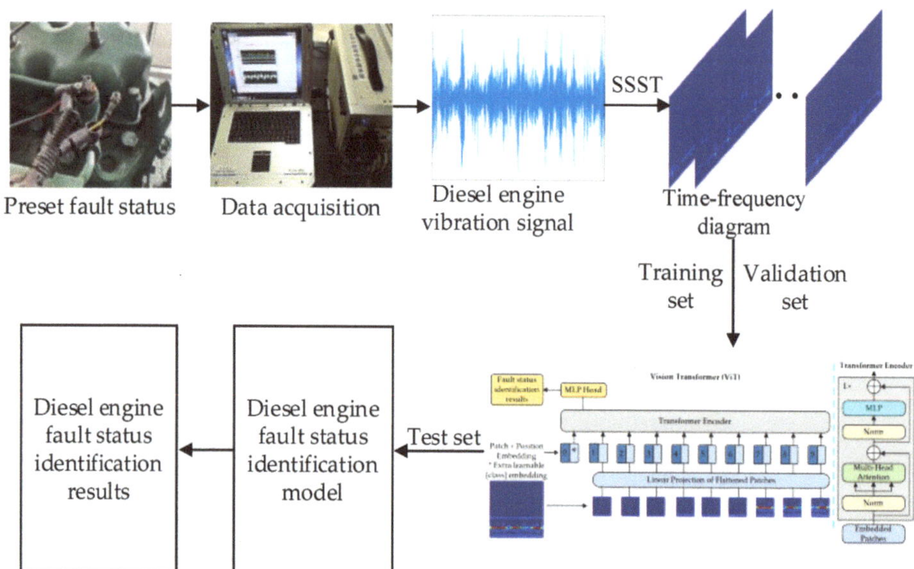

Figure 2. SSST-ViT fault status identification flow chart.

Step 1: The vibration signal of the diesel engine cylinder head is collected by the vibration sensor to obtain the raw dataset required for the experiment.

Step 2: The collected diesel engine vibration signals are transformed by SSST to obtain the time-frequency map. After pre-processing the time-frequency map, the required feature sample data are obtained. The sample data are divided into training set, validation set and test set according to a 7:2:1 ratio. The training set is used for model training, the validation set is used for initial evaluation of the accuracy of the model, and the test set is used for evaluating the performance of the model, which is not involved in the training of the model.

Step 3: The training set is used as the model input to the ViT network model and trained to obtain the desired diesel engine status recognition model.

Step 4: Use the test set as the trained model input to perform fault status recognition of the diesel engine.

3. Experimental Results and Comparative Analysis

The feasibility and effectiveness of the diesel engine condition identification method with SSST-ViT were validated using publicly available datasets and measured data from Case Western Reserve University (CWRU). The experiments were conducted using Windows 11 with a 12th Gen Intel(R) Core (TM) i7-12700H 2.30 GHz processor, a GeForce RTX 3060 Laptop GPU, 16G of RAM, Anaconda3, Python 3.9.13, and MATLAB2021b software environment. MATLAB2021b; deep learning framework is PyTorch1.11.0.

3.1. CWRU Dataset to Verify the Feasibility of SSST-ViT Method

Both rolling bearing and diesel engine vibration signals are characterized by time-varying, nonlinear non-smoothness [28]. Therefore, the feasibility of SSST with the ViT method was verified using the publicly available CWRU bearing vibration signal dataset. According to the literature, the publicly available dataset was obtained through a bearing failure simulation experiment bench.

The experiments were conducted using a deep groove ball bearing, model SKF6205, with a single point of failure of the bearing machined with electrical discharge machining (EDM), and the vibration acceleration signal of the bearing was collected using an accelerometer. The specific data used were the drive-end bearing data with a sampling frequency of 48 kHz, an approximate motor speed of 1797 r/min, and a load of 0 hp. The bearing statuses include normal, inner ring failure, outer ring failure, and rolling element failure, and each failure state can be classified into three types according to the depth of cut: 0.1778 mm, 0.3556 mm and 0.5334 mm. The status data of 10 bearings selected in this experiment are shown in Table 1.

Table 1. The 10 kinds of bearing data selected.

Serial Number	Fault Location	Fault Diameter (mm)	Load (hp)	Rotational Speed (r/min)
1	Normal	—	0	1797
2	Inner ring failure	0.1778	0	1797
3	Inner ring failure	0.3556	0	1797
4	Inner ring failure	0.5443	0	1797
5	Outer ring failure	0.1778	0	1797
6	Outer ring failure	0.3556	0	1797
7	Outer ring failure	0.5443	0	1797
8	Rolling body failure	0.1778	0	1797
9	Rolling body failure	0.3556	0	1797
10	Rolling body failure	0.5443	0	1797

The time-domain analysis of the vibration signals in each status of the bearing was performed. The data length of each fault status was intercepted to 5120 sampling points, and the time-domain waveforms of the bearing in 10 statuses were obtained, as shown in Figure 3.

As seen in the time-domain waveform diagram of the vibration signal, the time-domain waveforms of the status fluctuate widely, making it difficult to carry out effective fault status identification [29]. The signal waveforms of different status types are complex and do not differ greatly, and the individual status cannot be identified directly by hand. Therefore, it is difficult to carry out rolling bearing fault status identification from time-domain signal waveform analysis alone, and a more effective intelligent identification method is needed [30].

The SSST-ViT method proposed in this paper was applied to identify each status of the bearings. From each bearing status data point, 300 samples were randomly taken, and each sample length was 1024 sampling points, so a total of 3000 samples were obtained. By dividing the training set, validation set and test set according to the ratio of 7:2:1, 2100 training samples, 600 validation samples and 300 test samples were obtained for the feasibility verification experiments of SSST-ViT state recognition methods.

The SSST was performed on the original vibration signal to obtain the time-frequency diagram. To avoid the influence on the classification results, the coordinate system, legend and blank part were set not to be displayed, and the time-frequency diagram of the first sample in each state after processing is shown in Figure 4.

The warm and cold colors in Figure 4 represent energy values, the warmer the color the greater the energy, reflecting the energy magnitude of the signal at each frequency; the horizontal and vertical axes indicate time and frequency, respectively, showing the change of signal frequency components with time. The energy of the time frequency diagram of each state is more concentrated, with good time-frequency resolution, and the contained features are different, corresponding to different time-frequency diagrams, with the warm color part showing irregular block distribution. Although there are certain differences in expression, the similarity is high, and it is difficult to manually distinguish each fault status accurately. Therefore, each fault status was identified by the ViT network with a powerful image classification function.

The images were first set to not show the legend, coordinate system and blank parts. Then each time-frequency map was normalized to speed up the model convergence. Finally, without affecting the recognition rate, the grid was normalized and compressed to process the time-frequency map to improve the model training speed, and the image size was uniformly adjusted to $224 \times 224 \times 3$.

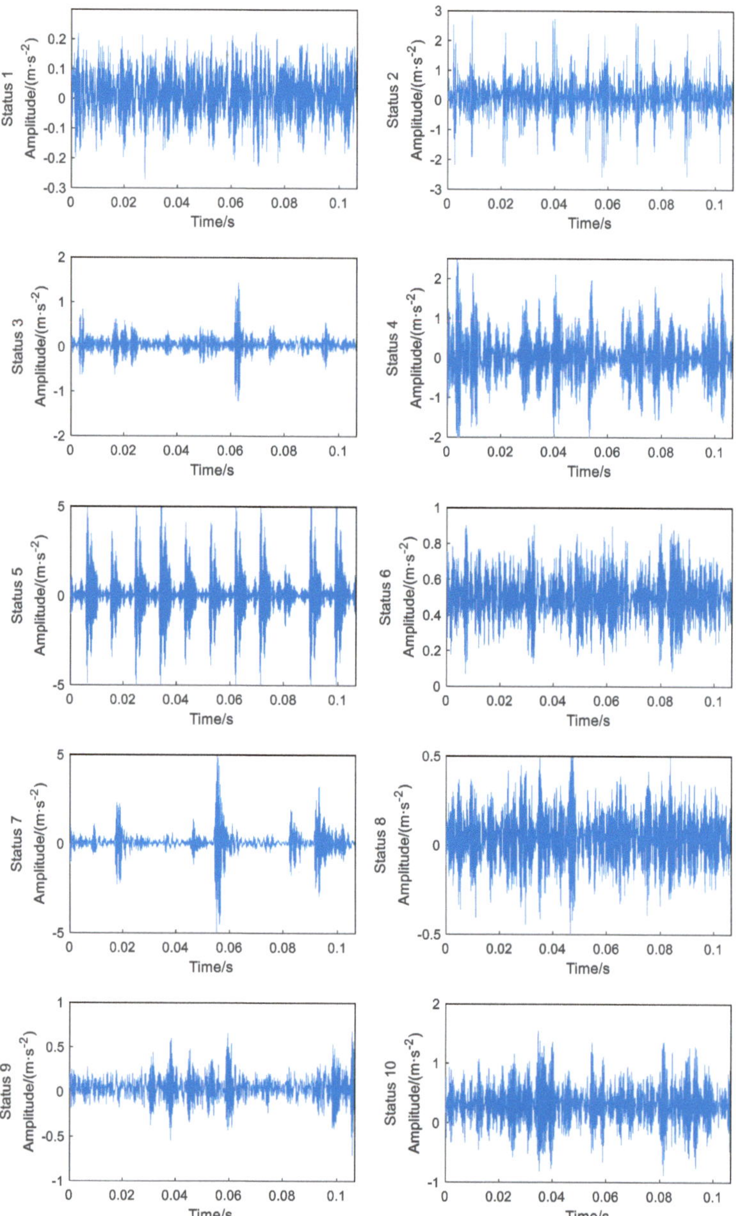

Figure 3. Time-domain waveforms of 10 failure statuses.

After considering the network structure, computer hardware level and sample characteristics and size, the parameters of the ViT network during training were configured as follows: batch processing size of 16; learning rate of 1×10^{-3}; weight decay of 1×10^{-5}; number of iterations—100; input image size of 224×224; number of classification categories—10;

optimizer—stochastic gradient descent; and loss function—cross-entropy loss function. The experimental results were extracted from the training log and plotted.

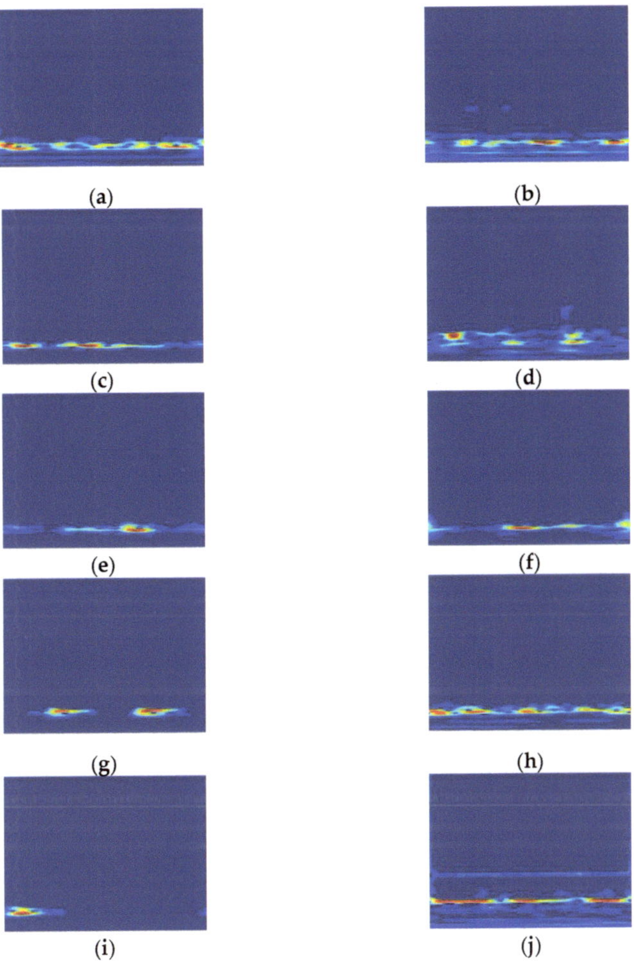

Figure 4. Time-frequency diagram of 10 fault statuses. (**a**) Status 1; (**b**) Status 2; (**c**) Status 3; (**d**) Status 4; (**e**) Status 5; (**f**) Status 6; (**g**) Status 7; (**h**) Status 8; (**i**) Status 9; and (**j**) Status 10.

ViT is the first transformer model used to replace the CNN and applied to image classification, which is able to achieve the desired results in the field of image classification. Therefore, in this paper, a ViT model is applied to diesel engine fault status identification. Since the network model is more suitable for extracting feature information from high-dimensional data, it is necessary to convert the one-dimensional vibration signals of diesel engines into two-dimensional images by some method. The S-transform combines the advantages of the continuous wavelet transform and the short-time Fourier transform, which has higher noise robustness and time-frequency analysis accuracy. However, the energy at a certain moment in the time-frequency map obtained by this method is distributed in a wider bandwidth near the instantaneous frequency, which causes instantaneous frequency energy leakage, leading to problems such as frequency band mixing and lower time-frequency resolution. Therefore, the synchro squeezing transform is combined with the S-transform to obtain the synchro squeezing S-transform method, which can effectively improve the time-frequency aggregation, time-frequency resolution and noise robustness compared with the traditional time-frequency analysis method. Therefore, the diesel engine fault status identification method proposed in this paper is obtained: SSST-ViT. The reason why ST-ViT is used as a comparison method in this paper is to verify that the SSST method can better characterize the feature information in the original signal and better retain the useful information. The reason why SSST-2DCNN is used as a comparison method in this paper is to verify that the ViT model has a more powerful image classification capability compared with the traditional CNN model, and is more suitable for identifying each fault status of the diesel engine. The reason why the FFT spectrum-1DCNN model is used as a comparison method in this paper is to verify that the 2D signal can better take into account the correlation of the signal in the time series compared to the 1D signal, i.e., to verify that the transformation of the original signal into a 2D image is more effective compared to the direct input of the 1D signal into the network model. In summary, it is completely reasonable to use the ST-ViT model, SSST-2DCNN model, and FFT spectrum-1DCNN model as comparisons in this paper.

In Reference [31], the authors proposed an intelligent fault diagnosis model for rolling bearings based on ViT, and achieved good results. Therefore, in this paper, the ViT model is applied to diesel engine fault status identification for the first time, and the feasibility and effectiveness of the proposed method are verified. Since ViT is the first transformer model used to replace the CNN and applied to image classification, and the network model is more suitable for extracting feature information from high-dimensional data, this paper proposes a diesel engine fault status recognition method based on SSST and ViT. Therefore, the ST-ViT model is improved on the basis of Reference [31], through which the model can be used to verify the superiority of the SSST time-frequency analysis method. In Reference [32], the authors proposed a 2DCNN-based fault diagnosis method for diesel engines by importing short-time Fourier transform (STFT) time-frequency maps into the 2DCNN model for training. However, the time-frequency map obtained using the STFT method suffers from low time-frequency resolution and weak energy aggregation. Therefore, the comparison method combines the SSST method with high time-frequency resolution and time-frequency aggregation with the 2DCNN, i.e., the SSST-2DCNN model is an improvement on Reference [32]. In Reference [33], the authors proposed a 1DCNN-based fault diagnosis method for diesel engines, where the features in the vibration signal are extracted and then input to the 1DCNN model for training. Since the method proposed in Reference [33] inputs multiple vibration signal features into the 1DCNN model for training, it tends to generate redundancy, which leads to a reduction in model efficiency. Therefore, the FFT spectrum is used as the fault feature in the comparison method of this paper. This is because different fault statuses of diesel engines generate different frequencies of fault features, and the FFT spectrum can well reflect the fault features of diesel engines at different fault states. Therefore, the source of the FFT spectrum-1DCNN model is Reference [33].

The training results of this model are compared with the training results of ST-ViT, SSST-2DCNN, and FFT spectrum-1DCNN models. The loss values and accuracy results of the training and validation sets of each model were obtained, as shown in Figure 5. The fault status identification results after 100 iterations are shown in Table 2 (Model 1: SSST-ViT; Model 2: ST-ViT; Model 3: SSST-2DCNN; Model 4: FFT spectrum-1DCNN).

Table 2. Accuracy and loss values for each model.

Models	Accuracy/ (%)		Loss Value	
	Training Set	**Validation Set**	**Training Set**	**Validation Set**
Model 1	100.00	97.33	2.08×10^{-4}	1.47×10^{-1}
Model 2	100.00	95.17	2.27×10^{-4}	1.74×10^{-1}
Model 3	90.09	92.50	2.89×10^{-1}	2.19×10^{-1}
Model 4	88.17	88.40	9.27×10^{-1}	9.62×10^{-1}

As seen in Figure 5 and Table 2, the different fault status identification models have converged after 100 iterations and all perform well on the CWRU public dataset. In terms of model accuracy and loss values, the SSST-ViT method proposed in this paper has the fastest convergence speed at iteration, the highest accuracy and the lowest loss values on both the training and validation sets, and the best performance compared to the other three methods. In terms of training stability, the method is optimal, and the accuracy and loss value curves are generally very stable, while the other three comparison methods all show different degrees of fluctuations. Therefore, compared with the comparison models, the SSST-ViT fault status identification method has better performance in terms of accuracy, loss value and stability, and the feasibility of the proposed fault status identification method has been verified.

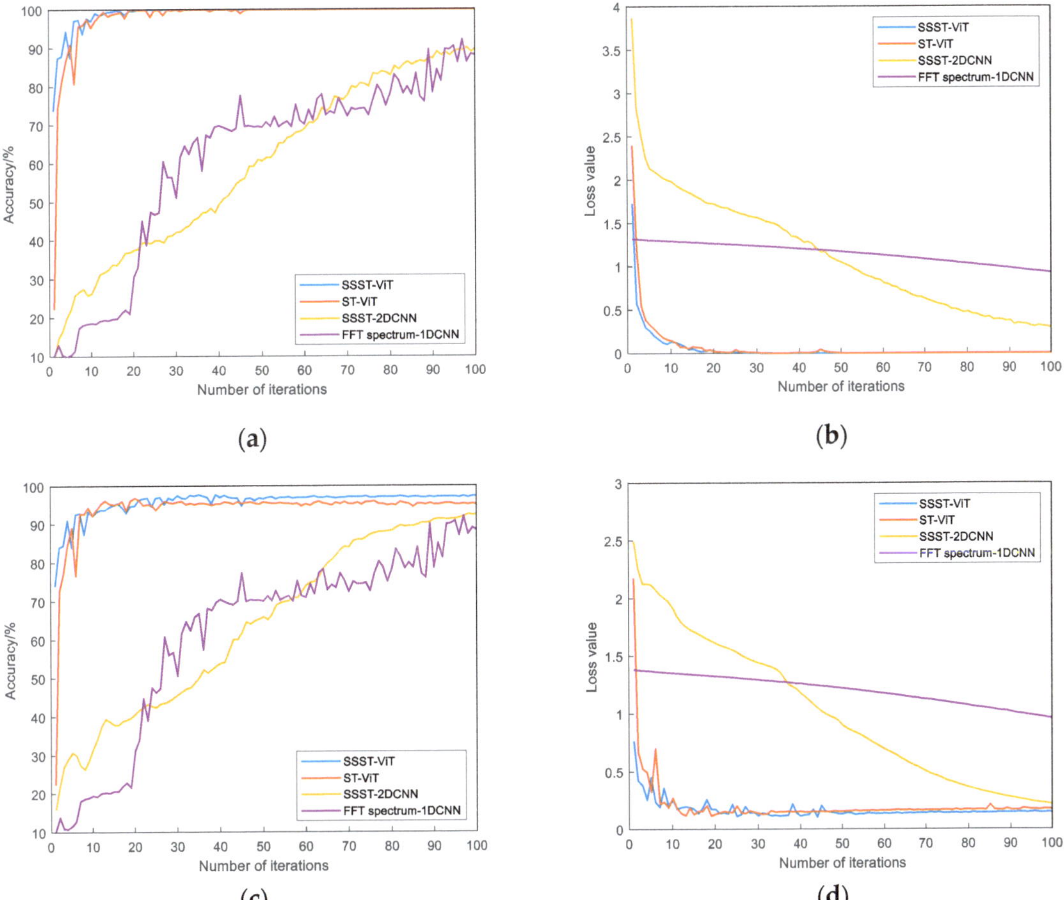

Figure 5. Comparison of training results of different models: (**a**) Accuracy of training set; (**b**) loss value of training set; (**c**) accuracy of validation set; and (**d**) loss value of validation set.

The accuracy and confusion matrix of different fault status identification models obtained under the test set are shown in Table 3 and Figure 6, respectively.

From Figure 6 and Table 3, it can be found that the proposed method has the optimal fault identification effect compared with other methods and can effectively distinguish the easily confused bearing fault types. To verify the feature extraction capability of SSST-ViT methods, the output of the classification layer network of the ViT model was extracted as discriminative features, and the identification results of bearing fault status were visualized in three dimensions by the t-SNE nonlinear dimensionality reduction technique, which is applicable to the visualization of high-dimensional data. The original data of the training set, the original data of the test set, the feature data of the training set and the feature data of the test set were obtained, as shown in Figure 7.

Table 3. Accuracy of each model under the test set.

Models	Accuracy
SSST-ViT	98.31%
ST-ViT	95.27%
SSST-2DCNN	92.33%
FFT spectrum-1DCNN	88.50%

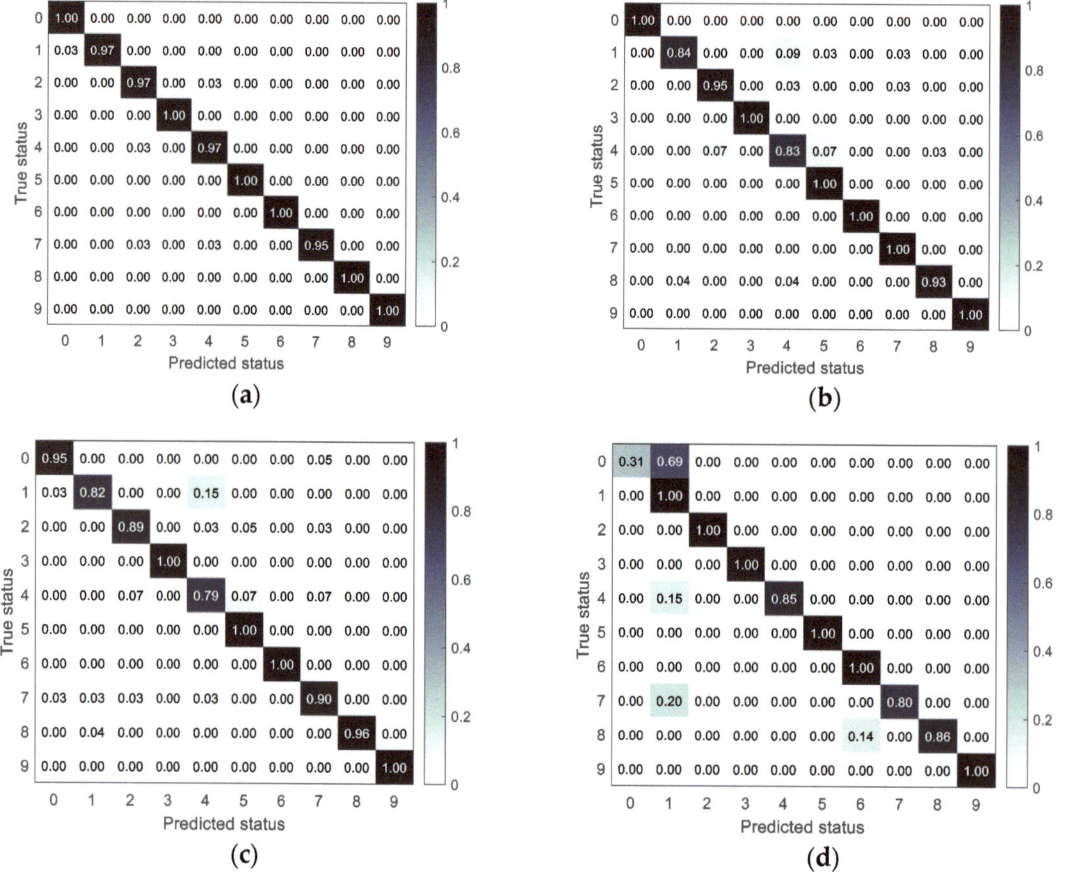

Figure 6. Confusion matrix for each model under the test set: (**a**) SSST-ViT; (**b**) ST-ViT; (**c**) SSST-2DCNN; and (**d**) FFT spectrum-1DCNN.

In Figure 7, using the test set feature data as an example, since none of the methods proposed in this paper achieved 100% accuracy under the test set, there must have been some points that did not fall within a cluster. In other words, it is because some features are identified as features of other fault statuses that some feature points are not in a cluster and therefore the accuracy is not 100%.

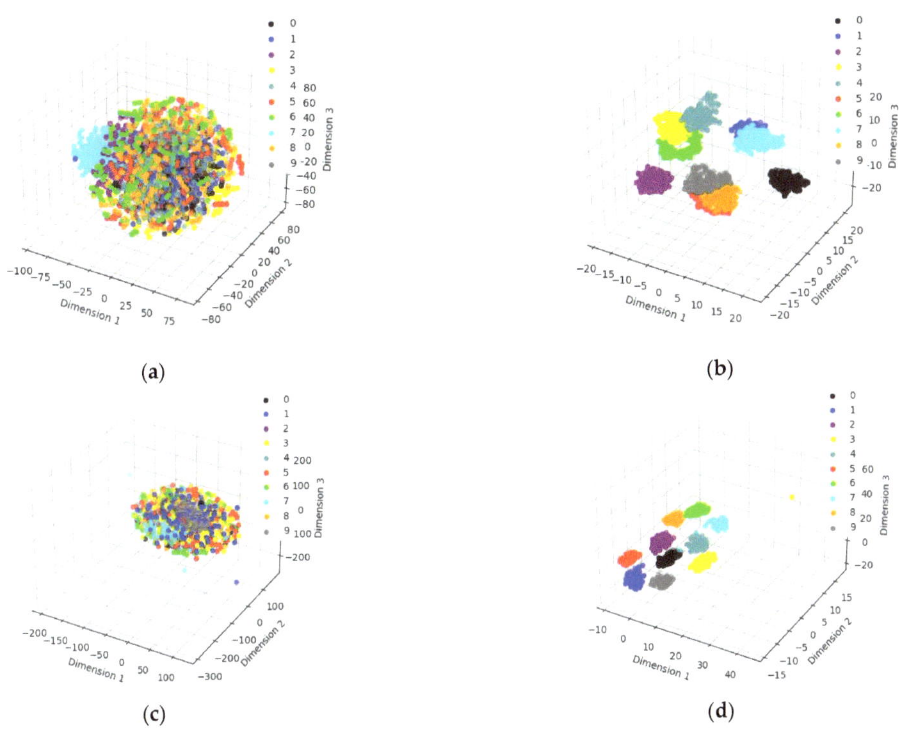

Figure 7. Three-dimensional visualization of status recognition results: (**a**) Training set raw data; (**b**) training set feature data; (**c**) test set raw data; and (**d**) test set feature data.

3.2. The Validity of the SSST-ViT Method Is Verified by the Measured Data

In order to verify the effectiveness of the SSST-ViT diesel engine fault status identification methods, this study relies on the high-pressure common rail diesel engine experimental bench in the laboratory, taking a CA6DF3-20E3 diesel engine as the research object to collect the state monitoring information during the operation of the diesel engine in different fault modes and provide data support for the research of diesel engine fault status identification methods. The experimental bench can be divided into two parts: the diesel engine system and the data acquisition system. The panoramic view of the experimental bench is shown in Figure 8, and the sensor installation is shown in Figure 9.

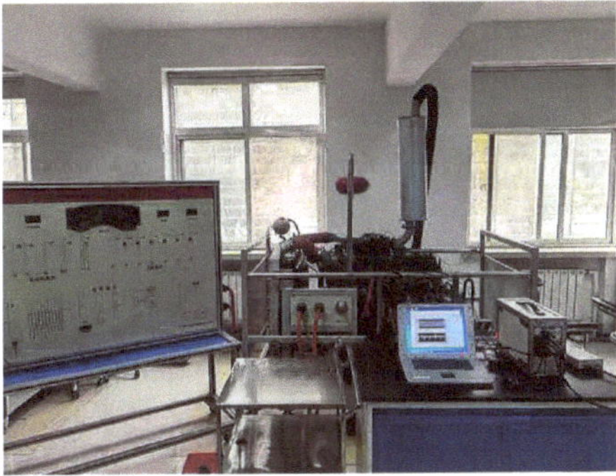

Figure 8. Diesel engine condition monitoring test bench.

Figure 9. Sensor mounting position.

By analyzing the composition structure and function of the diesel engine, and combining its typical failure modes in the process of use and maintenance, the pre-set failure experiment is carried out in the diesel engine condition-monitoring experimental bench (by artificially processing or replacing the faulty parts, the diesel engine components are pre-set to collect the data in the engine fault status and carry out research). Typical failure modes of diesel engines were set as shown in Table 4:

Table 4. Diesel engine preset fault mode.

Serial Number	Failure Mode
1	Normal
2	Fire in the first cylinder
3	Second cylinder fire
4	Clogged air filter
5	First cylinder and second cylinder misfire
6	Clogged air filter and first cylinder misfire
7	Clogged air filter and second cylinder misfire

In the actual equipment maintenance and repair process, due to the complex and harsh working environment, diesel engine failures may often be due to multiple, concurrent faults rather than a single failure mode. Therefore, when presetting the failure modes, a single failure mode is preset on the one hand and three mixed failure modes are preset on the other. As shown in Figure 10, the cylinder misfire fault is simulated by disconnecting the cylinder ignition power line, and the air intake outer cover is added to simulate the air filter blockage fault.

(a)

(b)

Figure 10. Pre-set diesel engine faults: (**a**) Disconnect the ignition power cord. (**b**) Install the air intake cover.

The diesel engine cylinder head vibration signal is acquired with a sampling frequency of 20 kHz, a single sampling time of 12 s, and a sample sampling interval of 30 s. After data acquisition experiments, there are 300 sets of data for each failure mode, six channels of data for each set, and a single sampling data volume of 240,000. In order to avoid the errors brought about by the process of the engine from start-up to steady status, the first 20 sets of data for each status are selected, and the last 20 sets of data from the 5th channel of each status are selected as the sample data.

Due to the simplicity, intuitiveness and clear physical meaning of the time-domain signal, the time-domain analysis of the vibration signal in each status of the diesel engine was performed. The length of the data intercepted for each failure mode sample is 5000 sampling points, and the time-domain waveforms of the diesel engine in each status are obtained as shown in Figure 11.

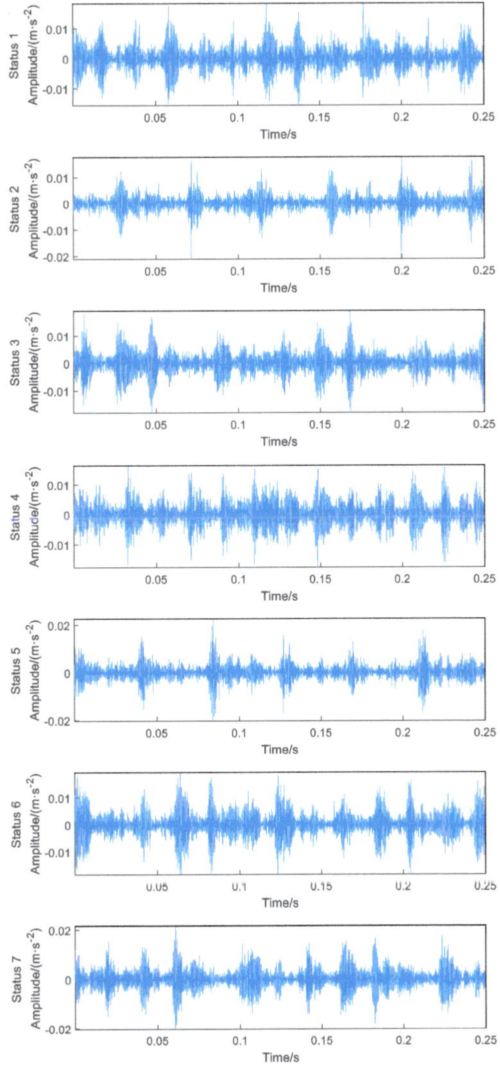

Figure 11. Time-domain wave forms of 7 failure statuses.

The engine head vibration signal presents a nonlinear and non-smooth status, and there are complex noise disturbances generated by the environment and the comprehensive action of each component during operation, so it is difficult to identify the fault status. From the time-domain waveform diagram in Figure 11, it can be seen that the vibration signal waveforms under different fault modes are complex and have basically the same amplitude change range, and there is no obvious difference from the time-domain waveform amplitude, so it is difficult to manually identify each status directly, so it is difficult to achieve effective identification of multiple engine faults from time-domain signal waveform analysis alone, and more effective fault information extraction and intelligent identification methods are needed.

The SSST-ViT method was applied to identify each status of the above diesel engine. A total of 2100 samples were obtained by taking 300 samples from each status of the diesel engine data, each with a sample length of 5000 sampling points. By dividing the training and validation sets according to the ratio of 7:2:1, 1470 training samples, 420 validation samples and 210 test samples were obtained, i.e., each status sample data included 210 training samples and 60 validation samples and 30 test samples. Processing of raw vibration signals by was carried out using the SSST and represented as a two-dimensional color time-frequency diagram, and the coordinate system, legend and blank part were set not to be displayed to avoid the influence on the classification results. The time-frequency diagram of the first sample in each status of the diesel engine after processing is shown in Figure 12.

Although each status in Figure 12 has some different expressions, the similarity is high, and it is difficult to distinguish each fault status only by hand. Therefore, each fault status is identified by the ViT network with a powerful image classification function.

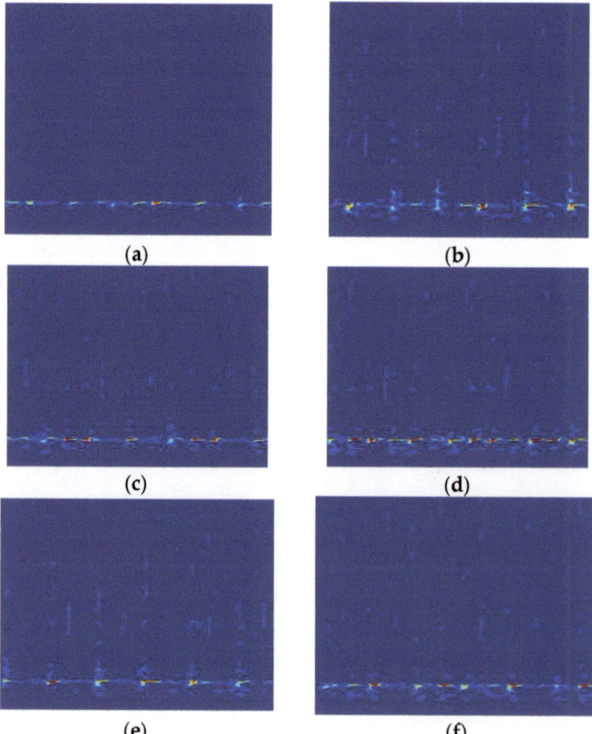

Figure 12. *Cont.*

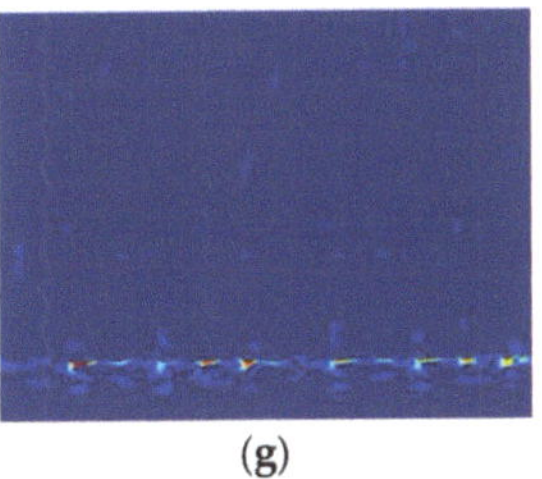
(**g**)

Figure 12. Time-frequency diagram of 7 fault status: (**a**) Status 1; (**b**) Status 2; (**c**) Status 3; (**d**) Status 4; (**e**) Status 5; (**f**) Status 6; and (**g**) Status 7.

The images were first set to not show the legend, coordinate system and blank parts. Then each time-frequency map was normalized to speed up the model convergence. Finally, the grid normalization compressed the time-frequency maps without affecting the recognition rate, and the image size was uniformly adjusted to $224 \times 224 \times 3$.

After considering the network structure, computer hardware level and sample characteristics and size, the parameters of the ViT network during training were configured as follows: batch processing size of 16; learning rate of 1×10^{-3}; weight decay of 1×10^{-5}; discard rate of 0.1; number of iterations—100; input image size of 224×224; number of classification categories—7; optimizer—stochastic gradient descent; loss function—cross entropy loss function. The experimental results are extracted from the training log and plotted.

The training results of this model are compared with those of the ST-Vision Transformer, SSST-2DCNN, and FFT spectrum-1DCNN models. The loss values and accuracy results of the training and validation sets of each model were obtained as shown in Figure 13. The fault status identification results after 100 iterations are shown in Table 5 (Model 1: SSST-ViT; Model 2: ST-ViT; Model 3: SSST-2DCNN; Model 4: FFT spectrum-1DCNN).

Table 5. Accuracy and loss values for each model.

Models	Accuracy/ (%)		Loss Value	
	Training Set	Validation Set	Training Set	Validation Set
Model 1	99.86	95.43	6.46×10^{-2}	1.69×10^{-1}
Model 2	96.70	91.47	4.10×10^{-2}	2.53×10^{-1}
Model 3	92.00	93.33	2.48×10^{-1}	2.25×10^{-1}
Model 4	90.68	88.33	9.85×10^{-1}	1.29

From Figure 13 and Table 5, it can be seen that in terms of model accuracy and loss values, the proposed SSST-ViT methods has the highest accuracy and lowest loss values in both training and validation sets with the best performance in terms of fast convergence during iterations compared with the other three compared methods. In terms of training stability, the accuracy and loss value curves of SSST-ViT methods is generally more stable. Therefore, compared with the comparison methods, SSST-ViT has better performance in terms of fault identification accuracy, loss value and stability.

The performance of the models was evaluated under the test set, and the accuracy and confusion matrix of different fault status identification models were obtained as shown in Table 6 and Figure 14, respectively.

Table 6. Accuracy of each model under the test set.

Models	Accuracy
SSST-ViT	95.67%
ST-ViT	94.23%
SSST-2DCNN	91.90%
FFT spectrum-1DCNN	87.62%

It can be found in Table 6 and Figure 14 that the proposed method in this paper has the optimal diesel engine fault identification effect compared with other methods, and can effectively distinguish the confusing fault types.

In order to test the feature extraction ability of the SSST-ViT method, the output of the classification layer network of the ViT model was extracted as the discriminative features, and the results of fault status recognition were visualized in three dimensions by the t-SNE nonlinear dimensionality reduction technique, which is suitable for visualizing high-dimensional data. The original data of the training set, the original data of the test set, the feature data of the training set and the feature data of the test set were obtained, as shown in Figure 15.

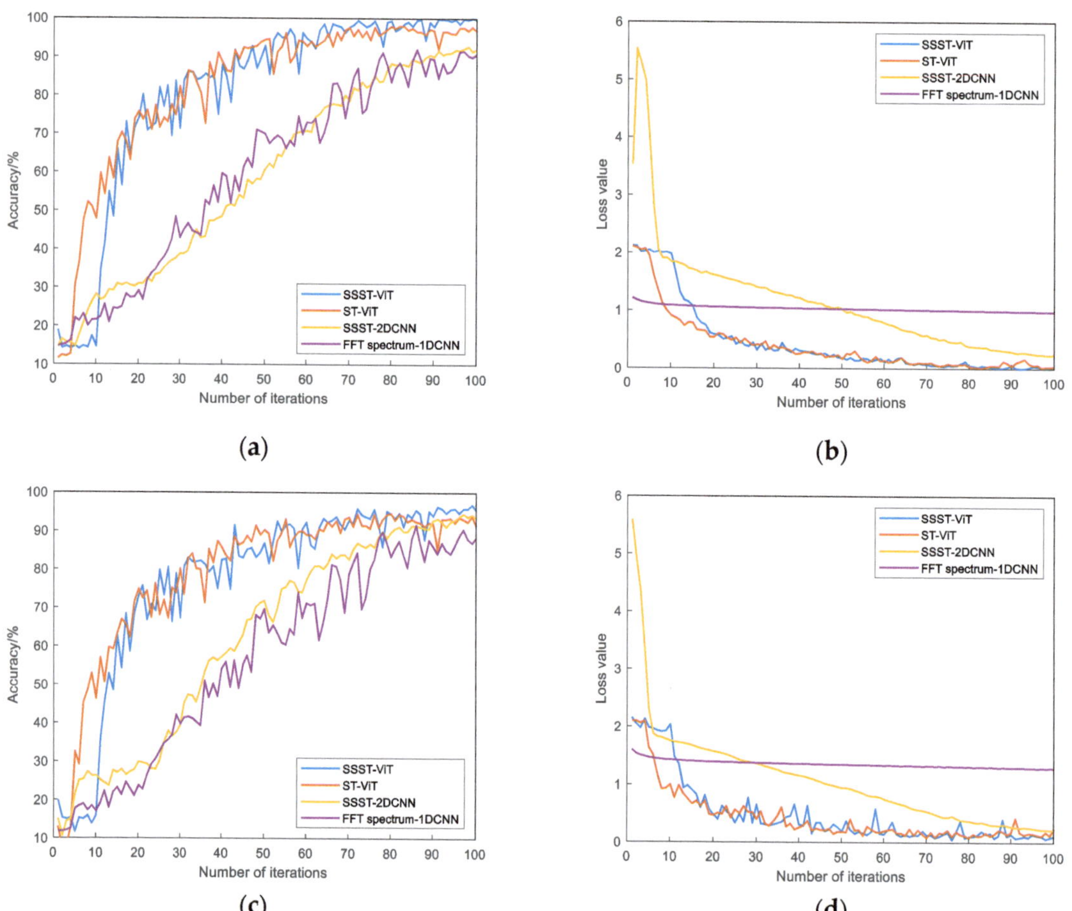

Figure 13. Comparison of the training results of each model: (**a**) Accuracy of training set; (**b**) loss value of training set; (**c**) accuracy of validation set; (**d**) loss value of validation set.

In Figure 15, using the test set feature data as an example, since none of the methods proposed in this paper achieved 100% accuracy under the test set, there must have been some points that did not fall within a cluster. In other words, it is because some features are identified as features of other fault statuses that some feature points are not in a cluster and therefore the accuracy is not 100%. As can be seen in Figure 15, the SSST-ViT method has excellent feature extraction performance, and the features of each fault status in the space have obvious differentiability. Different fault status types are distributed in different locations in the space and exhibit dense clustering.

In summary, the effectiveness and superiority of the proposed diesel engine fault status recognition method are verified. The SSST-ViT method can effectively extract fault features and has high recognition accuracy compared with other methods.

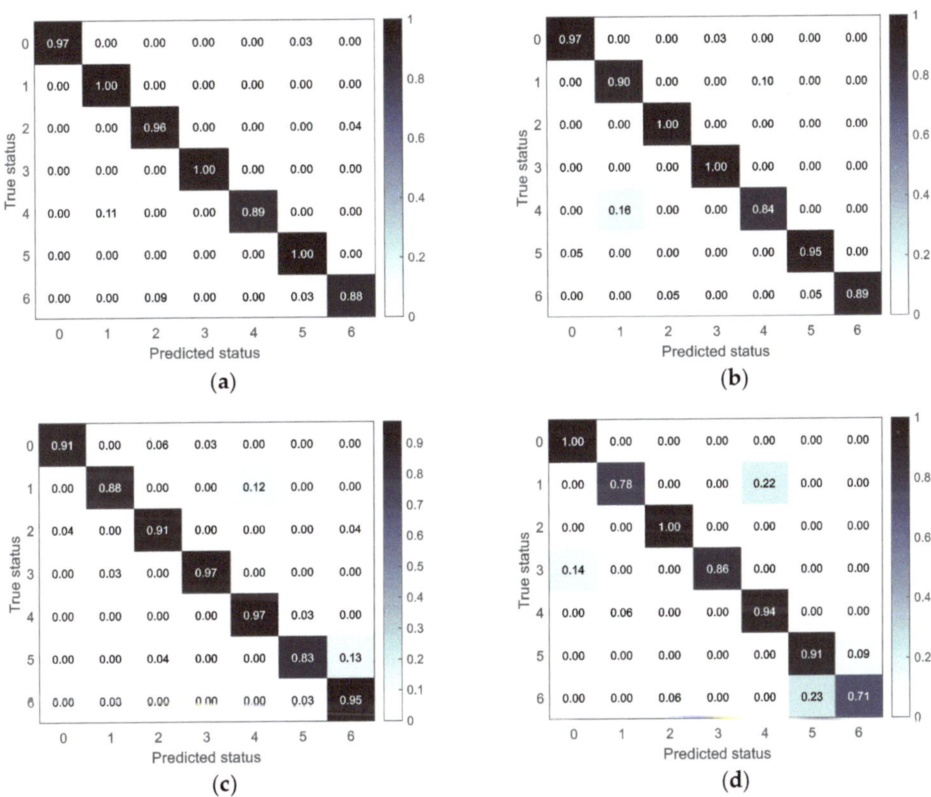

Figure 14. Confusion matrix for different fault status identification models: (**a**) SSST-ViT; (**b**) ST-ViT; (**c**) SSST-2DCNN; and (**d**) FFT spectrum-1DCNN.

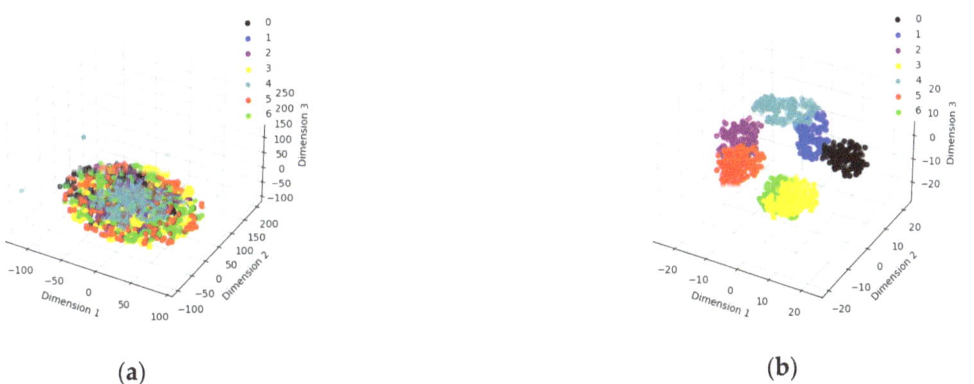

Figure 15. *Cont.*

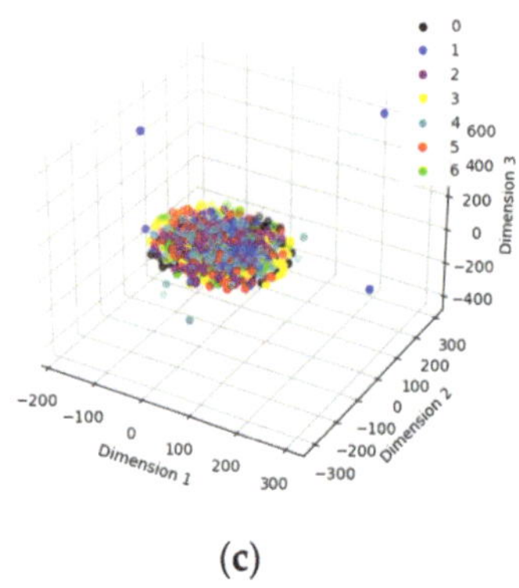

(**c**)

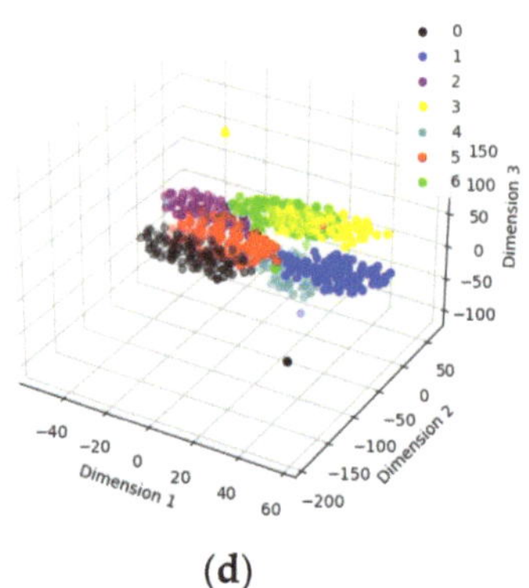

(**d**)

Figure 15. Three-dimensional visualization of status recognition results: (**a**) Training set raw data. (**b**) Training set feature data. (**c**) Test set raw data. (**d**) Test set feature data.

4. Conclusions

In this paper, a diesel engine fault status identification method based on SSST and ViT is proposed with a diesel engine as the engineering application background. Compared with the traditional method, the following conclusions can be drawn:

(1) SSST combines the high time-frequency aggregation of SST and the adaptive nature of ST, with better time-frequency aggregation and resolution.

(2) The method is the first to apply SSST, which can effectively characterize the original signal features, and the ViT model, which has excellent image classification capability, to the field of diesel engine fault status identification, and can effectively extract time-frequency image features.

(3) The method can provide theoretical and technical support for the research of diesel engine fault status recognition, which is of great military significance and realistic demand for improving the reliability and maintenance support capability of diesel engines.

Limited by the current experimental conditions, there are limitations in the development of diesel engine pre-set fault experiments. The focus of the next step is to carry out experimental research on diesel engine fault status identification in combination with diesel engine multi-part synthesis.

Author Contributions: Conceptualization, S.L. and Z.L.; methodology, Z.L.; software, Z.L. and S.L.; validation, Z.L., S.L. and R.W.; formal analysis, E.D. and Y.Y.; investigation, S.L. and Z.C.; resources, Z.C. and S.L.; data curation, Z.L.; writing—original draft preparation, S.L.; writing—review and editing, S.L.; visualization, Z.L. and S.L.; supervision, Z.C.; project administration, Z.C.; funding acquisition, Z.C and Z.L. All authors have read and agreed to the published version of the manuscript.

Funding: This research was funded by National Natural Science Foundation of China (Grant No. 71871219), National Defense Research Fund Project (Grant No. LJ20212C031173, LJ20222C020043).

Institutional Review Board Statement: Not applicable.

Informed Consent Statement: Not applicable.

Data Availability Statement: The authors confirm that the data supporting the findings of this study are available within the article.

Conflicts of Interest: The authors declare no conflict of interest.

References

1. Ke, Y.; Song, E.Z.; Yao, C.; Dong, Q. A review: Ship diesel engine prognostics and health management technology. *J. Harbin Eng. Univ.* **2020**, *41*, 125–131. [CrossRef]
2. Li, Y.F.; Han, T. Deep learning based industrial equipment prognostics and health management: A review. *J. Vib. Meas. Diagn.* **2022**, *42*, 835–847. [CrossRef]
3. Jiang, Z.N.; Wang, Z.J.; Zhang, J.J.; Huang, Y.F.; Mao, Z.W. Fault diagnosis method of EDG in nuclear power plants based on energy operator gradient neighborhood feature extraction. *Chin. Mech. Eng.* **2021**, *32*, 617–623. [CrossRef]
4. Liu, Z.; Zhang, C.; Dong, E.; Wang, R.; Li, S.; Han, Y. Research Progress and Development Trend of Prognostics and Health Management Key Technologies for Equipment Diesel Engine. *Processes* **2023**, *11*, 1972. [CrossRef]
5. Li, X.D. Research on Fault Diagnosis of Rotating Components Based on Deep Learning. Ph.D. Thesis, University of Chinese Academy of Sciences, Beijing, China, 2021. [CrossRef]
6. Liu, Z.; Bai, Y.S.; Li, S.Y.; Zhang, K.; Liu, M.; Jia, X.S. Assessment of diesel engine valve performance degradation status based on synchro extracting enhanced generalized S-transform. *Acta Armamentarii* **2023**, *in press*. Available online: http://kns.cnki.net/kcms/detail/11.2176.tj.20230629.1006.008.html (accessed on 12 July 2023).
7. He, Z.Y.; Shao, H.D.; Zhong, X.; Yang, Y.; Cheng, J.S. An intelligent fault diagnosis method for rotor-bearing system using small labeled infrared thermal images and enhanced CNN transferred from CAE. *Adv. Eng. Inform.* **2020**, *46*, 101150. [CrossRef]
8. Cai, Y.P.; Xu, G.H.; Zhang, H.; Fan, Y.; Li, A.H. Visual identification and diagnosis based on IC engine vibration signals. *J. Vib. Shock* **2019**, *38*, 150–157. [CrossRef]
9. Wei, K.Z. Research on Assessment Method of Diesel Engine Health Condition Based on Multi-Parameter Fusion. Master's Thesis, Beijing University of Chemical Technology, Beijing, China, 2019. [CrossRef]
10. Ke, Y.; Hu, Y.H.; Song, E.Z.; Yao, C.; Dong, Q. A method for degradation features extraction of diesel engine valve clearance based on modified complete ensemble empirical mode decomposition with adaptive noise and discriminant correlation analysis feature fusion. *J. Vib. Control* **2022**, *28*, 2570–2584. [CrossRef]
11. Liu, Y.S.; Kang, J.S.; Guo, C.M.; Bai, Y.J. Diesel engine small-sample transfer learning fault diagnosis algorithm based on STFT time–frequency image and hyperparameter autonomous optimization deep convolutional network improved by PSO–GWO–BPNN surrogate model. *Open Phys.* **2022**, *20*, 993–1018. [CrossRef]
12. Xi, W.K.; Li, Z.X.; Tian, Z.; Duan, Z.H. A feature extraction and visualization method for fault detection of marine diesel engines. *Measurement* **2018**, *116*, 429–437. [CrossRef]
13. Shen, H.; Zeng, R.L.; Yang, W.C.; Zhou, B.; Ma, W.P.; Zhang, L.L. Diesel engine fault diagnosis based on polar coordinate enhancement of time-frequency diagram. *J. Vib. Meas. Diagn.* **2018**, *38*, 27–33. [CrossRef]
14. Mu, W.J.; Shi, L.S.; Cai, Y.P.; Zheng, Y.; Liu, H. Diesel engine fault diagnosis based on the global and local features fusion of time-frequency image. *J. Vib. Shock* **2018**, *37*, 14–19. [CrossRef]
15. Zhao, Z.H.; Li, L.H.; Yang, S.P.; Zhao, J.J. Denoising method of stacked denoising auto-encoder for vibration signal. *J. Vib. Meas. Diagn.* **2022**, *42*, 315–321. [CrossRef]
16. He, X.Z.; Zhou, X.Q.; Yu, W.N.; Hou, Y.X.; Mechefske, C.K. Adaptive variational mode decomposition and its application to multi-fault detection using mechanical vibration signals. *ISA Trans.* **2021**, *111*, 360–375. [CrossRef]
17. Chen, Y.C. Health Assessment for Marine Medium-Speed Diesel Engine Based on Instantaneous Speed. Master's Thesis, Wuhan University of Technology, Wuhan, China, 2020. [CrossRef]
18. Bai, Y.J.; Jia, X.S.; Liang, Q.H.; Bai, Y.S. Evaluation of diesel engine valve health status based on deep learning. *Sci. Technol. Eng.* **2022**, *22*, 3941–3950.
19. Jiang, Z.; Lai, Y.; Zhang, J.; Zhao, H.; Mao, Z. Multi-Factor Operating Condition Recognition Using 1D Convolutional Long Short-Term Network. *Sensors* **2019**, *19*, 5488. [CrossRef]
20. Zhan, X.; Bai, H.; Yan, H.; Wang, R.; Guo, C.; Jia, X. Diesel Engine Fault Diagnosis Method Based on Optimized VMD and Improved CNN. *Processes* **2022**, *10*, 2162. [CrossRef]
21. Chen, S.; Lu, S.; Wang, S.; Ni, Y.; Zhang, Y. Shifted Window Vision Transformer for Blood Cell Classification. *Electronics* **2023**, *12*, 2442. [CrossRef]
22. Zhu, X.; Jia, Y.; Jian, S.; Gu, L.; Pu, Z. ViTT: Vision Transformer Tracker. *Sensors* **2021**, *21*, 5608. [CrossRef]
23. Xu, Y.G.; Deng, Y.J.; Zhao, J.Y.; Tian, W.K.; Ma, C.Y. A novel rolling bearing fault diagnosis method based on empirical wavelet transform and spectral trend. *IEEE Trans. Instrum. Meas.* **2020**, *69*, 2891–2904. [CrossRef]
24. Sandoval, S.; De Leon, P.L. Recasting the (Synchrosqueezed) Short-Time Fourier Transform as an Instantaneous Spectrum. *Entropy* **2022**, *24*, 518. [CrossRef] [PubMed]
25. Zhou, J. Research on the Intelligent Diagnosis Method of Rolling Bearing Faults Based on CYCBD and HRE. Master's Thesis, North University of China, Taiyuan, China, 2020. [CrossRef]
26. Lai, D.K.-H.; Yu, Z.-H.; Leung, T.Y.-N.; Lim, H.-J.; Tam, A.Y.-C.; So, B.P.-H.; Mao, Y.-J.; Cheung, D.S.K.; Wong, D.W.-C.; Cheung, J.C.-W. Vision Transformers (ViT) for Blanket-Penetrating Sleep Posture Recognition Using a Triple Ultra-Wideband (UWB) Radar System. *Sensors* **2023**, *23*, 2475. [CrossRef] [PubMed]
27. Sonain, J.; Arunabha, M.R. An efficient and robust Phonocardiography (PCG)-based Valvular Heart Diseases (VHD) detection framework using Vision Transformer (ViT). *IEEE Comput. Biol. Med.* **2023**, *158*, 106734. [CrossRef]

28. Cheng, Z.Y.; Wang, R.J. Enhanced symplectic characteristics mode decomposition method and its application in fault diagnosis of rolling bearing. *Measurement* **2020**, *166*, 108108. [CrossRef]
29. Ma, R.Z. Research on Wear Diagnosis of Piston-Cylinder Based on Block Surface Vibration Signal. Master's Thesis, Shandong University, Jinan, China, 2022. [CrossRef]
30. Bi, X.Y. Research on Intelligent Diagnosis Methods for Typical Diesel Engine Faults Based on Analytic Single Channel Vibration Signals. Ph.D. Thesis, Tianjin University, Tianjin, China, 2019. [CrossRef]
31. Du, K.N.; Ning, S.H.; Deng, G.Y. Intelligent fault diagnosis of rolling bearings based on Vision Transformer. *Modul. Mach. Tool Autom. Manuf. Tech.* **2023**, *4*, 96–99.
32. Liao, Y.C.; Zhao, J.H.; An, S.J.; Zhou, L. Fault diagnosis of abnormal diesel engine fuel supply advance angle based on 2DCNN. *J. Ordnance Equip. Eng.* **2021**, *42*, 250–255.
33. Bai, Y.J.; Jia, X.S.; Liang, Q.H.; Ma, Y.F.; Bai, H.J. Diesel engine fault diagnosis based on one-dimensional convolutional neural network and dual-channel information fusion. *Veh. Engine* **2021**, *6*, 76–81.

Article

Bayesian-Optimized Hybrid Kernel SVM for Rolling Bearing Fault Diagnosis

Xinmin Song [1], Weihua Wei [2], Junbo Zhou [1], Guojun Ji [3], Ghulam Hussain [4], Maohua Xiao [1,*] and Guosheng Geng [1,*]

1   College of Engineering, Nanjing Agricultural University, Nanjing 210031, China; 9203012015@stu.njau.edu.cn (X.S.); 2020112009@stu.njau.edu.cn (J.Z.)
2   College of Mechanical and Electronic Engineering, Nanjing Forestry University, Nanjing 210037, China; whwei@njfu.edu.cn
3   Essen Agricultural Machinery Changzhou Co., Ltd., Changzhou 213000, China; jx9816@163.com
4   Faculty of Mechanical Engineering, Ghulam Ishaq Khan Institute of Engineering Sciences & Technology, Topi 23460, Pakistan; gh_ghumman@hotmail.com
*   Correspondence: xiaomaohua@njau.edu.cn (M.X.); gsgeng@njau.edu.cn (G.G.)

Abstract: We propose a new fault diagnosis model for rolling bearings based on a hybrid kernel support vector machine (SVM) and Bayesian optimization (BO). The model uses discrete Fourier transform (DFT) to extract fifteen features from vibration signals in the time and frequency domains of four bearing failure forms, which addresses the issue of ambiguous fault identification caused by their nonlinearity and nonstationarity. The extracted feature vectors are then divided into training and test sets as SVM inputs for fault diagnosis. To optimize the SVM, we construct a hybrid kernel SVM using a polynomial kernel function and radial basis kernel function. BO is used to optimize the extreme values of the objective function and determine their weight coefficients. We create an objective function for the Gaussian regression process of BO using training and test data as inputs, respectively. The optimized parameters are used to rebuild the SVM, which is then trained for network classification prediction. We tested the proposed diagnostic model using the bearing dataset of the Case Western Reserve University. The verification results show that the fault diagnosis accuracy is improved from 85% to 100% compared with the direct input of vibration signal into the SVM, and the effect is significant. Compared with other diagnostic models, our Bayesian-optimized hybrid kernel SVM model has the highest accuracy. In laboratory verification, we took sixty sets of sample values for each of the four failure forms measured in the experiment, and the verification process was repeated. The experimental results showed that the accuracy of the Bayesian-optimized hybrid kernel SVM reached 100%, and the accuracy of five replicates reached 96.7%. These results demonstrate the feasibility and superiority of our proposed method for fault diagnosis in rolling bearings.

Keywords: Bayesian optimization; rolling bearing; fault diagnosis; hybrid kernel SVM

Citation: Song, X.; Wei, W.; Zhou, J.; Ji, G.; Hussain, G.; Xiao, M.; Geng, G. Bayesian-Optimized Hybrid Kernel SVM for Rolling Bearing Fault Diagnosis. *Sensors* **2023**, *23*, 5137. https://doi.org/10.3390/s23115137

Academic Editors: Dong Wang, Shilong Sun, Changqing Shen and Ka-Veng Yuen

Received: 31 March 2023
Revised: 11 May 2023
Accepted: 22 May 2023
Published: 28 May 2023

1. Introduction

With the development of the manufacturing industry, machine fault detection has become a very important field. Bearing, as a commonly used supporting part in machinery and equipment, has a great influence on the normal operation of the machine, while it also has a high incidence of failure because it often works under the condition of high speed and heavy load [1–4]. Statistics show that 30% of the failures of rotary machinery are related to bearings [5]. Therefore, fault diagnosis using vibration signals generated during its working process can reduce the probability of mechanical accidents and provide reliable decision support for later maintenance plans [6,7].

The most commonly used features for fault detection in rotating machines from vibration signals can be classified into three categories: time-domain, frequency-domain, and time-frequency domain features. Time-domain features include statistical features, such as mean, standard deviation, skewness, and kurtosis. Frequency-domain features include spectral features, such as power spectral density, frequency band energy, and frequency ratio. Time-frequency domain features include wavelet-based features, such as wavelet energy, wavelet entropy, and wavelet variance.

While these conventional features have been employed successfully in fault detection, there has been an increasing interest in the development of new methods for solving complex classification problems. One of these approaches is the Non-parallel Bounded Support Matrix Machine (NBSMM), which is a novel extension of SVMs that can effectively deal with non-linearly separable data by utilizing the concept of bounded support matrices. Another extension of SVMs is the multi-class fuzzy support matrix machine (MF-SMM), which is a robust and efficient method for multi-class classification problems. The Convolutional-Vector Fusion Network (CVFN) is a recent development in the field of deep learning that combines the strengths of convolutional neural networks (CNNs) and vector fusion networks. CVFN is particularly effective in handling complex and heterogeneous data by fusing information from multiple modalities [8,9].

In addition to these new methods, kurtosis and Kullback–Liebler divergence have also been employed successfully in fault detection. Kurtosis is sensitive to the presence of impulsive signals, which are often associated with faults in rotating machines. High kurtosis values indicate the presence of impulsive signals, which can be used to detect faults, such as bearing faults and gear faults. Kullback–Liebler divergence has been used for fault detection in rotating machines by comparing the probability distribution functions of healthy and faulty signals. The Kullback–Liebler divergence between the two distributions can be used as a feature to detect faults [10,11].

On the other hand, with the continuous development of the Bayesian optimization (BO) algorithm, more and more researchers have begun to apply it to fault detection [12,13]. BO is a method used to optimize "black box" function which is defined as a function whose analytic expression is unknown. Therefore, we do not have access to their gradients. Hence, their evaluation, in terms of computing time and other resources, is costly. In addition, the evaluation of these functions may be subjected to noise pollution, which means that two evaluations at the same input location may yield varying results [14]. On the one hand, SVM is a kind of machine learning algorithm for classification and regression. BO can achieve the efficient optimization of "black box" functions by constructing a Gaussian process model to predict the value of unknown functions and selecting the next point for evaluation according to Bayes' theorem. Additionally, SVM divides the data into two or more categories by finding the optimal decision hyperplane.

In relevant literature, we can see that many researchers have discussed the application of BO and SVM in the field of fault diagnosis. For example, Orhan et al. [15] employed BO and SVM algorithms for the diagnosis of motor faults. They used the BO algorithm to select the optimal SVM parameters and subsequently performed feature selection using the SVM algorithm. The SVM algorithm was then applied for motor fault diagnosis. Their results indicated that this approach exhibited high accuracy and robustness in diagnosing motor faults. Similarly, Li et al. [16] utilized BO and SVM algorithms for detecting rolling bearing faults. They applied the BO algorithm to optimize the SVM parameters and utilized statistics-based methods to extract vibration signal features from rolling bearings. Experimental results demonstrated that their approach effectively detected rolling bearing faults. Additionally, Xiong et al. [17] utilized the BO algorithm to improve the accuracy of bearing fault diagnosis by selecting the optimal SVM parameters. Their method automatically searched for the SVM parameters with the highest classification results. The approach yielded favorable results in bearing fault diagnosis. Furthermore, James Bergstra and Yoshua Bengio [18] proposed a technique based on random search and BO to optimize SVM hyperparameters and achieved excellent outcomes. Some researchers have also explored the combination of SVM with other machine learning techniques for fault diagnosis. For instance, Tian Han [19] proposed a method that combined improved SVM and convolutional neural networks for diagnosing rolling bearing faults and achieved favorable outcomes.

Overall, relevant studies have demonstrated promising results when combining SVM with other machine learning techniques for fault diagnosis.

Although BO has advantages in fault diagnosis, it has some limitations. Firstly, the optimization of the Gaussian process in each iteration requires significant computing resources and time, particularly for large datasets. Secondly, BO is a global optimization algorithm based on probability and may fall into local optima, particularly for complex non-convex function problems. Thirdly, BO's performance is highly reliant on parameter settings, such as the kernel function and hyperparameters of the Gaussian process. These settings can significantly impact the algorithm's performance. Finally, BO requires prior knowledge to guide the search process, which can lead to decreased performance if the prior knowledge is insufficient or inaccurate. When using BO for fault diagnosis, it is essential to be aware of these limitations to better use the algorithm's advantages and address the existing issues [20,21].

Our proposed theory focuses on the development of a Bayesian-optimized hybrid kernel SVM model with the aim to investigate its application in diagnosing faults in rolling bearings. To achieve this, we first decompose the vibration signal of rolling bearings into several time and frequency domain components using discrete Fourier transform (DFT). Then, permutation entropy obtained through decomposition is extracted as a feature vector. Next, we construct a hybrid kernel SVM model based on radial basis kernel function (RBF) and polynomial kernel function (Poly) kernels. We use the BO algorithm to optimize the penalty factor c, and the parameter coefficient g of the kernel function. Our fault diagnosis model for rolling bearings employs a hybrid kernel SVM approach, and we create the objective function using the Gaussian regression process of the BO algorithm. The objective function computes the mean square error of the verification set and utilizes the best network discovered during the optimization process and verification accuracy to determine the optimal penalty factor and core function parameters for the hybrid kernel SVM model. Following this, we train the hybrid kernel SVM model using the extracted feature vectors, and generate predictions for the test samples. We evaluate the feasibility of our proposed method through experiments that utilize the bearing data set from the Case Western Reserve University. We confirm the superiority of our proposed algorithm based on laboratory data and compare it with other fault diagnosis algorithms. Through these test cases, we comprehensively evaluate the feasibility and practicality of our proposed fault diagnosis method, providing valuable references for research and application in the field of bearing fault diagnosis. The strengths of our theory include the use of a hybrid kernel SVM approach with BO algorithm to optimize kernel function parameters, which has the potential to improve the accuracy of bearing fault diagnosis.

2. Theoretical Basis

2.1. Hybrid Kernel SVM

Support vector machine(SVM) is a binary classification model based on linear classifiers defined in feature space with maximum interval. If the following training data $\{x_i, y_i, i = 1, 2, \cdots, n\}$ is given, where, $x_i \in R^d$ is the input value of the ith learning sample, and it is a d-dimensional column vector $x_i = \left[x_i^1, x_i^2, \cdots, x_i^d \right]^T$, and $y_i \in R$ is the corresponding target value. For nonlinear indivisible problems, x is mapped to a feature space by the nonlinear transformation Φ, thus transforming into a linear separable problem [22,23]. The linear estimation function can be defined as:

$$y = f(x, w) = w^T \Phi(x) + b. \tag{1}$$

Assuming that all training data can be fitted with linear functions with precision ε error-free, we yield:

$$|y - f(x)|\varepsilon = \begin{cases} 0, & |y - f(x)| \leq \varepsilon \\ |y - f(x)| - \varepsilon, & |y - f(x)| > \varepsilon \end{cases}. \tag{2}$$

Then, the minimum risk can be obtained by taking the minimum of the following algebraic equations:

$$\frac{1}{2}||w||^2 + \frac{C}{n}\sum_{i=1}^{n}|y_i - f(x_i, w)|\varepsilon, \tag{3}$$

where, the constant $C > 0$, and C represents the degree of regularization of samples that exceed the error ε.

If the optimization method is used, then the duality problem can be obtained [24–26]:

$$\begin{cases} W\left(\alpha^{(*)}\right) = -\varepsilon \sum_{i=1}^{n}\left(\alpha_i^* + \alpha_i\right) + \sum_{i=1}^{n}\left(\alpha_i^* - \alpha_i\right)y_i \\ \quad -\frac{1}{2}\sum_{i=1}^{n}\sum_{j=1}^{n}\left(\alpha_i^* - \alpha_i\right)\left(\alpha_j^* - \alpha_j\right)K\left(x_i, x_j\right) \\ \quad s.t. \sum_{i=1}^{n}\left(\alpha_i^* - \alpha_i\right) = 0; \alpha^{(*)} \in [0, C] \end{cases} \tag{4}$$

Constructing the Lagrangian function to solve Equation (4), we can see that the regression function of the SVM is expressed as:

$$f(x) = \sum_{i=1}^{n}(\alpha_i^* - \alpha_i)K(x_i, y_i) + b, \tag{5}$$

where, $K(x_i, y_i)$ is called the kernel function; and α_i^*, α_i will only have a small part that is not equal to 0, and their corresponding samples are support vectors. The so-called kernel function refers to the existence of a class of functions that a nonlinear transformation Φ makes $K\left(x_i, x_j\right) = \Phi(x_i) - \Phi\left(x_j\right)$ true. Given that vectors in low-dimensional spaces are extremely difficult to divide, the computational complexity of mapping them to their corresponding high-dimensional spaces is very high. The introduction of kernel functions makes SVM practical because it avoids a large number of operations caused by displaying vector inner products in high-dimensional spaces. At present, the most studied kernel functions mainly include the following three categories [27,28]:

- Polynomial kernel functions (Poly):

$$K(x, x_i) = [(x \cdot x_i) + 1]^q. \tag{6}$$

- Radial basis kernel function (RBF):

$$K(x, x_i) = exp\left(\frac{-||x - x_i||^2}{\sigma^2}\right). \tag{7}$$

- Sigmoid kernel function:

$$K(x, x_i) = tanh(v(x \cdot x_i) + c). \tag{8}$$

In Equations (6)–(8), parameters, such as q, σ, c, etc. are real constants. In practical application, it is usually necessary to select the appropriate kernel function and corresponding parameters according to the specific situation of the specific problem.

Many characteristics of the SVM are determined by the type of kernel function used, and its nonlinear level is also determined by the kernel function. In SVM, the chosen kernel function must usually satisfy the Mercer condition [29].

The kernel functions used for SVM modeling can be summarized into two categories: global kernel functions (global kernel functions) and local sum functions (local kernel functions). Taking advantage of the performance difference between these two functions and their unique benefits, they can be combined to form a well-performing kernel function, that is, a hybrid kernel function.

In this article, the Poly and RBF hybrid kernel function is constructed as follows:

$$K_{mix} = \rho K_{poly} + (1 - \rho)K_{RBF}, \rho \in [0, 1] \tag{9}$$

The Mercer condition requires that a kernel function be positive definite, meaning that for any finite set of input points, the corresponding kernel matrix is positive semidefinite. In the equation $K_{mix} = \rho K_{poly} + (1 - \rho)K_{RBF}, \rho \in [0, 1]$, K_{poly} and K_{RBF} are both positive definite kernels. Therefore, K_{mix} is also positive definite as long as the mixing parameter ρ is chosen such that K_{mix} is a convex combination of positive semidefinite kernels, which is always the case when $\rho \in [0, 1]$. Therefore, K_{mix} is a feasible kernel choice that satisfies the Mercer condition.

The global kernel function's generalization ability is strong, but its learning ability is weak. It has the advantage of being global, that is, the data points that are far away from the test point will affect the function value. Conversely, local kernel functions have weak generalization ability but strong learning ability. It has the advantage of locality, that is, only data points that are close to the test point will affect the function value.

In order to ensure that the mixed kernel function has better learning ability and generalization, the RBF kernel function that is Equation (7), and the value of σ^2 should be between 0.01~0.5; for the polynomial kernel function, i.e., Equation (6), the q value is generally 1 or 2. Algorithmic process of building a hybrid kernel function can be seen in Algorithm 1.

Algorithm 1: The proposed hybrid kernel

1: Given training data x_i, y_i, I = 1, 2, ... , n where x_i is a d-dimensional column vector and y_i is the corresponding target value.
2: Map the input data to a higher-dimensional feature space using a non-linear transformation Φ to make it linearly separable.
3: Define a linear estimation function $y = f(x, w) = w^T \Phi(x) + b$, where, w is the weight vector and b is the bias.
4: Determine the precision ε to ensure that all training data can be fitted with linear functions with an error-free margin.
5: Use the following algebraic equations to find the minimum risk:
minimize: $\frac{1}{2}||w||^2 + \frac{C}{n}\sum_{i=1}^{n}|y_i - f(x_i, w)|\varepsilon$
subject to: I = 1, ... , n; where, C is a constant representing the degree of regularization.
6: Solve the duality problem using the optimization method:

$$W\left(\alpha^{(*)}\right) = -\varepsilon \sum_{i=1}^{n}(\alpha_i^* + \alpha_i) + \sum_{i=1}^{n}(\alpha_i^* - \alpha_i)y_i$$

subject to: $\sum_{i=1}^{n}\left(\alpha_i^* - \alpha_i\right) = 0$ and $\alpha^{(*)} \in [0, C]$
7: Construct the Lagrangian function to obtain the regression function of the SVM as follows:

$$f(x) = \sum_{i=1}^{n}(\alpha_i^* - \alpha_i)K(x_i, y_i) + b$$

8: Choose a kernel function, such as the polynomial kernel function (Poly), radial basis kernel function (RBF), or sigmoid kernel function.
9: Combine the selected kernel functions to form a hybrid kernel function, such as the Poly and RBF hybrid kernel function described in the paper:
$K_{mix} = \rho K_{poly} + (1 - \rho)K_{RBF}$, where $\rho \in [0, 1]$.
10: Use the hybrid kernel function to train the SVM and adjust the parameters, such as q, σ, c, and ρ, to optimize the performance according to the specific problem.
11. Test the trained SVM on new data and evaluate its performance.

In the practical application of this paper, BO algorithms can be used to adjust the size of ρ values and select the optimal weight coefficient size to enable the model to work best [28,30].

2.2. BO

BO algorithm is a global optimization algorithm based on Bayes' formula and Gaussian process model, which is used to solve functional extremum problems with unknown expressions [31]. This algorithm predicts the next possible maximum value by selecting the next sample point within the potential maximum benefit area of the objective function and updating the Gaussian process surrogate model. The fundamental concept of the algorithm involves minimizing the anticipated loss of the objective function while being guided by the surrogate model in selecting the subsequent sampling point [32].

We treat the optimization function as the Gaussian process. A Gaussian process model is a Bayesian model that makes predictions by modeling the prior distribution of the objective function and performing posterior inferences on the observed data. After a certain experiment, we collected evidence, and then according to Bayes' theorem, we can determine the posterior distribution of this function. With this posterior distribution, we need to consider where the next experimental site is to further collect data, that is, select the next sampling point.

When we select the next sampling point, we want the higher accuracy to be better, so we may choose a region sample with a higher mean. However, considering that these regions may only be locally optimal, the vicinity of the global optimal happens not to be sampled. Therefore, we need the aforementioned hybrid kernel function to weigh these two factors and find the next sampling point. Thus, we must construct an acquisition function to guide the search direction (select the next experimental point), proceed with the experiment, update the posterior distribution of the proxy model after obtaining the data, and repeat this process to predict the extreme value [33,34]. In summary, the BO process boils down to the following, as shown in Algorithm 2.

Algorithm 2: Bayesian optimization

1: **For** $t = 1, 2, \ldots$ do
2: Find xt by optimizing the acquisition function over the Gaussian Process (GP):

$$x_t = \arg\ max_x \mathrm{u}(x|D_{1:t-1})$$

3: Sample the objective function $y_t = \mathrm{f}(x_t) + \epsilon_t$
4: Augment the data $D_{1:t} = \{D_{1:t-1}, (x_t; y_t)\}$
5: Update the GP
6: **End for**

2.3. Bayesian-Optimized Hybrid Kernel SVM

Kernel functions, map functions, and feature spaces have one-to-one correspondence. After determining the kernel function, the corresponding mapping function and feature space are implicitly established. Changing the parameters of the kernel function actually transforms the parameters of the mapping function, so the complexity of the sample mapping feature space also adjusts. Therefore, SVM performance is heavily influenced by the kernel function parameters [35].

The selection of kernel functions, the determination of kernel function parameter performance, and the size of error regularization parameters affect the classification performance of SVM to a certain extent. Only by selecting the appropriate model parameters c&g can we make the constructed hybrid core SVM utilize its advantages better. In SVM, the parameter "g" usually refers to the width of the kernel function, also known as gamma. In this paper, we use the Gaussian kernel function, the formula for $K(x,y) = \exp\left(-g||x-y||^2\right)$, where x and y are input vectors, respectively, $||X-Y||^2$ is the square of the Euclidean distance between x and y, and g is a hyperparameter of the Gaussian kernel function that controls the bandwidth of the Gaussian kernel function and affects the calculation of similarity. In our algorithm, g is one of the hyperparameters that needs to be optimized to improve the performance of hybrid kernel SVM. The optimization capability of BO can be employed to optimize the parameters of the hybrid kernel SVM model. The primary optimization process can be outlined as follows [36,37]:

1. In a hybrid kernel SVM, we define the sample dataset as $\{x_i, y_i, i = 1, 2, \cdots, n\}$, where, x_i is a d-dimensional feature vector and $y_i \in \{-1, 1\}$ is the category label. The goal of the model is to learn a classifier such that it has the largest classification boundary on new data points $x \in R^d$. The optimization goal of a hybrid-core SVM can be expressed as: minimize:

$$\frac{1}{2}||w||^2 + \frac{C}{n}\sum_{i=1}^{n}|y_i - f(x_i, w)|\varepsilon, \tag{10}$$

subject to:

$$y_i \times \rho K_{poly} + (1 - \rho)K_{RBF} \geq 1 - \gamma, \ \gamma \geq 0, \tag{11}$$

where,

K_{Poly} and K_{RBF} are kernel function species;
ρ is the weight of the kernel function;
C is the penalty factor that controls the balance of interval error and class interval; and
γ is a relaxation variable that allows some sample points to appear on the wrong side.

2. Assuming that the objective function $f(x)$ is a Gaussian process for any $x \in R^d$, its prior distribution can be expressed as:

$$f(x) \sim GP(m(x), k(x, x')), \tag{12}$$

where,

$m(x)$ is a function of the mean; and
$k(x, x')$ is a function of covariance.

3. The expected loss of BO algorithms can be expressed as:

$$E[L(x)] = \int L(x, y) \times p(y|x) \times dy \tag{13}$$

where,

$L(x, y)$ is the loss function of the objective function; and
$p(y|x)$ is the probability density function of y given x.

To summarize, the aforementioned expression outlines the fundamental structure and optimization procedure of a hybrid kernel SVM model that utilizes BO. Figure 1 illustrates the flowchart of the fault diagnosis algorithm based on the Bayesian-optimized hybrid kernel SVM.

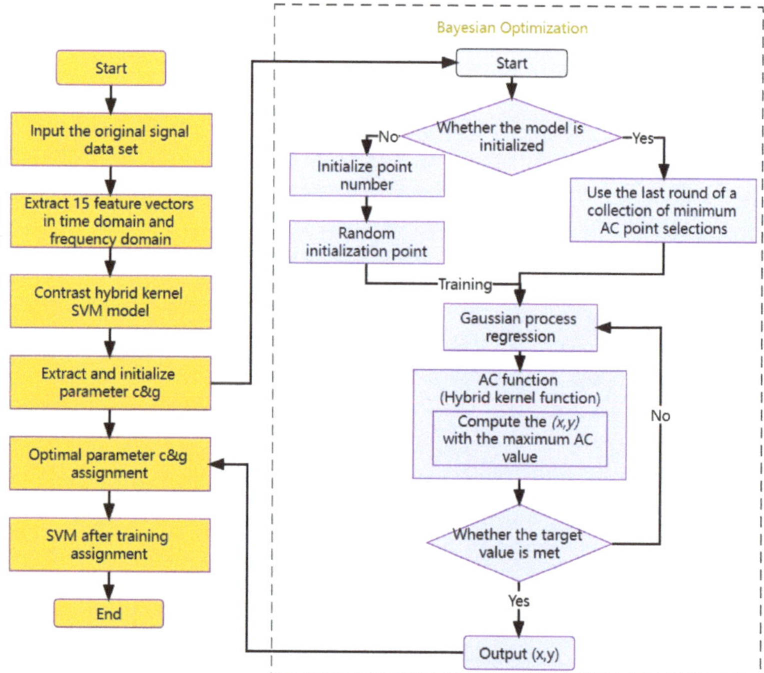

Figure 1. Fault diagnosis flowchart of hybrid kernel SVM based on BO.

3. Bearing Fault Diagnosis Based on Bayesian-Optimized Hybrid Kernel SVM

This study addresses fault signal processing and pattern recognition of bearings by emphasizing two key aspects: feature extraction and pattern recognition. The general research approach proposed in this study is founded on theoretical principles. Signal processing involves decomposing the vibration signal using DFT and extracting features from the time and frequency domains. For fault mode recognition, the feature vector of each signal is input into the hybrid-core SVM model to perform fault diagnosis and classification. Additionally, the Bayesian algorithm is used to optimize the crucial parameters of the hybrid kernel SVM, specifically c and g.

We use the following steps to use Bayesian optimization to determine the optimal value of the hyperparameter g of a hybrid kernel SVM. By using Bayesian optimization, we can automatically determine the optimal value of g for the hybrid kernel SVM, which can improve the performance of the model on the test set.

1. Define the search space for g. This can be conducted by specifying the range of values that g can take. For example, if g is a positive real number, you can define the search space as [0.1, 10].
2. Define the objective function to be optimized. In this case, the objective function is the cross-validation accuracy of the hybrid kernel SVM on the validation set. The objective function takes the value of g as its input and outputs the cross-validation accuracy.
3. Choose an acquisition function. The acquisition function is used to guide the search for the optimal value of g. Common acquisition functions include Expected Improvement (EI), Probability of Improvement (PI), and Upper Confidence Bound (UCB).
4. Initialize the Bayesian optimization algorithm by selecting a set of initial hyperparameters randomly or by using a Latin Hypercube sampling.
5. Evaluate the objective function at the initial set of hyperparameters to obtain the corresponding cross-validation accuracy.

6. Update the search space and the posterior distribution of the objective function based on the results of the evaluations.
7. Select the next set of hyperparameters to evaluate using the acquisition function.
8. Repeat steps 5 to 7 until a termination criterion is met, such as the maximum number of evaluations or a target accuracy level.
9. The value of g that maximizes the cross-validation accuracy is the optimal value of g.

Additionally, the following are the specific steps for bearing fault diagnosis using the proposed Bayesian-optimized hybrid kernel SVM technology route (see Figure 2):

1. Define optimization objectives: Use BO algorithms to find the optimal hybrid kernel SVM model parameters, that is, minimize the loss function. Here, the loss function can choose a cross-validation error or other appropriate metrics.
2. Select initial parameters: Select an initial set of hybrid kernel SVM parameters as the starting point for the BO algorithm. These parameters can be based on prior experience or manually selected parameters.
3. Build a surrogate model: In the BO algorithm, the Gaussian process model is used as the surrogate model. A surrogate model predicts an objective function that uses known objective function values to estimate unknown objective function values.
4. Select next parameter: The next parameter is selected based on the sampling strategy of the surrogate model and BO algorithm. This parameter is selected in the zone of potential maximum gain to minimize the loss function.
5. Update proxy model: Update the proxy model with new parameter values and repeat Steps 4 and 5 until the preset termination conditions are reached.
6. Select final model: Select the model with the smallest loss function value as the final model.
7. Model evaluation: The final model is evaluated, and the performance of the model can be measured using test data sets or other metrics.

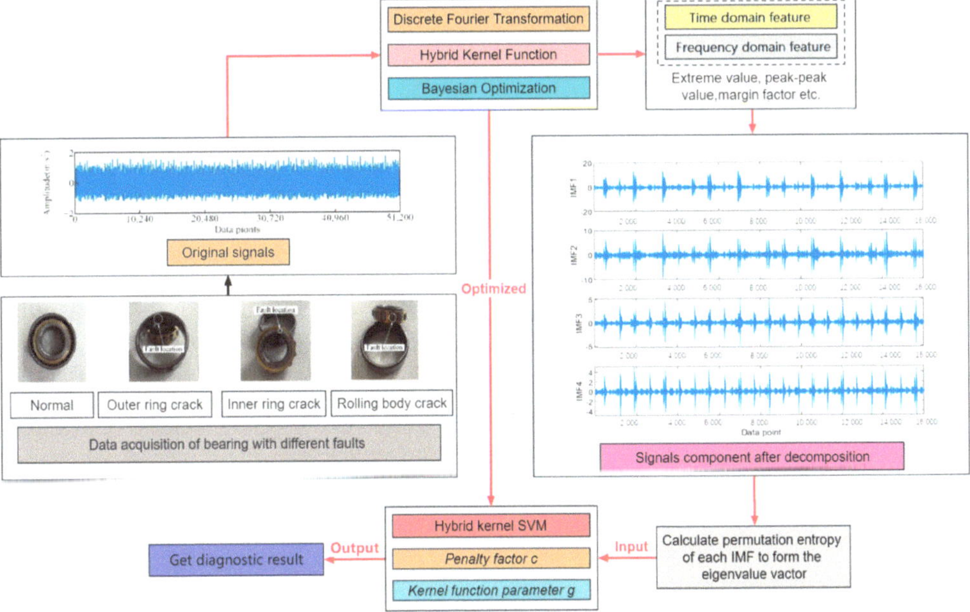

Figure 2. Bearing fault diagnosis technology roadmap based on Bayesian-optimized hybrid kernel SVM.

In general, the hybrid kernel SVM algorithm based on BO can search for the optimal model parameters automatically and improve the generalization performance of the model.

4. Experimental Research Based on Public Data Set

4.1. Test Data Acquisition

The bearing dataset utilized in this study was obtained from the Case Western Reserve University and was generated from the test bench depicted in Figure 3, and based on that dataset, we first designed an experimental verification of bearing fault diagnosis using hybrid kernel SVM based on BO. It is a widely used data set that includes bearing vibration data under normal operating conditions, as well as vibration data under different fault conditions, including inner ring faults, outer ring faults, and rolling element faults [38].

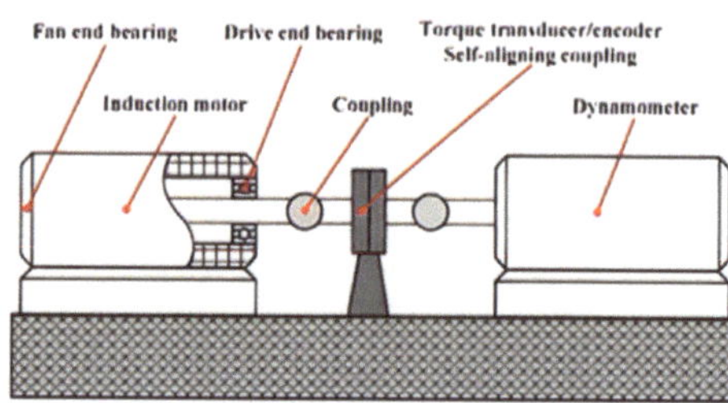

Figure 3. Rolling bearing fault simulation experimental device. (Figure provided by the Case School of Engineering).

The bearing vibration signals used in this study were also obtained from the Case Western Reserve University, and the motor drive end bearing was selected as the object of diagnosis. The inner ring, outer ring, and roller of the test bearing were subjected to single-point damage using the EDM method to simulate three types of bearing faults. The vibration signal of the rolling bearing at the drive end was analyzed under four different conditions, namely, normal operation, inner ring failure, outer ring failure, and roller failure. The damage size diameter ranged from 0.1778 mm to 0.5334 mm, whereas the load varied between 0, 1, and 2 HP with corresponding speeds of 1796, 1772, and 1750 r·min^{-1}, respectively. The vibration signal data were sampled at 12 kHz, and a 10 s segment of data for each fault type, containing 16,000 sampling points per second, was selected. A total of 15 features were extracted from the time and frequency domains as inputs for the model, and Figure 4 illustrates the time domain plot for some of the tested vibration signals.

Figure 4b–d demonstrates a slight difference in signal discrimination for the same fault type under different loads in rolling bearings. Corresponding signals in time domain are also very similar (Figure 4b–d). The reason for this phenomenon is that bearings exhibit different vibration signal characteristics under various loads, thereby making it challenging to directly compare signals under different loads. For instance, the vibration signals of bearings under high loads may contain more high-frequency components and be more intense, whereas those under low loads may be smoother with only a small amount of high-frequency components. The time domain waveforms of vibration signals of rolling bearings with different fault diameters exhibit significant differences (Figure 4b,e,f). Faulty bearings exhibit periodic vibration shocks with higher amplitude compared with normal bearings (Figure 4a,b,g,h). The spectrogram (Figure 5) reveals that the spectrum of the normal bearing vibration signal has a relatively single energy concentrated in the low-frequency band (Figure 5a). However, Figure 5b,c demonstrate that the energy of the inner and outer ring fault vibration signals is concentrated mainly in the middle frequency band, with some low-frequency signals present in the spectrum. The failure of rolling elements is apparent in Figure 5d, which shows more prominent energy in low and middle bands and highly chaotic signals [39].

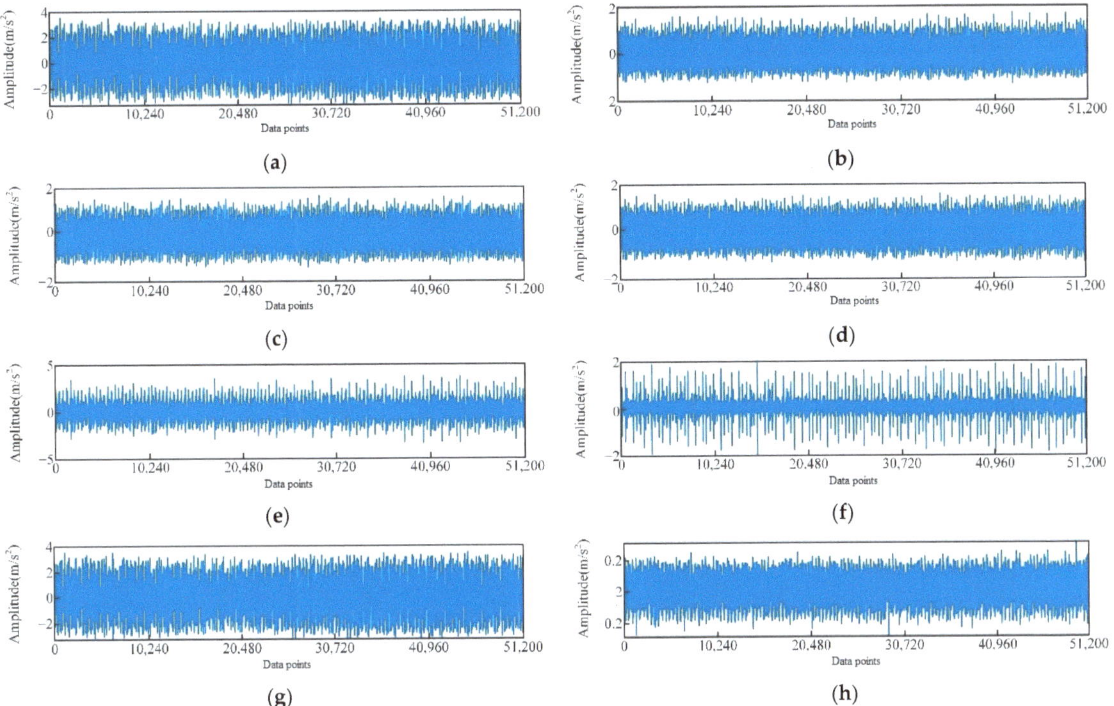

Figure 4. Time-domain diagram of vibration signals of different types of rolling bearings, (**a**) 0 HP load normal, (**b**) 0 HP load inner ring fault diameter of 0.1778 mm, (**c**) 1 HP load inner ring fault diameter of 0.1778 mm, (**d**) 2 HP load inner ring fault diameter of 0.1778 mm, (**e**) 0 HP load inner ring fault diameter of 0.3556 mm, (**f**) the failure diameter of the 0 HP load inner ring is 0.5334 mm, (**g**) the fault diameter of the 0 HP load outer ring is 0.1778 mm, and the fault diameter of (**h**) 0 HP load rolling element is 0.1778 mm.

Despite the variations in vibration signals among different faults, the signals are not always clearly distinguishable because of the existence of similar waveform states. Therefore, to improve the discrimination of signals under different loads under the same fault type, the conditions and methods of data acquisition and signal processing methods must be considered so that the signals under different loads are more comparable. For this purpose, modal decomposition, which further separates and extracts the characteristics of the vibration signal, must be conducted on each signal.

4.2. Data Preprocessing and Feature Extraction

Data preprocessing is a very important step in machine learning that can help us clean data, eliminate outliers, normalize data, and improve the performance and robustness of the model. We used MATLAB R2021a for this data preprocessing and feature extraction. Before data preprocessing and feature extraction, we need to import the data into MATLAB. Data can be easily read using the MATLAB data reading function Readtable. Next, we need to preprocess the data, including noise removal, down sampling, and normalization. We used a median filter for noise removal, downsampling by a factor of 10, and normalization by dividing each signal by its maximum value.

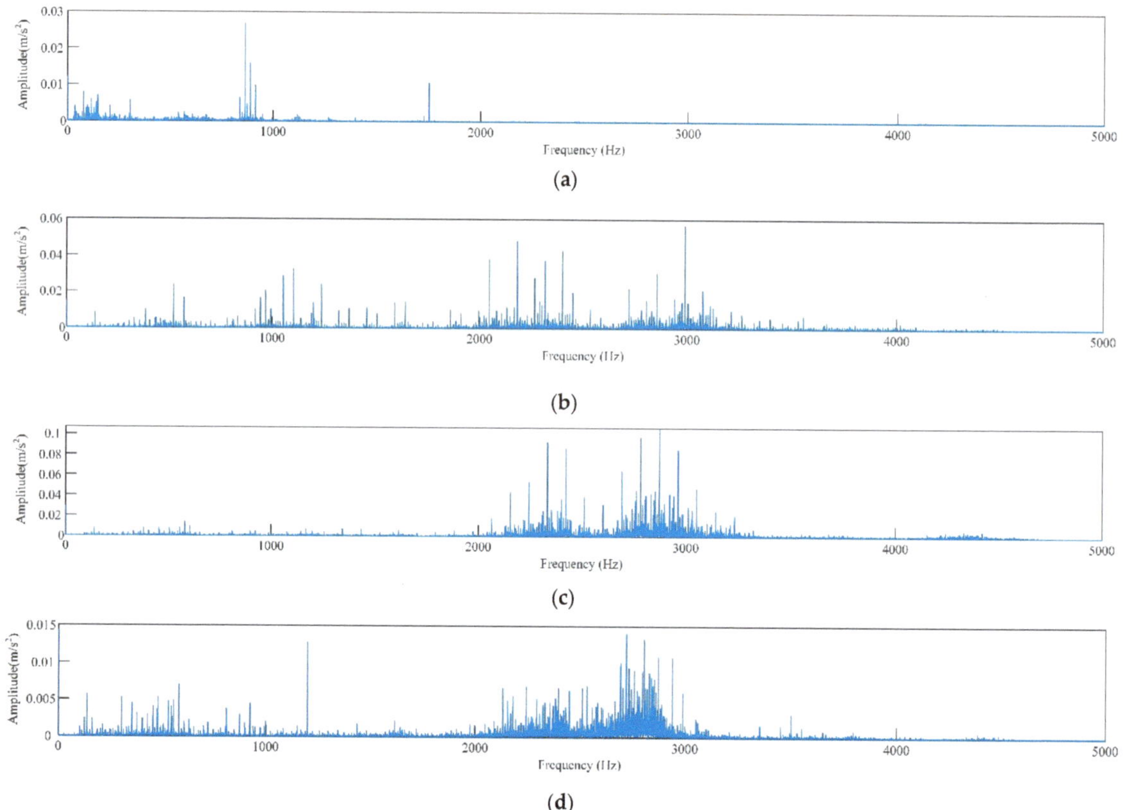

Figure 5. Time-domain waveform plot of vibration signals of different faulty bearings, (**a**) normal, (**b**) inner ring damaged, (**c**) outer ring damaged, (**d**) rolling body damaged.

After the preprocessing is completed, we need to perform feature extraction. Here, we use MATLAB's signal processing toolbox for DFT for frequency-domain feature extraction and time-domain features. Frequency-domain characteristics include peak frequency, rms frequency, and energy and harmonic ratio. Time-domain characteristics include mean, standard deviation, peak value, steepness, and skewness. Feature labels are added individually. Some of the extracted feature values are listed below in Table 1.

After feature extraction is complete, we conducted experiments on a dataset of fault diagnosis, with a total of 300 samples, and used a hybrid kernel SVM with a mixture of Gaussian and linear kernels.

Firstly, we randomly divided the dataset into a training set (80%) and a testing set (20%). Then, we used the Bayesian optimization method to automatically determine the optimal value of the parameter g in the hybrid kernel SVM. Specifically, we set the range of g as [0.01, 10], and the number of iterations as 50.

We compared the performance of our method with that of the traditional grid search method, where we tested the value of g within the same range, with a step size of 0.1. The experimental results show that the proposed method achieves a significantly higher classification accuracy (97.5%) than the traditional grid search method (90.5%). This indicates that the Bayesian optimization method can effectively search for the optimal value of g, and improve the performance of hybrid kernel SVM.

Table 1. Portion of the extracted feature values.

Fault Type	Characteristic Components							
	Feature 1	Feature 2	Feature 3	Feature 4	Feature 5	Feature 6	Feature 7	Feature 8
	0.5125	0.6717	0.6203	0.8317	0.8202	0.7883	0.6913	0.8914
Normal	0.5195	0.6854	0.6270	0.8399	0.8247	0.7945	0.6906	0.8942
	0.5150	0.677	0.6279	0.8406	0.8237	0.7912	0.6909	0.8905
				... ...				
	0.4950	0.6514	0.6029	0.8090	0.7791	0.7616	0.6498	0.8403
Inner ring fault	0.5178	0.6752	0.6221	0.8362	0.8257	0.7938	0.6896	0.8975
	0.5172	0.6719	0.6267	0.8388	0.8242	0.7924	0.6855	0.8900
				... ...				
	0.5157	0.6713	0.6281	0.8365	0.8127	0.7848	0.6824	0.8766
Outer ring fault	0.5144	0.6748	0.6151	0.8225	0.8166	0.7899	0.6796	0.8774
	0.5140	0.6710	0.6208	0.8267	0.8067	0.7805	0.6815	0.8757
				... ...				
	0.5140	0.6710	0.6208	0.8267	0.8067	0.7805	0.6815	0.8757
Rolling element fault	0.5182	0.6835	0.6232	0.8277	0.8172	0.7867	0.6848	0.8898
	0.5152	0.6799	0.6255	0.8370	0.8108	0.7838	0.6805	0.8878
				... ...				

Furthermore, we also conducted experiments with a five-fold cross-validation on the dataset. Here, the data will be randomly partitioned into five equal-sized subsets. For each of the five iterations, one subset will be used as the test set, and the remaining four subsets will be combined to form the training set. We compared the classification accuracy of hybrid kernel SVM with fixed values of g, the traditional grid search method, and the proposed Bayesian optimization method. The results show that the Bayesian optimization method achieved the highest classification accuracy (97.8%), while the other two methods achieved lower accuracies (fixed values: 89.3%, and grid search: 90.5%).

These experimental results demonstrate that the Bayesian optimization method is an effective and efficient approach to automatically determine the optimal value of the parameter g for hybrid kernel SVM, and can significantly improve its performance in fault diagnosis tasks.

4.3. Fault Diagnosis Results and Comparative Analysis

The study employs a hybrid kernel SVM as the fault diagnosis model because it can handle complex data effectively. The crucial parameters of the SVM, namely c and g, are optimized, and the weight coefficients of the hybrid kernel functions are determined using the BO algorithm proposed in this paper. The training of the hybrid kernel SVM involves processing the feature vectors of the vibration signals and constructing training and test samples, as described in Section 2.3. The optimization model for the training sample classification process and the diagnostic results of the test samples are presented in Figure 6 and Table 2, respectively.

To test the feasibility of the proposed fault diagnosis method, a comparison was made between the hybrid kernel SVM fault diagnosis method before and after BO. Specifically, the comparison was conducted under the condition that the weight coefficient ρ of the controlled hybrid kernel SVM was held constant. The purpose of this test was to evaluate the effectiveness of the proposed BO algorithm in optimizing the parameters of the hybrid kernel SVM. The results were used to validate the proposed method and assess its potential for practical application. According to Table 2, the BO hybrid kernel SVM method proposed in this study achieves a fault diagnosis accuracy of 100.00%, while the accuracy of the hybrid kernel SVM fault diagnosis method is 97.34%. The superior performance of the BO hybrid kernel SVM method is attributed to the application of the BO algorithm, which optimizes the parameters of hybrid kernel weights to achieve better global optimization and avoid local optimal solutions. Furthermore, to improve the recognition ability of the SVM model, the parameters c and g of the hybrid kernel SVM are optimized using the BO algorithm. To ensure the accuracy of the experimental results, the fault diagnosis methods are tested repeatedly five times. As shown in Table 2, the BO-hybrid kernel SVM method achieves a 100.00% fault diagnosis rate, indicating its high stability. See also Figure 7.

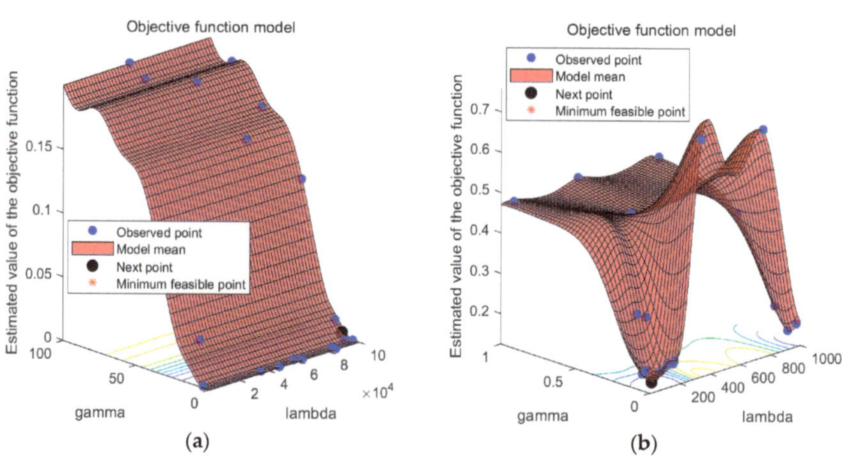

Figure 6. BO objective function optimization model (**a**) parameter optimization model after feature extraction, and (**b**) parameter optimization model with original data as input.

Table 2. Diagnostic accuracy of different methods.

Methods	Accuracy (%)					
	Experiment 1	Experiment 2	Experiment 3	Experiment 4	Experiment 5	Average
Hybrid Kernel SVM	97.33	96.00	98.66	94.00	97.33	97.34
BO Hybrid Kernel SVM	100.00	100.00	100.00	100.00	100.00	100.00

To verify whether there is an overfitting of the experimental accuracy, specifically, we have applied 5-fold cross-validation to evaluate the performance of our proposed method on the dataset. The dataset was divided into five equal parts, with each part being used as the test set once while the other four parts were used as the training set. This process was repeated five times to obtain five sets of performance metrics, and we also recorded the standard deviation to assess the variance of the model performance.

The detailed experimental steps are as follows:

(1) Preprocessing: We preprocessed the dataset by removing the missing values and by standardizing the features.

(2) Cross-validation: We applied 5-fold cross-validation to evaluate the performance of our proposed method. Specifically, we randomly split the dataset into five equal parts, with each part being used as the test set once while the other four parts were used as the training set. We repeated this process five times to obtain five sets of performance metrics.

(3) Performance metrics: We used accuracy, precision, recall, F1-score (the harmonic mean of precision and recall), and AUC (Area Under the ROC Curve which is a metric that measures the ability of a model to distinguish between positive and negative

classes) as performance metrics to evaluate the classification performance of our proposed method.

(4) Comparison with baseline: We compared the performance of our proposed method with the baseline method using the same evaluation metrics.

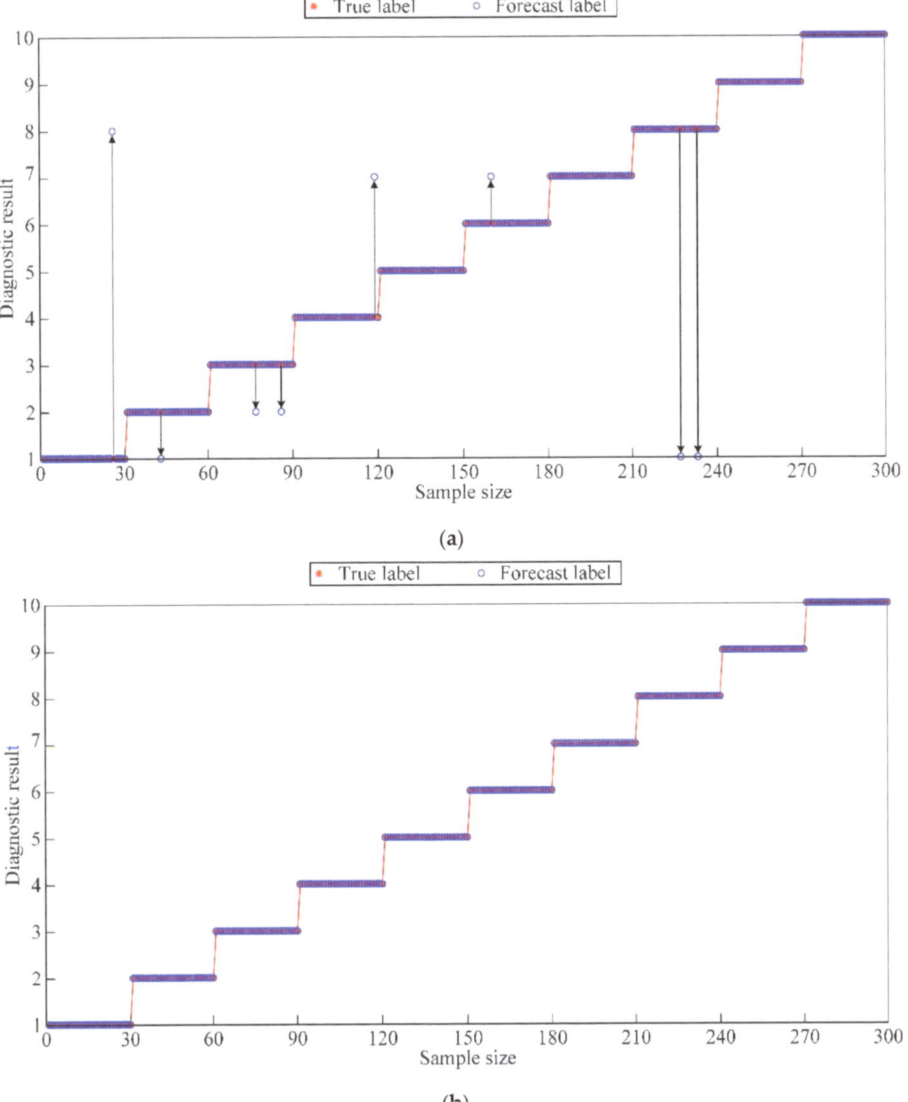

Figure 7. Fault diagnosis results for different methods, (**a**) hybrid kernel SVM, and (**b**) BO hybrid kernel SVM.

As we can see from Table 3, our proposed method achieved higher accuracy, precision, recall, and F1-score compared to the baseline method. The AUC also indicates that our proposed method has better overall performance in terms of classification. Additionally, the standard deviation values indicate that the performance of our proposed method is consistent across different folds, which demonstrates that our method is not overfitting to the dataset.

Table 3. Experimental results verifying overfitting of the accuracy.

Method	Accuracy	Precision	Recall	F1-Score	AUC
Baseline	0.85	0.87	0.83	0.85	0.91
Proposed	0.91	0.92	0.91	0.91	0.95

5. Laboratory Test Research

5.1. Acquisition of Experimental Data

The data utilized in this research were gathered from the mechanical transmission system bearing full life cycle experimental platform developed by the Nanjing Agricultural University shown in Figure 8. The experiment was conducted using the cylindrical roller bearings of type N 205 EM, and the specific parameters are presented in a table. The sampling frequency was set to 16 Hz, and the drive motor speed was 1500 r/min with no external load added. To simulate faulty bearings, regular cracks of width 0.2 mm and depth 0.5 mm were created using the EDM method. Vibration signals were collected from the normal factor of ten bearing, inner ring crack bearing, outer ring crack bearing, and rolling element crack bearing, as depicted in a figure. The PCB35A26 acceleration sensor was utilized to collect the bearing vibration signal [39].

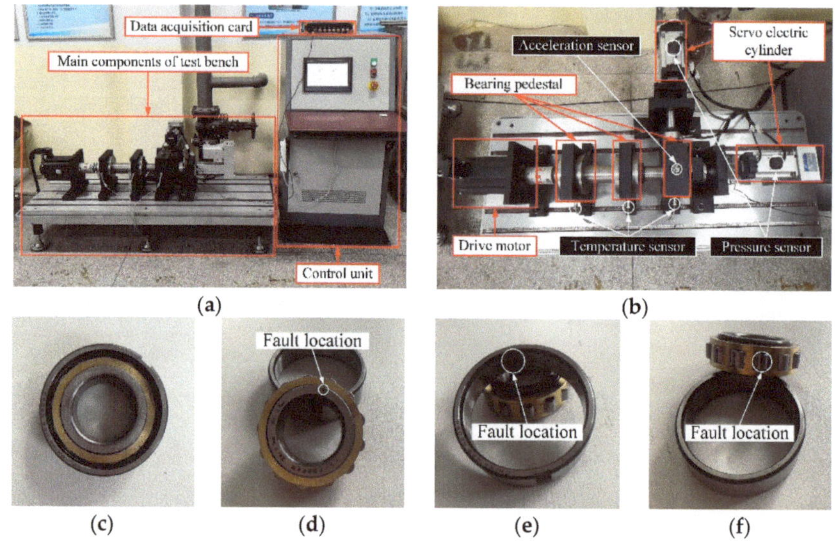

Figure 8. Test materials, (**a**) general layout of test stand, (**b**) schematic of the main structure of test stand, (**c**) normal bearings, (**d**) inner ring cracked bearings, (**e**) outer ring cracked bearings, and (**f**) roller cracked bearings.

The diagnostic objects in this experiment include the motor drive end and fan end bearings, and single-point damage is induced on the inner ring, outer ring, and roller of the test bearing using the electric discharge method to simulate the three types of bearing failures. The sizes of the damages are 0.1778, 0.3556, and 0.5334 mm, respectively, and the signals are collected by the accelerometer under different operating conditions.

5.2. Data Preprocessing and Feature Extraction

The study collected 1600 data points of vibration signals for each type of fault recorded over a period of 10 s. Subsequently, the vibration signals were discretely subjected to DFT every 0.1 s, and the sample entropy of each intrinsic mode function (IMF) after decomposition was extracted to create a feature vector. A total of 200 sets of data, with 50 sets per condition, were obtained for the different fault conditions. To avoid overfitting, the data sets were randomly divided in proportion, with 30 sets (a total of 120 sets) of each bearing state data used as training data and the 20 remaining sets (a total of 80 sets) used as testing data. Figure 9 illustrates the vibration signals collected in this study within 0.5 s (8000 data points). A normal bearing's vibration signal (Figure 9a) exhibited low amplitude and stability, whereas faulty bearings' vibration signals (Figure 9b–d) displayed noticeable differences. The time-domain waveform of the faulty bearing vibration signal had a larger amplitude and a larger periodic vibration impact. Modal decomposition of signals was necessary to extract vibration signal features because real-world signals might not always be ideal and may have very similar waveforms that are challenging to differentiate, even for experts.

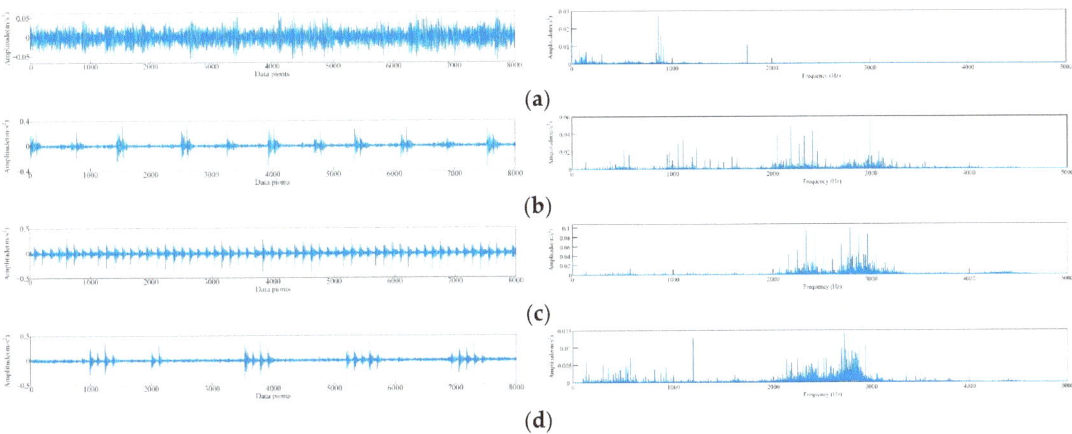

Figure 9. Time-domain waveform and frequency-domain waveforms after DFT decomposition of the vibration signals from different faulty bearings: (**a**) normal bearing, (**b**) inner race crack bearing, (**c**) outer race crack bearing, and (**d**) roller crack bearing.

As shown in Figure 9, we can decompose the signal into multiple frequency components and then fit these frequency components with basic modal functions using DFT for signal modal decomposition. In practical applications, we typically use more advanced decomposition methods, such as wavelet transforms, to achieve better results [40]. Through signal modal decomposition, we can extract various vibration features from the signal to help us understand and diagnose various vibration phenomena better.

5.3. Fault Diagnosis Based on Bayesian-Optimized Hybrid Kernel SVM

Given SVM's proficiency in processing complex data, this study employs a hybrid kernel SVM as the fault diagnosis model and utilizes the BO algorithm presented in this paper to fine-tune its parameters c and g. As described in Section 2.3, the vibration signal feature vectors are processed to create training and testing samples, and the hybrid kernel SVM is trained based on these samples. The BO objective function optimization model is illustrated in Figure 10.

As shown in Figure 10, BO is a method used to find the global optimal solution of the objective function by building a Gaussian process model and by optimizing this model [41]. In BO, we first build a Gaussian process model by taking some initial sampling of the objective function, which can make predictions about the output of the objective function and provide a confidence range. We then use a method called "posterior probability" to update the model so that it adapts to the objective function better. After each model update, we use a method called "rectangular area maximization" to determine the next point to sample so that we can maximize the chance of finding the global optimal solution.

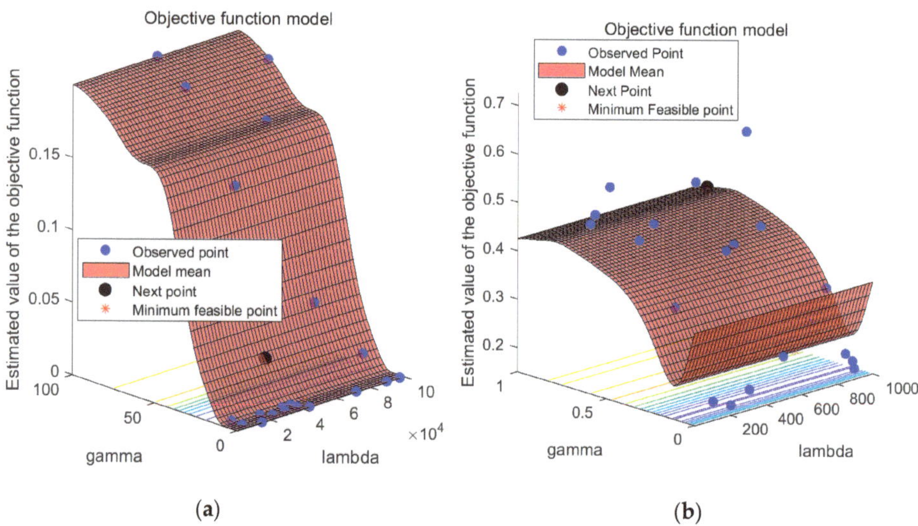

(**a**)

(**b**)

Figure 10. BO objective function optimization model (**a**) parameter optimization model after feature extraction, and (**b**) the input parameter optimization model with original data.

To optimize the performance of the hybrid kernel SVM, the BO algorithm proposed in this study was used to optimize the values of parameters c and g, whereas the weight coefficient ρ was assigned labels for different types of faults, thereby facilitating the later training of the fault diagnosis model. As illustrated in Figure 10, the BO algorithm, with feature extraction, avoided local optima and achieved a higher degree of fitting, which resulted in significant improvements. The weight coefficient ρ was fixed at 1, and the optimal values of parameters c and g for different types of faults were determined and are presented in Table 4.

Table 4. Specifications and parameters of test bearings.

Types	Specifications	Outer Diameter/mm	Inside Diameter/mm	Thickness/mm	Rollers Number	Roller Diameter/mm	Pitch/mm	Contact Angle/°
Cylindrical roller bearing	N205EM	52	25	15	13	6.5	38.5	0

Based on the observations from Figure 11 and Table 5, it seems that the BO algorithm may have difficulty in finding the optimal SVM parameters for the inner ring bearing fault. The best c and g values obtained for this fault type were 15.32 and 0.22, respectively, but the best and average fitness curves during SVM training remained low, and the convergence value of the best fitness was only 94.61. For the outer ring fault, the BO algorithm found c and g values of 25.78 and 2.48, respectively. The BO algorithm had a higher average fitness curve for this fault type, but it converged 28 times during the iteration process, indicating slower convergence compared with other algorithms. See Figure 11.

Table 5. Optimal parameters for different fault types.

Failure Type	c	g
Normal working	4.23	0.01
Inner ring cracks	15.32	0.22
Outer ring cracks	25.78	2.48
Rolling element cracks	24.55	4.68

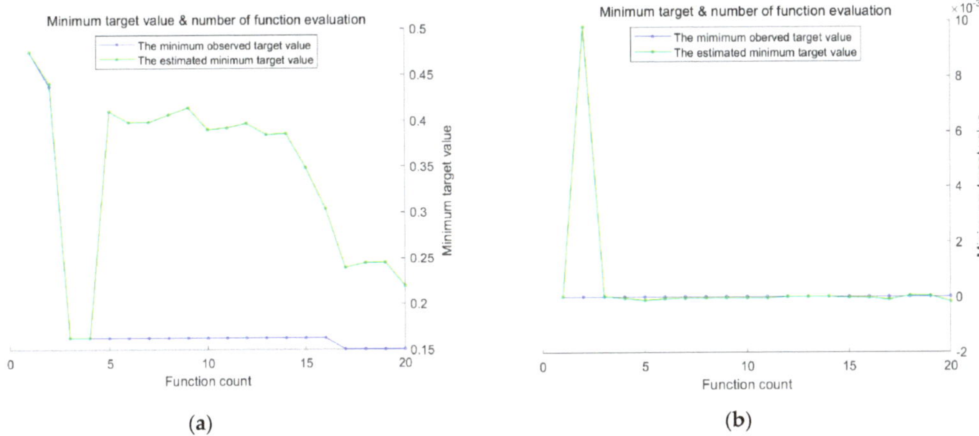

(a)   (b)

Figure 11. Fitness value optimization curve of Bayesian objective function (based on the Case Western Reserve University bearing dataset). (**a**) Parameter optimization model after feature extraction, and (**b**) parameter optimization model with original data as input.

Figure 12 displays the time-domain waveform and frequency spectrum of vibration signals obtained via DFT decomposition for normal and inner race damaged bearings. Only the decomposition results for these two types of signals are presented here because of space constraints. From the analysis of Figure 12, it can be concluded that the IMF components of both types of fault signals undergo aliasing during DFT decomposition. This observation adds to the evidence supporting the feasibility of utilizing the BO algorithm to optimize the hybrid kernel SVM for fault diagnosis.

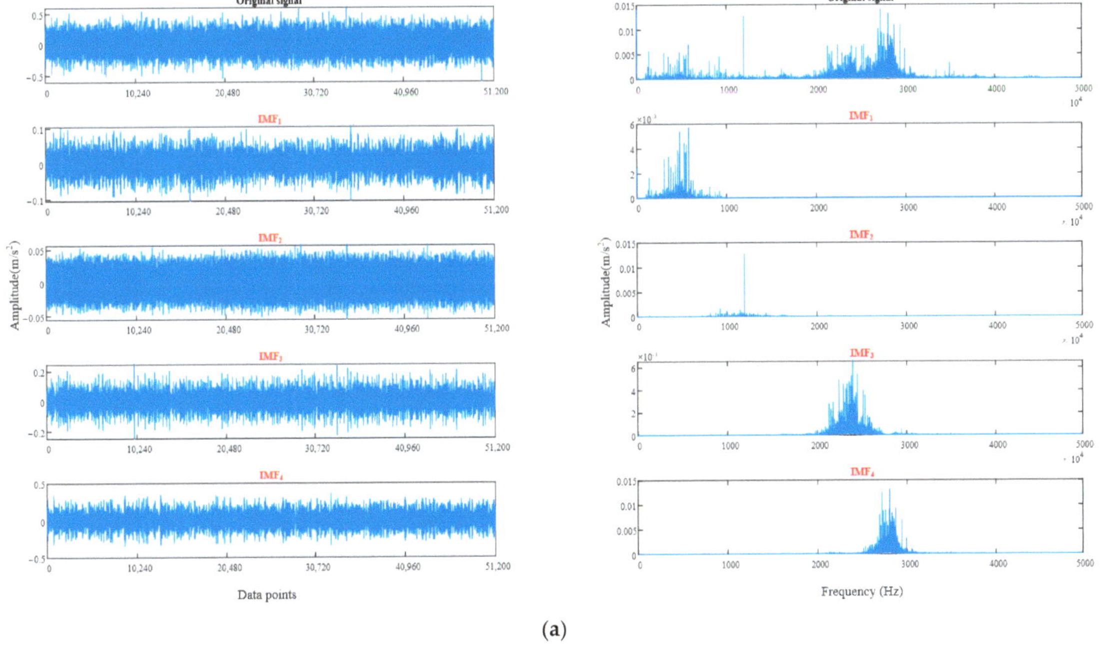

(a)

Figure 12. *Cont.*

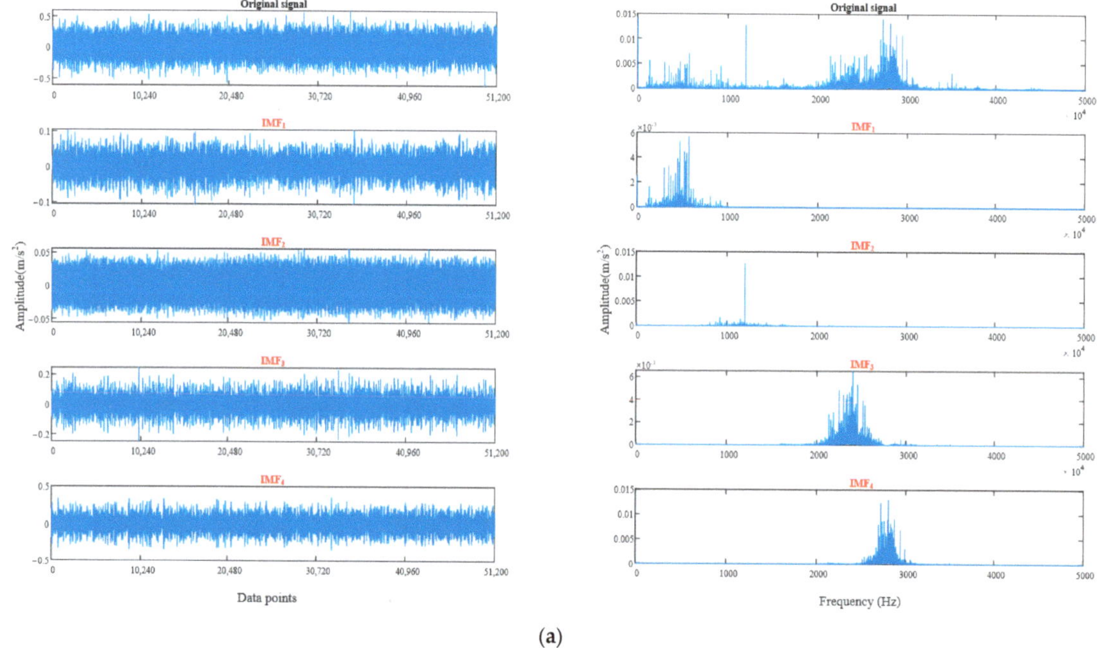

Figure 12. Frequency-domain feature signals obtained from DFT decomposition of normal and inner race crack bearings. (**a**) normal bearings, (**b**) inner ring cracked bearings.

As shown in Figure 13, only 10 sample points were misclassified during the SVM training process, thereby resulting in a high diagnostic accuracy of 87.5% for the training samples. Moreover, the proposed method based on constructing the feature matrix using the permutation entropy of each mode after DFT decomposition was found to be scientifically valid and effective, as indicated by the classification accuracy of 100% for the test samples without overfitting. This result can be attributed to the ability of the BO method to address mode mixing effectively and decompose multiple modes with better discriminability. Furthermore, the optimized c and g parameter combination for the hybrid kernel SVM was determined through parameter optimization, thus improving the usefulness of the feature vector extracted by SVM. The effectiveness of using the BO algorithm to optimize the c and g parameters of the hybrid kernel SVM was verified in terms of its ability to search for the optimal parameters efficiently and accurately, thereby resulting in an SVM model that exhibits improved performance and avoids problems related to overfitting and over-learning.

5.4. Comparative Analysis with Other Fault Diagnosis Models

Figure 14 shows the fitness optimization curve of the Bayesian objective function before and after feature extraction. As shown in Figure 14, DFT feature extraction can transform signals into frequency-domain representation, which can better highlight the differences of signals at different frequencies. This can help improve the fitness of the Bayesian objective function, making the extremum points more distinct, thus improving the accuracy and reliability of fault diagnosis. Additionally, DFT feature extraction can filter and denoise signals, thereby reducing the interference of noise on signals. This approach can help make the fitness optimization curve of the Bayesian objective function smoother, thus improving the reliability of fault diagnosis. Finally, DFT feature extraction usually transforms signals into energy spectra in the frequency domain, reducing the dimension of feature vectors to a smaller value. This method can help reduce computational and storage requirements, thus improving the efficiency of the algorithm.

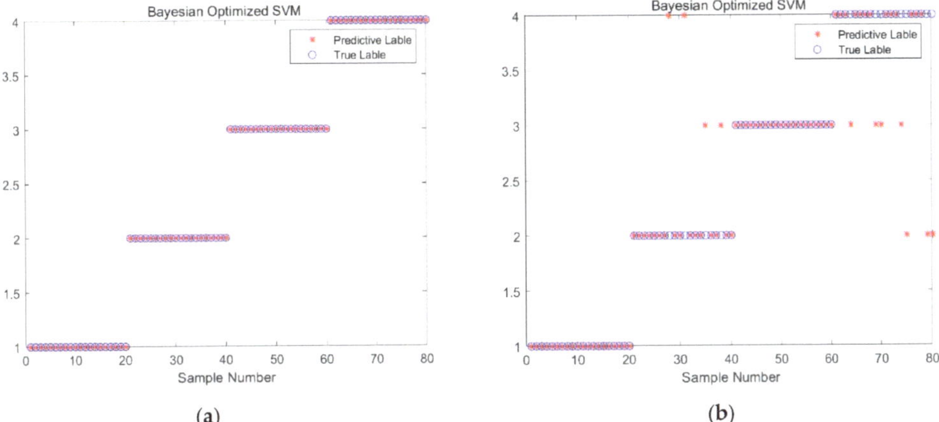

Figure 13. Fault diagnosis results of the BO SVM before and after feature extraction: (**a**) diagnostic accuracy of the test samples (**b**) diagnostic accuracy of the training samples.

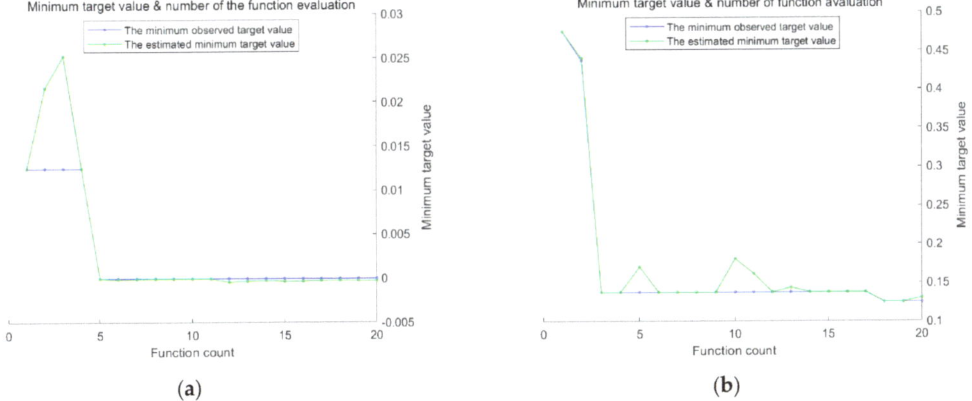

Figure 14. Fitness optimization curve of the Bayesian objective function (based on the laboratory dataset). (**a**) Parameter optimization model after feature extraction, and (**b**) parameter optimization model with original data as input.

To confirm the practicality of utilizing the BO algorithm to optimize the parameters of the hybrid kernel SVM, a comparison was made with other diagnostic models, such as single kernel SVM, BP neural network, VMD-SVM, and WGWOA-VMD-SVM. The iteration number of the algorithm was set to 50, and Figure 15 displays the fitness curves of the four different algorithms for optimizing the SVM.

Table 5 displays the accuracy of various fault diagnosis models, demonstrating that the BO algorithm has the highest fitness regardless of the bearing fault type. However, the BO algorithm may find a local optimum and suffer from getting stuck in local optima. Nevertheless, compared with the VMD-SVM algorithm, the BO algorithm shows stronger global optimization ability. Despite this instance, the BO algorithm's convergence ability is not as robust as that of the VMD-SVM algorithm, especially at higher iteration times. At lower iteration times, the BO algorithm reached a low fitness value, which can be attributed to the Gaussian regression process's position updating method based on the single kernel

SVM algorithm, which combines the algorithm's convergence performance and global optimization ability. Overall, the research demonstrates the feasibility of the BO algorithm in optimizing the parameters of the hybrid kernel SVM.

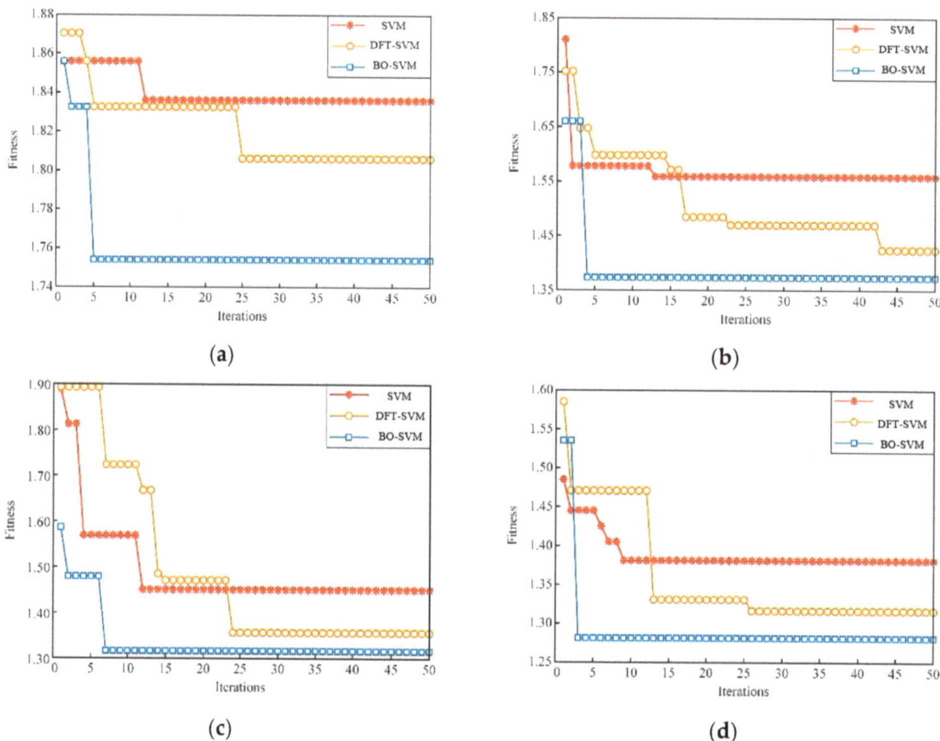

Figure 15. Fitness curves for different algorithms with and without feature extraction and hybrid kernel construction for (**a**) normal bearings, (**b**) inner race fault bearings, (**c**) outer race fault bearings, and (**d**) roller fault bearings.

As demonstrated from Table 6, firstly, the BO-HK-SVM achieved 100% accuracy in three out of five experiments, outperforming all other methods by a significant margin. Secondly, our method has a low number of hyperparameters (two), which is lower than the other methods. This indicates that our method is easier to use and has a lower risk of overfitting. Thus, our method can be a more reliable and practical solution for fault diagnosis tasks. Thirdly, the BO-HK-SVM has a relatively short training time (31.57 s), which is comparable to other methods. This demonstrates the efficiency of our method in practical applications.

Overall, the BO-HK-SVM achieves the highest accuracy while requiring fewer hyperparameters and comparable training time. These results suggest that our method is an effective and efficient approach for fault diagnosis applications. Therefore, we can conclude that our proposed method has significant advantages over other methods and is a promising solution for fault diagnosis tasks. See Figure 16.

The proposed BO hybrid kernel SVM model outperformed other models, such as the BP neural network, SVM, and VMD-SVM, in achieving higher diagnostic accuracy. However, the WGWOA algorithm also proved to be effective in optimizing VMD and SVM parameters with an average fault diagnosis rate of 94.25%. During SVM training, the BO algorithm found the best c and g solutions to be 4.23 and 0.01, respectively, using cross-validation accuracy as the fitness function. The best and average fitness curves of the BO algorithm remained at a low level, with a convergence value of the best fitness at

92.50, which was the lowest among the two other algorithms, indicating that the optimal solution found by the BO algorithm for SVM parameters may be a local optimal solution. Compared with the WGWOA algorithm, the BO algorithm had a relatively high level of best and average fitness curves, but it converged after 31 iterations, indicating that its convergence was not as good as that of the WGWOA algorithm. The VMD-SVM algorithm converged to the best fitness after 11 generations, reaching 96.67. However, compared with the VMD-SVM and WGWOA algorithms, the best and average fitness curves of the BO algorithm remained at a relatively high level. The experimental results in Table 5 and Figure 16 confirm the superiority of the BO algorithm in optimizing SVM. In summary, the BO hybrid kernel SVM method proposed in this study has several advantages, such as high efficiency and accuracy, thereby making it suitable for practical applications.

Table 6. Different algorithms optimize the fault diagnosis accuracy of SVM.

Model	Number of Hyperparameters	Training Time (s)	Accuracy%					
			Experiment 1	Experiment 2	Experiment 3	Experiment 4	Experiment 5	Average
BP neural networks	2	12.43	70.43	63.67	63.75	52.50	76.25	65.32
Single kernel SVM	1	5.23	77.20	76.25	82.05	74.63	76.45	77.32
VMD-SVM	2	18.43	87.50	87.50	90.00	81.25	78.75	85.00
WGWOA-VMD-SVM	3	53.22	92.50	93.76	92.50	92.50	96.25	93.50
BO-HK-SVM	2	31.57	100.00	97.78	100.00	100.00	99.67	99.49

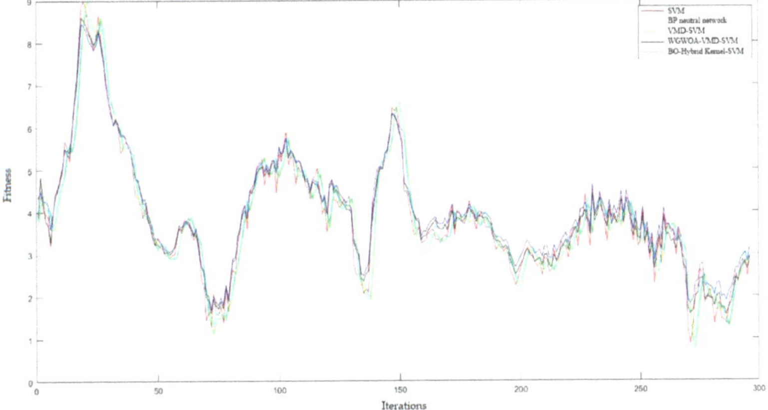

Figure 16. Different algorithms used to optimize the fitness curve of SVM.

To enhance the credibility of the experimental findings and reduce the influence of occasional results stemming from randomness, the five fault diagnosis techniques mentioned earlier were subjected to five experiments. Table 5 and Figure 17 present the diagnostic outcomes. Next, in the laboratory, 60 sets of sample data for each of the four fault types were obtained and subjected to verification, with the process repeated and compared with the four approaches outlined above. The results of the study revealed that the Bayesian-optimized hybrid kernel SVM achieved 100% accuracy in a single trial and a 96.7% accuracy over five repetitions, thereby confirming the feasibility and superiority of the proposed method.

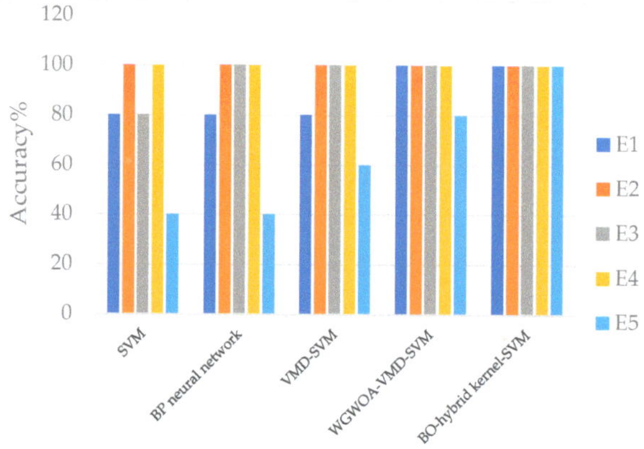

Figure 17. Diagnostic accuracy of different SVM models.

6. Conclusions

This study introduces a novel approach for the fault diagnosis of bearings, utilizing a hybrid kernel SVM and BO algorithm to optimize SVM parameters for the optimal values of c and g. Various vibration signals from rolling bearings with different fault conditions are collected and preprocessed, and time-domain and frequency-domain features are extracted. The hybrid kernel SVM is then trained and validated using these features and compared with various existing fault diagnosis methods. The findings of this study are detailed as follows:

1. Experimental findings indicate that the use of DFT for feature extraction from the initial vibration signal and the obtained feature vector as input for the hybrid kernel SVM yields an average accuracy rate of 96.75% across five iterations. This technique offers notable benefits over alternative fault diagnosis methods, including high accuracy and consistent performance, thereby providing a promising novel approach for existing fault diagnosis procedures;

2. Experimental results demonstrate that the combination of Poly and RBF kernel functions in the hybrid kernel SVM, optimized by the BO algorithm, can suppress mode mixing successfully. Moreover, the use of permutation entropy as the feature vector and sample entropy as the fitness value allows for a more efficient feature extraction of fault samples. Gaussian regression process is then utilized to optimize the parameters c and g of hybrid kernel SVM, leading to increased accuracy and adaptability of the model classification. Impressively, this method has achieved a 100% single fault diagnosis rate; and

3. In comparison with the alternative optimization algorithms, the BO approach presented in this study exhibits favorable performance in terms of optimization accuracy, algorithmic efficiency, and convergence. This method offers the added benefits of streamlined model training and efficient processing, thereby resulting in excellent diagnostic accuracy following training.

In summary, the experimental outcomes suggest that the proposed hybrid kernel SVM method for fault diagnosis of bearings is feasible and superior, providing a new direction for the advancement of fault diagnosis techniques in this area.

Author Contributions: Conceptualization, X.S. and M.X.; methodology, X.S.; software, X.S.; validation, W.W. and M.X.; formal analysis, G.G. and X.S.; investigation, J.Z. and X.S.; resources, M.X. and G.J.; data curation, X.S.; writing—original draft preparation, X.S. and J.Z.; writing—review and editing, G.G., G.H. and G.J.; visualization, G.H. and W.W.; supervision, M.X. and X.S.; project administration, M.X. and G.J.; funding acquisition, M.X. and G.J. All authors have read and agreed to the published version of the manuscript.

Funding: This research was supported by the Jiangsu International Science and Technology Cooperation Project (No. BZ2022002), in part by the Agricultural Science and Technology Independent Innovation Fund of Jiangsu Province (No. CX(22)3101) and in part by the National key research and development program (No. 2022YFD2001805).

Institutional Review Board Statement: Not applicable.

Informed Consent Statement: Not applicable.

Data Availability Statement: Not applicable.

Conflicts of Interest: The authors declare no conflict of interest.

References

1. He, C.; Li, H.; Zhao, X. Weak characteristic determination for blade crack of centrifugal compressors based on underdetermined blind source separation. *Measurement* **2018**, *128*, 545–557. [CrossRef]
2. Duan, L.; Ren, Y.; Duan, F. Adaptive stochastic resonance based convolutional neural network for image classification. *Chaos Solitons Fractals* **2022**, *162*, 112429. [CrossRef]
3. Wang, G.; Xiang, J. Remain useful life prediction of rolling bearings based on exponential model optimized by gradient method. *Measurement* **2021**, *176*, 109161. [CrossRef]
4. Islam, M.M.; Prosvirin, A.E.; Kim, J.-M. Data-driven prognostic scheme for rolling-element bearings using a new health index and variants of least-square support vector machines. *Mech. Syst. Signal Process.* **2021**, *160*, 107853. [CrossRef]
5. Nirwan, N.W.; Ramani, H.B. Condition monitoring and fault detection in roller bearing used in rolling mill by acoustic emission and vibration analysis. *Mater. Today Proc.* **2021**, *51*, 344–354. [CrossRef]
6. Wang, Z.; Yao, L.; Chen, G.; Ding, J. Modified multiscale weighted permutation entropy and optimized support vector machine method for rolling bearing fault diagnosis with complex signals. *ISA Trans.* **2021**, *114*, 470–484. [CrossRef]
7. Zeng, F.; Li, Y.; Jiang, Y.; Song, G. An online transfer learning-based remaining useful life prediction method of ball bearings. *Measurement* **2021**, *176*, 109201. [CrossRef]
8. Pan, H.; Xu, H.; Zheng, J.; Tong, J. Non-parallel bounded support matrix machine and its application in roller bearing fault diagnosis. *Inf. Sci.* **2023**, *624*, 395–415. [CrossRef]
9. Pan, H.; Xu, H.; Zheng, J.; Su, J.; Tong, J. Multi-class fuzzy support matrix machine for classification in roller bearing fault diagnosis. *Adv. Eng. Inform.* **2021**, *51*, 101445. [CrossRef]
10. Liang, L.; Ding, X.; Wen, H.; Liu, F. Impulsive components separation using minimum-determinant KL-divergence NMF of bi-variable map for bearing diagnosis. *Mech. Syst. Signal Process.* **2022**, *175*, 109129. [CrossRef]
11. Qin, A.-S.; Mao, H.-L.; Hu, Q. Cross-domain fault diagnosis of rolling bearing using similar features-based transfer approach. *Measurement* **2020**, *172*, 108900. [CrossRef]
12. Basha, N.; Kravaris, C.; Nounou, H.; Nounou, M. Bayesian-optimized Gaussian process-based fault classification in industrial processes. *Comput. Chem. Eng.* **2023**, *170*, 108126. [CrossRef]
13. Tang, S.; Zhu, Y.; Yuan, S. Intelligent fault diagnosis of hydraulic piston pump based on deep learning and Bayesian optimization. *ISA Trans.* **2022**, *129*, 555–563. [CrossRef]
14. Garrido-Merchán, E.C.; Fernández-Sánchez, D.; Hernández-Lobato, D. Parallel predictive entropy search for multi-objective Bayesian optimization with constraints applied to the tuning of machine learning algorithms. *Expert Syst. Appl.* **2023**, *215*, 119328. [CrossRef]
15. Yaman, O.; Yol, F.; Altinors, A. A Fault Detection Method Based on Embedded Feature Extraction and SVM Classification for UAV Motors. *Microprocess. Microsyst.* **2022**, *94*, 104683. [CrossRef]
16. Li, C.; Ledo, L.; Delgado, M.; Cerrada, M.; Pacheco, F.; Cabrera, D.; Sánchez, R.-V.; de Oliveira, J.V. A Bayesian approach to consequent parameter estimation in probabilistic fuzzy systems and its application to bearing fault classification. *Knowl. Based Syst.* **2017**, *129*, 39–60. [CrossRef]
17. Xiong, H.; Szedmak, S.; Piater, J. Scalable, accurate image annotation with joint SVMs and output kernels. *Neurocomputing* **2015**, *169*, 205–214. [CrossRef]
18. Goodfellow, I.J.; Erhan, D.; Carrier, P.L.; Courville, A.; Mirza, M.; Hamner, B.; Cukierski, W.; Tang, Y.; Thaler, D.; Lee, D.-H.; et al. Challenges in representation learning: A report on three machine learning contests. *Neural Netw.* **2015**, *64*, 59–63. [CrossRef]
19. Han, T.; Zhang, L.; Yin, Z.; Tan, A.C. Rolling bearing fault diagnosis with combined convolutional neural networks and support vector machine. *Measurement* **2021**, *177*, 109022. [CrossRef]
20. Perrone, V.; Donini, M.; Zafar, M.B.; Schmucker, R.; Kenthapadi, K.; Archambeau, C. Fair Bayesian Optimization. In Proceedings of the 2021 AAAI/ACM Conference on AI, Ethics, and Society (AIES '21), Association for Computing Machinery, New York, NY, USA, 19–21 May 2021; pp. 854–863. [CrossRef]
21. Shahriari, B.; Swersky, K.; Wang, Z.; Adams, R.P.; de Freitas, N. Taking the Human Out of the Loop: A Review of Bayesian Optimization. *Proc. IEEE* **2015**, *104*, 148–175. [CrossRef]
22. Tan, Y.; Wang, J. A support vector machine with a hybrid kernel and minimal vapnik-chervonenkis dimension. *IEEE Trans. Knowl. Data Eng.* **2004**, *16*, 385–395. [CrossRef]

23. Sangeetha, R.; Kalpana, B. A Comparative Study and Choice of an Appropriate Kernel for Support Vector Machines. In *Information and Communication Technologies. ICT 2010. Communications in Computer and Information Science*; Das, V.V., Vijaykumar, R., Eds.; Springer: Berlin/Heidelberg, Germany, 2010; Volume 101. [CrossRef]

24. Riazi, A.; Saraeian, S. Sustainable production using a hybrid IPSO optimized SVM-based technique: Fashion industry. *Sustain. Comput. Inform. Syst.* **2023**, *37*, 100838. [CrossRef]

25. Nieto, P.G.; Fernández, J.A.; Suárez, V.G.; Muñiz, C.D.; García-Gonzalo, E.; Bayón, R.M. A hybrid PSO optimized SVM-based method for predicting of the cyanotoxin content from experimental cyanobacteria concentrations in the Trasona reservoir: A case study in Northern Spain. *Appl. Math. Comput.* **2015**, *260*, 170–187. [CrossRef]

26. Wang, C.; Zhang, Y.; Song, J.; Liu, Q.; Dong, H. A novel optimized SVM algorithm based on PSO with saturation and mixed time-delays for classification of oil pipeline leak detection. *Syst. Sci. Control. Eng.* **2019**, *7*, 75–88. [CrossRef]

27. Song, H.; Ding, Z.; Guo, C.; Li, Z.; Xia, H. Research on Combination Kernel Function of Support Vector Machine. In Proceedings of the 2008 International Conference on Computer Science and Software Engineering, Wuhan, China, 12–14 December 2008; pp. 838–841. [CrossRef]

28. Wu, X.; Tang, W.; Wu, X. Support Vector Machine Based on Hybrid Kernel Function. In *Information Engineering and Applications*; Information Engineering and Applications; Lecture Notes in Electrical Engineering Book Series; Zhu, R., Ma, Y., Eds.; Springer: London, UK, 2012; Volume 154. [CrossRef]

29. Figuera, C.; Barquero-Pérez, Ó.; Rojo-Álvarez, J.L.; Martínez-Ramón, M.; Guerrero-Curieses, A.; Caamaño, A.J. Spectrally adapted Mercer kernels for support vector nonuniform interpolation. *Signal Process.* **2014**, *94*, 421–433. [CrossRef]

30. Zhou, X.; Jiang, P.; Wang, X. Recognition of control chart patterns using fuzzy SVM with a hybrid kernel function. *J. Intell. Manuf.* **2015**, *29*, 51–67. [CrossRef]

31. Zeng, Y.; Cheng, Y.; Liu, J. An efficient global optimization algorithm for expensive constrained black-box problems by reducing candidate infilling region. *Inf. Sci.* **2022**, *609*, 1641–1669. [CrossRef]

32. Snoek, J.; Rippel, O.; Swersky, K.; Kiros, R.; Satish, N.; Sundaram, N.; Patwary, M.; Prabhat, M.; Adams, R. Scalable Bayesian Optimization Using Deep Neural Networks. In Proceedings of the 32nd International Conference on Machine Learning, Lille, France, 6–11 July 2015; Volume 37, pp. 2171–2180.

33. Folch, J.P.; Lee, R.M.; Shafei, B.; Walz, D.; Tsay, C.; van der Wilk, M.; Misener, R. Combining multi-fidelity modelling and asynchronous batch Bayesian Optimization. *Comput. Chem. Eng.* **2023**, *172*, 108194. [CrossRef]

34. Anh, D.T.; Pandey, M.; Mishra, V.N.; Singh, K.K.; Ahmadi, K.; Janizadeh, S.; Tran, T.T.; Linh, N.T.T.; Dang, N.M. Assessment of groundwater potential modeling using support vector machine optimization based on Bayesian multi-objective hyperparameter algorithm. *Appl. Soft Comput.* **2023**, *132*, 109848. [CrossRef]

35. Zuo, X. Rolling Bearing Fault Diagnosis Based on Gaussian Dimensionality Reduction and Hybrid Core SVM Fusion. Master's thsis, Wuhan University of Technology, Wuhan, China, 2017.

36. Kouziokas, G.N. SVM kernel based on particle swarm optimized vector and Bayesian optimized SVM in atmospheric particulate matter forecasting. *Appl. Soft Comput.* **2020**, *93*, 106410. [CrossRef]

37. Elsayad, A.M.; Nassef, A.M.; Al-Dhaifallah, M. Bayesian optimization of multiclass SVM for efficient diagnosis of erythemato-squamous diseases. *Biomed. Signal Process. Control.* **2021**, *71*, 103223. [CrossRef]

38. He, C.; Wu, T.; Gu, R.; Jin, Z.; Ma, R.; Qu, H. Rolling bearing fault diagnosis based on composite multiscale permutation entropy and reverse cognitive fruit fly optimization algorithm—Extreme learning machine. *Measurement* **2020**, *173*, 108636. [CrossRef]

39. Zhou, J.; Xiao, M.; Niu, Y.; Ji, G. Rolling Bearing Fault Diagnosis Based on WGWOA-VMD-SVM. *Sensors* **2022**, *22*, 6281. [CrossRef] [PubMed]

40. Tek, Y.I.; Tuna, E.B.; Savaşcıhabeş, A.; Özen, A. A new PAPR and BER enhancement technique based on lifting wavelet transform and selected mapping method for the next generation waveforms. *AEU—Int. J. Electron. Commun.* **2021**, *138*, 153871. [CrossRef]

41. Pelikan, M. Bayesian Optimization Algorithm. In *Hierarchical Bayesian Optimization Algorithm: Toward a New Generation of Evolutionary Algorithms*; Springer: Berlin/Heidelberg, Germany, 2005; pp. 31–48. [CrossRef]

Article

Convolutional Neural Network-Based Transformer Fault Diagnosis Using Vibration Signals

Chao Li [1], Jie Chen [1,*], Cheng Yang [2], Jingjian Yang [1], Zhigang Liu [3] and Pooya Davari [4,*]

[1] School of Electrical Engineering, Beijing Jiaotong University, Beijing 100044, China; 19117012@bjtu.edu.cn (C.L.); 21117023@bjtu.edu.cn (J.Y.)
[2] China Institute of Marine Technology and Economy, Beijing 100081, China; yangcheng@cimtec.net.cn
[3] Beijing Rail Transit Electrical Engineering Technology Research Center, Beijing 100044, China; zhgliu@bjtu.edu.cn
[4] AAU Energy, Aalborg University, 9220 Aalborg, Denmark
* Correspondence: jiechen@bjtu.edu.cn (J.C.); pda@energy.aau.dk (P.D.)

Abstract: Fast and accurate fault diagnosis is crucial to transformer safety and cost-effectiveness. Recently, vibration analysis for transformer fault diagnosis is attracting increasing attention due to its ease of implementation and low cost, while the complex operating environment and loads of transformers also pose challenges. This study proposed a novel deep-learning-enabled method for fault diagnosis of dry-type transformers using vibration signals. An experimental setup is designed to simulate different faults and collect the corresponding vibration signals. To find out the fault information hidden in the vibration signals, the continuous wavelet transform (CWT) is applied for feature extraction, which can convert vibration signals to red-green-blue (RGB) images with the time–frequency relationship. Then, an improved convolutional neural network (CNN) model is proposed to complete the image recognition task of transformer fault diagnosis. Finally, the proposed CNN model is trained and tested with the collected data, and its optimal structure and hyperparameters are determined. The results show that the proposed intelligent diagnosis method achieves an overall accuracy of 99.95%, which is superior to other compared machine learning methods.

Keywords: fault diagnosis; vibration analysis; deep learning; convolutional neural network (CNN); power transformer

Citation: Li, C.; Chen, J.; Yang, C.; Yang, J.; Liu, Z.; Davari, P. Convolutional Neural Network-Based Transformer Fault Diagnosis Using Vibration Signals. *Sensors* **2023**, *23*, 4781. https://doi.org/10.3390/s23104781

Academic Editors: Dong Wang, Shilong Sun and Changqing Shen

Received: 18 April 2023
Revised: 10 May 2023
Accepted: 14 May 2023
Published: 16 May 2023

1. Introduction

As one of the most important and expensive piece of equipment in a power system, the power transformer plays a vital role in power conversion and delivery [1]. Power transformers are generally designed to have a lifetime of 20 to 35 years, and can actually last up to 60 years with proper maintenance [2]. However, occasional in-service faults of a transformer can cause catastrophic consequences for the power system and even endanger personal safety; moreover, it is very costly to repair or replace transformers. With the increase in operation time, under the long-term influence of mechanical stress, thermal stress, etc., more and more transformers begin to deteriorate, which brings a great potential threat to the power system and puts forward higher requirements for fault diagnosis technology. In general, transformer faults can be classified as electrical, mechanical, and thermal; how to prevent these faults and ensure a healthy working condition of the transformer is a significant topic. Traditionally, scheduled maintenance makes its plans for inspection and testing based on experience, trying to find a balance between low-risk and low-cost, which can easily result in over-maintenance or under-maintenance. Alternatively, by monitoring the characteristic parameters of a transformer in real-time, condition-based maintenance (CBM) can detect the abnormal state of the equipment and make a diagnosis at the first time, which can minimize the damage to the equipment by failure [3]. Thus, transformer condition monitoring and fault diagnosis techniques have recently attracted extensive attention from researchers and engineers.

Generally, transformer fault diagnosis methods can be classified as offline and online according to the working state of the transformer. The offline methods, due to their simple principle and accurate results, are commonly used for annual maintenance and fault analysis. For instance, frequency response analysis (FRA) can determine the condition of the winding by measuring the impedance or admittance of the winding [4–6]. Short-circuit impedance (SCI) is available to evaluate the transformer operating condition [7]. Similarly, the winding resistance measurement is used to evaluate the contact condition of the winding conductors and the tap changer, and the winding ratio test can determine if there are shorted turns or open winding circuits. However, these methods require transformer shutdown during implementation.

By contrast, the online methods can be implemented while the transformer is in operation. Dissolved gas analysis (DGA) can be used to diagnose latent transformer faults by continuously detecting and analyzing the components of different gases dissolved in the insulating oil [8,9]. Similarly, insulating oil quality (IOQ) tests can be used to analyze the condition of the transformer-insulating oil [10]. However, the above approach is only applicable to oil-immersed transformers but not to dry-type transformers. Recently, with the rapid development of sensor technology and signal processing, some non-traditional diagnostic methods are rapidly evolving, such as partial discharge (PD) testing which is utilized to detect whether the partial discharge is occurring in the transformer [11,12]. Ultra-wideband (UWB) signals are used to diagnose mechanical faults in the transformer winding [13]. In addition, the thermal imaging monitoring can detect abnormal thermal faults in a transformer [14]. Nevertheless, some of these methods are expensive or not accurate enough.

Alternatively, vibration analysis provides a new online diagnosis method for transformers with easy and low-cost implementation, which has attracted increasing attention in the recent years. The authors of [15] proved that the vibration intensity of a transformer is related to its location and load current by investigating the distribution characteristics of vibration signals. Different short-circuited turn conditions of the transformer can be recognized by classifying the indicators extracted from vibration signals using support vector machines (SVM), as reported in [16]. Similarly, using the total harmonic distortion (THD) from vibration signals as a fault feature, ref. [17] effectively diagnosed the transformer short-circuit faults. Based on vibration and reactance information, the loose state and deformation of the transformer winding can be monitored, as reported in [18]. An effective feature extraction method from transformer vibration signals was introduced in [19], which decomposed the vibrations into multiple modes using variational mode decomposition (VMD); then, they extracted the feature vector from those modes by wavelet transform. However, most of the above methods require detailed parameters or information about the transformer, which are highly dependent on the expertise and limits their development.

Recent research has shown that fault diagnosis methods with deep learning (DL) can overcome the expertise dependence issue [20]; furthermore, they can also achieve higher accuracy [21]. Typically, there are three main types in DL, which are deep belief network (DBN), recurrent neural network (RNN), and CNN. Since the problem of gradient extinction has been solved and the performance of the graphics processing unit (GPU) has improved, DL has made remarkable progress, especially in the fields of speech recognition [22], image recognition [23], and automatic driving [24]. Meanwhile, some achievements have also been made in transformer fault diagnosis with DL. For instance, RNN was adopted in [3] to capture the hidden patterns of vibration time series directly, which can diagnose the abnormal excitation voltage and turn-to-turn short-circuit faults of the transformer. The authors of [25] recognized converted vibrating images using CNN to identify three working conditions of transformers. Similarly, a multi-scale fusion feature extraction model based on CNN with attention mechanism was designed in [26], which can recognize the operating conditions of the transformer with different voltages and loads. However, the types of faults they can identify are relatively limited; also, most of the current research has focused on oil-immersed transformers, while little research has been done on dry-type transformers. Therefore, it needs further research on how to quickly and effectively implement online multiple fault diagnosis for dry-type transformers.

The main contributions of this study are summarized in the following.

(1)   An intelligent fault diagnosis method for dry-type transformers using vibration signals is proposed, which can quickly identify different faults under various loads of the transformer with high accuracy.

(2)   A CWT method is adopted to convert the raw vibration signals of the transformer to RGB images, which could adequately extract fault features from the different conditions.

(3)   An improved CNN model is designed to accurately classify the RGB images for transformer fault diagnosis, and its optimal structure and parameters are determined.

The rest of this article is organized as follows. Section 2 introduces the theoretical background. Section 3 describes the experimental setup and data. Section 4 presents the proposed method in detail, including the feature extraction and proposed CNN structure. In Section 5, experimental and test results are presented to validate the performance of the proposed method. Finally, the conclusion is drawn in Section 6.

2. Theoretical Background

2.1. Mechanism of Transformer Vibration

The transformer vibrates all the time in service with or without load, and the vibrations are mainly caused by core vibration and winding vibration. Core vibrations are mainly generated by magnetostriction since the geometry of magnetic material changes slightly when it is in a magnetic field, and the vibration occurs when the strength of the magnetic field varies considerably [16]. The fundamental frequency of the core vibration is twice the source. It should be noted that the core vibration will also contain high-frequency harmonics because of the nonlinear property of magnetostriction. The amplitude of core vibrations is basically proportional to the voltage squared, which can be represented by

$$\alpha_{core} \propto U^2, \tag{1}$$

where α_{core} is the amplitude of core vibrations, U is the voltage.

The winding vibrations are mainly generated by electromagnetic forces due to the interaction between the current in winding and the leakage flux field. Those electromagnetic forces are proportional to the current squared [15]; since the current waveform is practically sinusoidal, the fundamental frequency of the winding vibration is 100 Hz (in the case of a 50 Hz grid). The amplitude of winding vibration is basically proportional to the current squared, which can be represented by

$$\alpha_{winding} \propto I^2, \tag{2}$$

where $\alpha_{winding}$ is the amplitude of winding vibrations, I is the current.

The vibration of a transformer is highly correlated with its condition [27]; therefore, the vibration is employed in transformer fault diagnosis as a fault feature in this study.

2.2. Wavelet Transform

Wavelet transform is a popular tool for extracting time–frequency information from time-domain signals [28]. It inherits and develops the localization idea of short-time Fourier transform (STFT), and overcomes its shortcomings of a non-changing window size with frequency [29]. The wavelet transform can provide a "time–frequency" window that changes with frequency. Then, the time subdivision at high frequency and frequency subdivision at low frequency can be realized. There are two main types of the wavelet transform, CWT [30] and discrete wavelet transform (DWT) [31]. The difference between them is that CWT operates on all possible combinations of shifting and compression, while the DWT only operates on a specific subset of shifting and compression.

CWT is defined by the wavelet coefficients which are produced by the convolution of the original signal $x(t)$ with the mother wavelet function $\psi(t)$. Through the translation (shift in time) and dilation (compression in time) by the mother wavelet function $\psi(t)$, a multi-scale refinement of the original signal $x(t)$ is gradually carried out. The transformation process can be described by

$$WC(a,b) = \frac{1}{\sqrt{|a|}} \int_{-\infty}^{\infty} x(t)\psi^* \left(\frac{t-b}{a} \right) dt, \tag{3}$$

where WC is the wavelet coefficient, a is the scale of the mother wavelet, and b is the translation of the mother wavelet. DWT can transform the discrete input data sequence $f = \{f_n\} = \{f_0, f_1, \ldots, f_{N-1}\}$ to a vector matrix form as

$$\alpha = \mathbf{W}f, \tag{4}$$

where α is composed of N wavelet coefficients, and $\mathbf{W}$ is an orthogonal matrix.

Wavelet decomposition is implemented through two filters: the low-pass filter (scaling filter) and the high-pass filter (wavelet filter) [32]. They share the same set of wavelet filter coefficients, but with alternating signs and in reversed order, which means they complement each other. After the signal down-sampling operation for each decomposition level, the signal reconstruction process is done by applying the inverse way to the decomposition process. Each reconstruction level is followed by a signal up-sampling operation, which is known as the Mallat algorithm, and the procedure is illustrated in Figure 1.

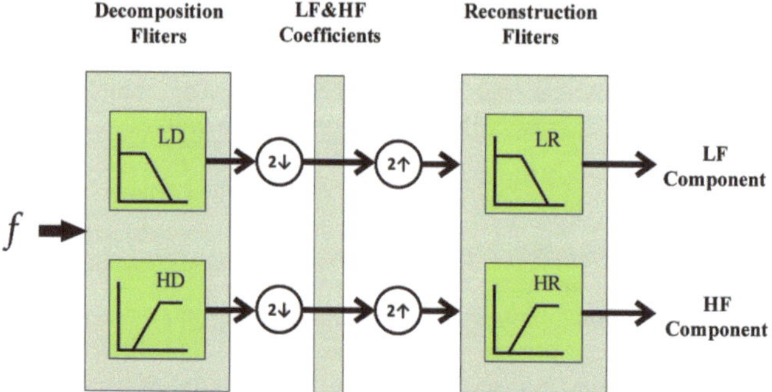

Figure 1. Mallat algorithm of wavelet decomposition and reconstruction.

2.3. CNN

CNN is a typical deep learning algorithm, inspired by the concept of the visual nervous system [33], which can reduce image dimensionality and improve the efficiency and accuracy of image processing. It has made great achievements in computer vision [34], natural language processing [35], etc.

The typical CNN structure consists of three types of layers, which are the convolutional layer, pooling layer, and fully connected layer. The process of pooling operation is illustrated in Figure 2. According to task requirements, these layers are combined in different ways to form different CNN models, such as LeNet-5 [36], ResNet [37], EfficientNet [38], and 1-D CNN [39].

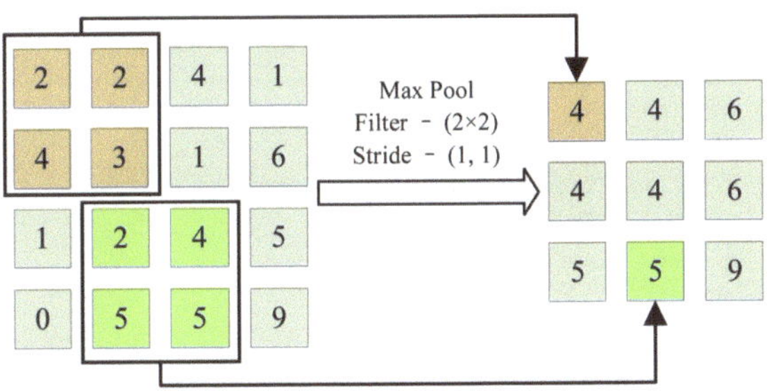

Figure 2. Process of the pooling operation.

3. Experimental Setup and Data

3.1. Experimental Setup

The transformer under study is a customized 50 kVA dry-type transformer with two terminals A and B, which can easily simulate turn-to-turn short circuit faults. Its main parameters are shown in Table 1. The output terminal of the transformer was connected to an adjustable load cabinet, whose power ranges from 0 to 200 kW.

Two accelerometers with the sensitivity of 500 mV/g of type CA-YD-188T were used to collect vibration signals of the transformer. Then, the collected raw signals are processed by the SIRIUSm-4xACC data acquisition instrument with a sampling rate of 8000 Hz, and saved by the Devesoft X3 software. Considering the structural characteristics and insulation safety of the studied transformer, as shown in Figure 3, the above accelerometers were fixed in the vertical direction (CH1) and horizontal direction (CH2) of the core clamp, respectively. The whole experimental system is shown in Figure 4.

The loosening faults of the core, winding, and connection bar were simulated by adjusting the tightness of the clamp bolts from 50 to 80 Nm using a torque wrench, the turn-to-turn short circuit fault was simulated by connecting a resistor between terminals A and B. It is worth mentioning that all fault types have multiple load levels to represent changing loads.

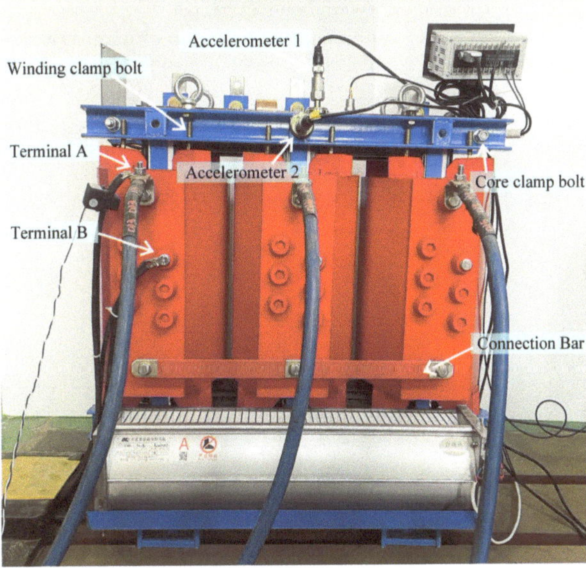

Figure 3. Position of the accelerometer on the studied transformer.

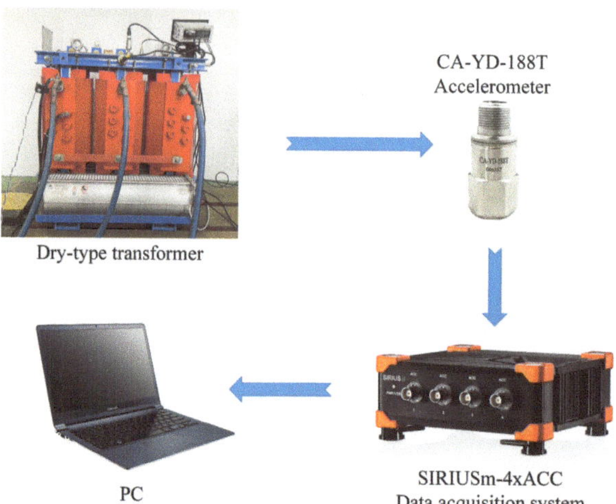

Figure 4. Experimental system of transformer fault diagnosis.

Table 1. Main parameters of the studied transformer.

Categories	Parameters
Rated power	50 kVA
Rated frequency	50 Hz
Type of cooling	air natural cooling
Service condition	Indoor
Host weight	330 kg
Shape size	$740 \times 460 \times 790$ mm
Rated voltage (primary)	10 kV
Rated voltage (secondary)	0.4 kV

3.2. Data Description and Preprocessing

As shown in Table 2, there are four different transformer faults, respectively, core clamp looseness (CC), winding clamp looseness (WC), connection bar looseness (CB), and turn-to-turn short circuit (TT), which were simulated in this study. Meanwhile, two different load levels are applied for each fault, along with the normal state (NO), and a total of 10 different working conditions are obtained.

Table 2. Working states of the studied transformer.

Working States	Loads (kW)	Categories
Normal state	20	NO20
	40	NO40
Core clamp looseness	20	CC20
	40	CC40
Winding clamp looseness	20	WC20
	40	WC40
Connection bar looseness	20	CB20
	40	CB40
Turn-to-turn short circuit	20	TT20
	40	TT40

In order to train the proposed diagnosis model, 400 segments of the vibration signal were collected for each working condition, which eventually constituted a total dataset of

4000 samples, of which 70% were selected as the training dataset, 20% as the validation dataset, and the remaining 10% as the test dataset. It should be noted that each sample can only be assigned to one dataset, which means that the samples of the testing dataset are completely different from the training dataset and validation dataset.

Figure 5 illustrates the converted RGB image of the normal state with load of 20 kW (NO20), and the remaining 9 cases are shown in Figure 6. It is obvious that the RGB pictures of different conditions have unique features in both the time domain and frequency domain, which demonstrates that the proposed feature extraction method works effectively.

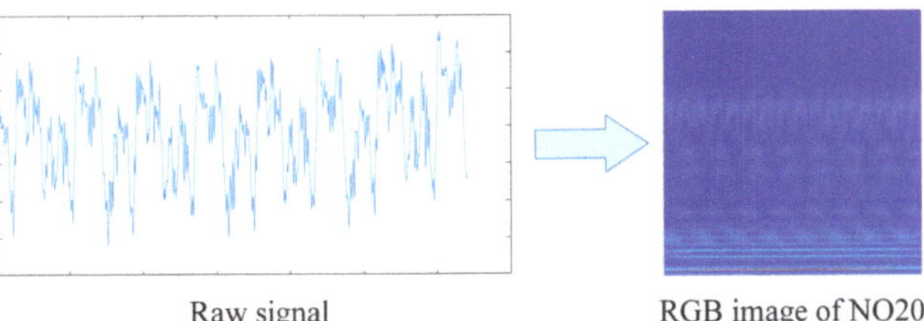

Figure 5. CWT conversion image of the normal state.

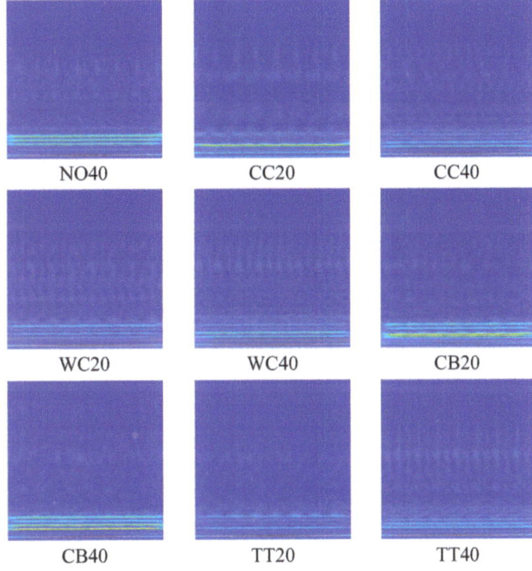

Figure 6. Converted RGB images of nine conditions.

4. Proposed Fault Diagnosis Method

The proposed transformer fault diagnosis method is presented in this section. After the vibration signals are acquired from the transformer, they are converted into RGB images by the CWT method described in Section 2.2. Then, the RGB images are classified by the proposed diagnosis model.

4.1. Feature Extraction

Vibration signals are collected by the high-frequency accelerometers. In order to fully collect transformer vibration characteristics, the sampling rate is usually around 10 kHz. The collected time-domain signals contain rich characteristic information; however, it can hardly be used directly for fault diagnosis. Therefore, a proper feature extraction method is essential.

For the purpose of extracting sufficient feature information from the original vibration signal, CWT is used to process the vibration signal in this study. The length of the selected raw signal segment is 1280 (i.e., 160 ms), and the cmor3-3 (Morlet wavelet) is employed as the mother wavelet with a total scale of 256. It is worth mentioning that the sampling rate is set to 8000 Hz since the vibration frequency of the transformer in this case is basically below 4000 Hz. As shown in Figure 7, the time-domain vibration signals is converted to RGB images after translation and dilation by the mother wavelet. Meanwhile, the images are labeled and proportionally divided into training, validation, and testing datasets.

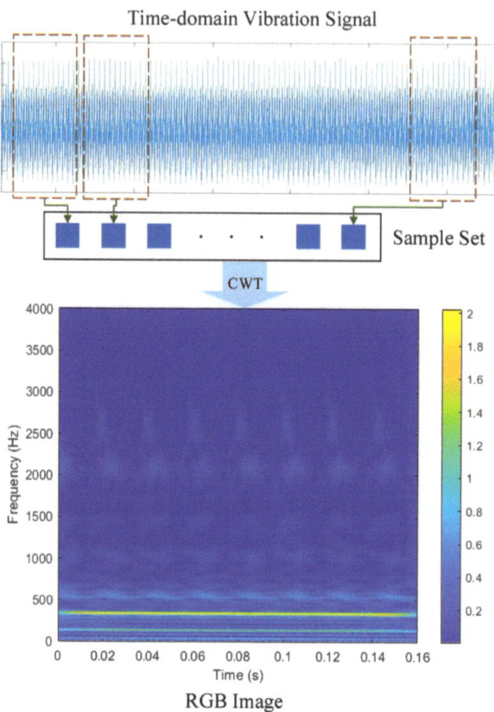

Figure 7. Feature extraction procedure.

4.2. Proposed CNN Structure

After converting the raw signals to RGB images, there are n classes of images corresponding to n transformer working conditions. The RGB image can be divided into 3 monochrome layers to meet the requirements of the input format. In order to improve the accuracy of image recognition, the input size of proposed model is set to 64 × 64 in this study.

Based on experience and comparison, the proposed CNN structure was finally determined as shown in Figure 8. There are two alternating convolutional and pooling layers in the proposed CNN structure. The size of the convolution kernels (filter) in the first and second convolutional layers is 6@5 × 5 and 16@5 × 5, respectively, which determines the number and dimensionality of the feature maps. The process of pooling operation can reduce the size of the image by selecting the dominant pixels on the feature map, and the kernel size of both pooling layers is 2 × 2. Meanwhile, to fully capture the features of the images and control the size of feature maps, in this study, the strides of convolutional kernels and pooling kernels are set to 1 and 2, respectively. In addition, three successive fully connected layers are designed to calculate the final feature information by converting the pooled feature maps to the 1-D vector. Eventually, the image classification is implemented by a softmax process.

Some other initial hyperparameters of the structure are set as follows: learning rate = 0.015, batch size = 12. The optimal combination of the above parameters will be discussed in Section 4. Finally, the flowchart of the proposed method is shown in Figure 9.

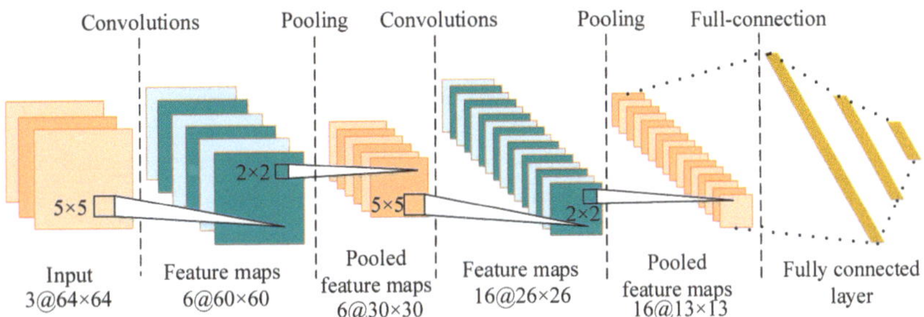

Figure 8. The structure of the proposed diagnosis model.

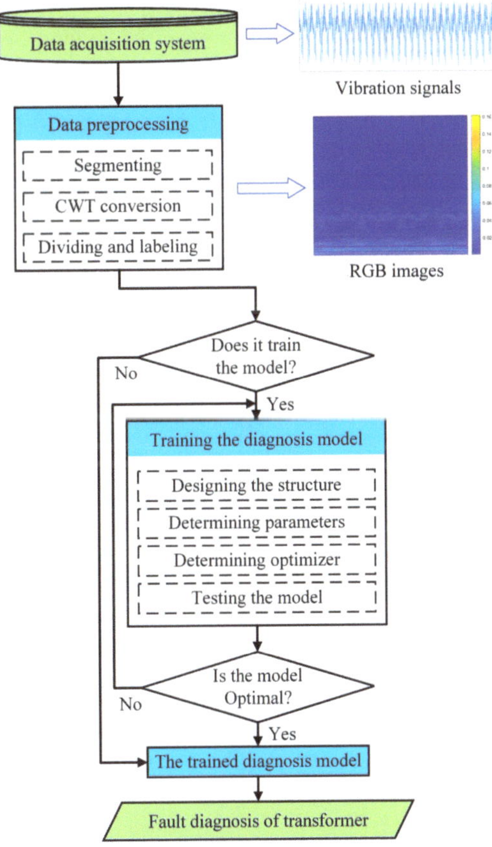

Figure 9. Flowchart of the proposed diagnosis method.

5. Experimental Verification and Discussion

In this section, an experimental setup was designed to simulate different faults, and the corresponding vibration signals were collected to train and test the proposed diagnosis model. Moreover, the performances of different parameters in the proposed model were compared to select the optimal combination. The CNN model is written in Python 3.7 with PyTorch and runs on windows 10 with two Nvidia RTX 2080Ti GPUs.

5.1. Comparison of Different Structures

The structure of the proposed model has a crucial impact on diagnosis accuracy. In order to find the best combination of structures, the performances of different structures were compared, and the results are shown in Table 3, where CNN-*x*-*y*-*z* means that there are *x*, *y*, and *z* neurons in the first, second, and third fully connected layer, respectively. For example, CNN-2704-126 means that there are 2704 neurons in the first layer, 126 neurons in the second layer, and there is no third layer in this structure.

Each model was run ten times, and the maximum, minimum, mean, and standard deviation (SD) of the testing accuracy were employed as criteria to evaluate the performance of diagnostic models. From the results shown in Table 3, it can be concluded that the model of CNN-2704-126-64 achieves the best performance on CH2. Its maximum, minimum, mean, and SD of testing accuracy are 100%, 97.5%, 98%, and 1.96%, respectively. All of those criteria are superior to the other structures compared. It should be noted that all six models performed better on CH2 than CH1, which indicates that the horizontal component of the transformer vibration signal contains richer fault characteristics than the vertical component in this study.

Figure 10 shows the training process of CNN-2704-126-64. It can be seen that when the epoch was around 70, the accuracy of the training dataset is close to 100%, and the training loss is minimized accordingly, which indicates that the structure has good fitting performance.

Table 3. Result of CNN models with different structures.

Structures	Testing Accuracy (%)							
	Max		**Min**		**Mean**		**SD**	
	CH1	CH2	CH1	CH2	CH1	CH2	CH1	CH2
CNN-2704-126	96.5	97.5	58.5	63	93.95	95.3	12.31	6.30
CNN-2704-256	95	98	65.5	87	92.3	94.15	14.92	9.11
CNN-2704-126-32	100	99.5	84	79.5	94.55	96.35	4.81	4.39
CNN-2704-126-64	**99**	**100**	**95.5**	**97.5**	**95.15**	**98**	**2.94**	**1.96**
CNN-2704-126-128	100	100	87.5	93.5	93.85	95.3	5.19	3.03

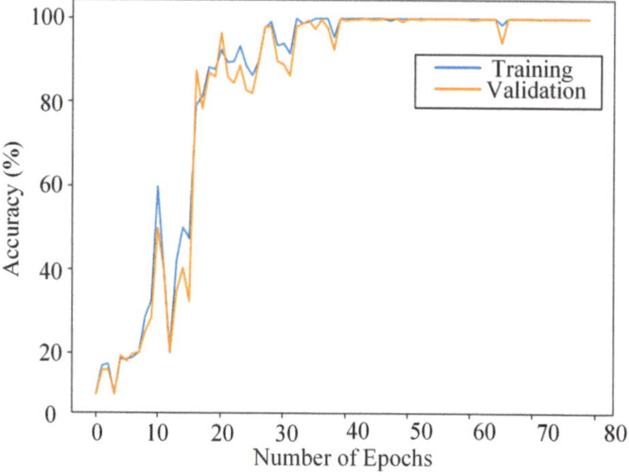

Figure 10. Training process of the proposed structure.

5.2. Comparison of Different Hyperparameters

The batch size (BS) is one of the most important hyperparameters in deep learning, which represents the number of samples picked for a training session. It affects the degree of model optimization as well as the speed of optimization by changing the GPU memory usage. In order to select the most suitable BS, the diagnosis performances of different BS are compared, which are shown in Figure 11. The results show that the model achieves the best performance when BS = 20; its maximum, minimum, mean, and SD of testing accuracy are 100%, 97%, 99.2%, and 0.95%, respectively.

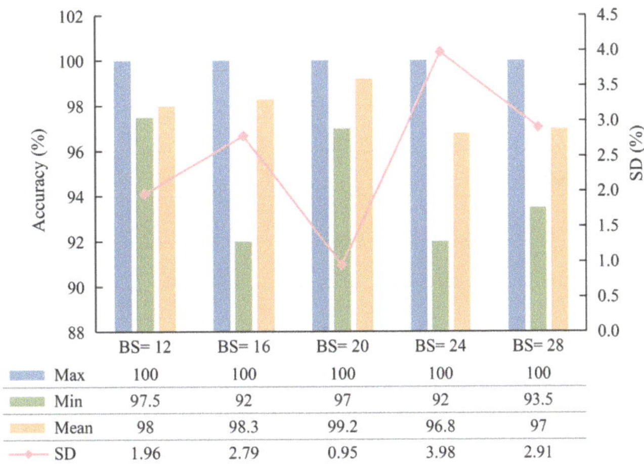

	BS= 12	BS= 16	BS= 20	BS= 24	BS= 28
Max	100	100	100	100	100
Min	97.5	92	97	92	93.5
Mean	98	98.3	99.2	96.8	97
SD	1.96	2.79	0.95	3.98	2.91

Figure 11. Diagnosis result of different batch sizes.

The learning rate (LR) determines whether and when the objective function can converge to a local minimum. A suitable LR can make the objective function converge fast and efficiently. To this end, the diagnostic performances of different LR are compared, and the results are shown in Figure 12, from which it can be seen that the best performance with a mean accuracy of 99.95% is achieved when LR = 0.02. In addition, it has a low SD of 0.32%, which indicates that the proposed parameter combination has very stable performance.

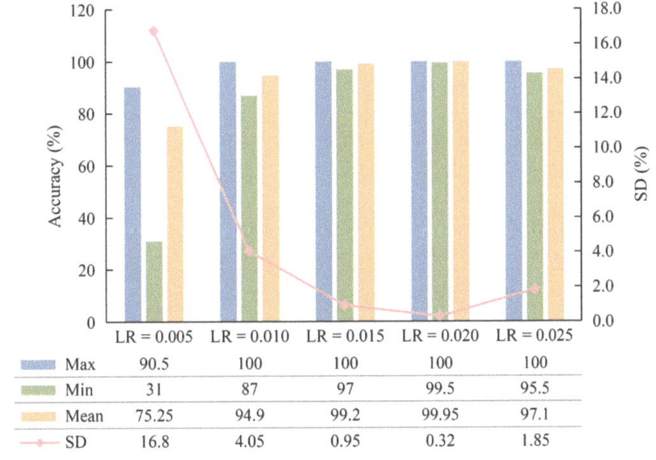

	LR = 0.005	LR = 0.010	LR = 0.015	LR = 0.020	LR = 0.025
Max	90.5	100	100	100	100
Min	31	87	97	99.5	95.5
Mean	75.25	94.9	99.2	99.95	97.1
SD	16.8	4.05	0.95	0.32	1.85

Figure 12. Diagnosis result of different learning rates.

Based on the above comparison and analysis, the hyperparameters of the proposed diagnosis model are finally determined as BS = 20 and LR = 0.02. The confusion matrix of diagnosis results is illustrated in Figure 13, where the columns represent prediction labels and the rows represent actual labels, and the intersection of them represents that the predicted conditions are consistent with the actual conditions. As shown in Figure 13, all the 400 testing samples, divided into 10 conditions, are matched with an accuracy rate of 100%, which demonstrates that the proposed method is quite effective in transformer fault diagnosis.

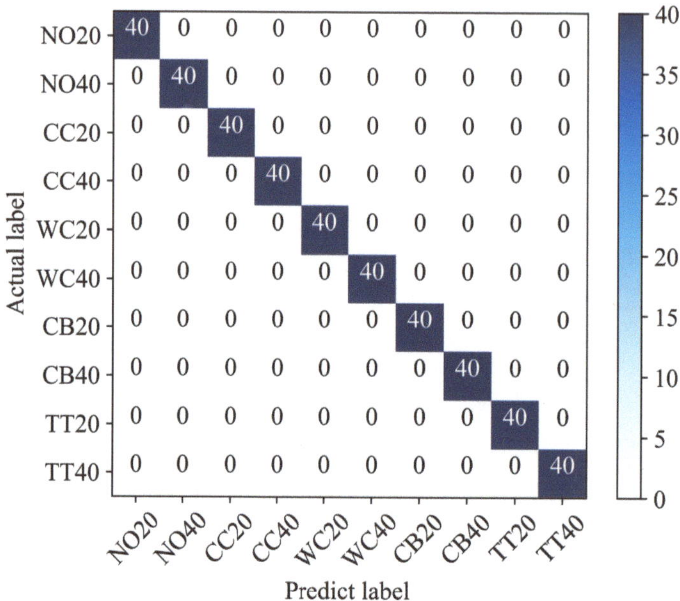

Figure 13. Confusion matrix of the proposed method.

5.3. Verification of Superiority

To verify the superiority of the proposed diagnosis method in this study, the performances of different methods are compared, including ANN [40], DBN [41], 1D-CNN, Hilbert–Huang Transform (HHT)-CNN, short-time Fourier transform (STFT)-CNN, and CWT-CNN. It is worth mentioning that the vibration signals used in all methods are collected by CH2, and each method was run ten times. The results are shown in Table 4. It can be seen that the proposed CWT-CNN method achieves the best performance, and the maximum, minimum, mean, and SD of its prediction accuracy are 100%, 99.5%, 99.95%, and 0.32%, respectively. Compared with other methods, CWT-CNN can perform better feature extraction and identification from the raw vibration signal in this study.

Table 4. Diagnosis performance of different methods.

Methods	Testing Accuracy (%)			
	Max	Min	Mean	SD
ANN	84.5	55.5	71.73	9.25
DBN	87.5	68	82.1	8.9
1D-CNN	92.5	84.5	91.52	5.47
HHT-CNN	95.5	89	93.25	2.84
STFT-CNN	95	87.5	94.14	3.93
CWT-CNN	**100**	**99.5**	**99.95**	**0.32**

6. Conclusions

This study proposed a deep learning-based fault diagnosis method for transformers, which converted vibration signals into RGB images to extract the corresponding fault features using CWT and then achieved fault diagnosis through an improved CNN model. In order to train and validate the proposed model, an experimental setup was designed to simulate transformer faults, including core clamp looseness, winding clamp looseness, connection bar looseness, and turn-to-turn short circuit. The optimal structural and hyperparameters of the proposed model were determined by comparing their diagnostic performances. Compared with other methods, the proposed diagnosis method can achieve the highest mean accuracy of 99.95% and the lowest SD of 0.32%. Moreover, due to the offline training strategy, the feature extraction and diagnosis process took less than 7 s, which can provide fast and accurate online fault diagnosis for the transformer. This study can expand the field of transformer fault diagnosis and offer technical support for condition-based maintenance of operating transformers.

Author Contributions: Conceptualization, C.L. and J.C.; methodology, C.L. and P.D.; software, C.L. and J.Y.; validation, C.L. and J.Y.; formal analysis, C.L.; investigation, C.L.; resources, C.L.; data curation, C.L. and C.Y.; writing—original draft preparation, C.L.; writing—review and editing, C.L. and P.D.; visualization, C.L.; supervision, J.C. and P.D.; project administration, C.L. and J.C.; funding acquisition, Z.L. and J.C. All authors have read and agreed to the published version of the manuscript.

Funding: This research was funded by the Fundamental Research Funds for the Central Universities (2018JBZ004).

Institutional Review Board Statement: Not applicable.

Informed Consent Statement: Not applicable.

Data Availability Statement: All data and codes are available at https://github.com/ldcrchao/chao.

Conflicts of Interest: The authors declare no conflicts of interest.

References

1. Tightiz, L.; Nasab, M.A.; Yang, H.; Addeh, A. An intelligent system based on optimized ANFIS and association rules for power transformer fault diagnosis. *ISA Trans.* **2020**, *103*, 63–74. [CrossRef] [PubMed]
2. Wang, M.; Vandermaar, A.J.; Srivastava, K.D. Review of condition assessment of power transformers in service. *IEEE Electr. Insul. Mag.* **2002**, *18*, 12–25. [CrossRef]
3. Zollanvari, A.; Kunanbayev, K.; Akhavan Bitaghsir, S.; Bagheri, M. Transformer Fault Prognosis Using Deep Recurrent Neural Network over Vibration Signals. *IEEE Trans. Instrum. Meas.* **2020**, *70*, 1–11. [CrossRef]
4. Akhmetov, Y.; Nurmanova, V.; Bagheri, M.; Zollanvari, A.; Gharehpetian, G.B. A new diagnostic technique for reliable decision-making on transformer FRA data in interturn short-circuit condition. *IEEE Trans. Ind. Inform.* **2020**, *17*, 3020–3031. [CrossRef]
5. Wu, Z.; Zhou, L.; Wang, D.; Zhou, M.; Jiang, F.; Yu, X.; Tang, H.; Zhao, H. Feature Analysis of Oscillating Wave Signal for Axial Displacement in Autotransformer. *IEEE Trans. Instrum. Meas.* **2021**, *70*, 1–13. [CrossRef]
6. Abbasi, A.R.; Mahmoudi, M.R.; Arefi, M.M. Transformer Winding Faults Detection Based on Time Series Analysis. *IEEE Trans. Instrum. Meas.* **2021**, *70*, 1–10. [CrossRef]
7. Ye, Z.; Yu, W.; Gou, J.; Tan, K.; Zeng, W.; An, B.; Li, Y. A Calculation Method to Adjust the Short-Circuit Impedance of a Transformer. *IEEE Access* **2020**, *8*, 223848–223858. [CrossRef]
8. Wang, L.; Littler, T.; Liu, X. Gaussian Process Multi-Class Classification for Transformer Fault Diagnosis Using Dissolved Gas Analysis. *IEEE Trans. Dielectr. Electr. Insul.* **2021**, *28*, 1703–1712. [CrossRef]
9. Ma, X.; Hu, H.; Shang, Y. A New Method for Transformer Fault Prediction Based on Multifeature Enhancement and Refined Long Short-Term Memory. *IEEE Trans. Instrum. Meas.* **2021**, *70*, 1–11. [CrossRef]
10. Soni, R.; Chakrabarti, P.; Leonowicz, Z.; Jasiński, M.; Wieczorek, K.; Bolshev, V. Estimation of Life Cycle of Distribution Transformer in Context to Furan Content Formation, Pollution Index, and Dielectric Strength. *IEEE Access* **2021**, *9*, 37456–37465. [CrossRef]
11. Gao, C.; Yu, L.; Xu, Y.; Wang, W.; Wang, S.; Wang, P. Partial discharge localization inside transformer windings via fiber-optic acoustic sensor array. *IEEE Trans. Power Deliv.* **2019**, *34*, 1251–1260. [CrossRef]
12. Sharifinia, S.; Allahbakhshi, M.; Ghanbari, T.; Akbari, A.; Mirzaei, H.R. A New Application of Rogowski Coil Sensor for Partial Discharge Localization in Power Transformers. *IEEE Sens. J.* **2021**, *21*, 10743–10751. [CrossRef]

13. Alehosseini, A.; Hejazi, M.A.; Mokhtari, G.; Gharehpetian, G.B.; Mohammadi, M. Detection and classification of transformer winding mechanical faults using UWB sensors and Bayesian classifier. *Int. J. Emerg. Electr. Power Syst.* **2015**, *16*, 207–215. [CrossRef]
14. Mariprasath, T.; Kirubakaran, V. A real time study on condition monitoring of distribution transformer using thermal imager. *Infrared Phys. Technol.* **2018**, *90*, 78–86. [CrossRef]
15. Jiang, P.; Zhang, Z.; Dong, Z.; Wu, Y.; Xiao, R.; Deng, J.; Pan, Z. Research on distribution characteristics of vibration signals of ±500 kV HVDC converter transformer winding based on load test. *Int. J. Electr. Power Energy Syst.* **2021**, *132*, 107200–107210. [CrossRef]
16. Huerta-Rosales, J.R.; Granados-Lieberman, D.; Garcia-Perez, A.; Camarena-Martinez, D.; Amezquita-Sanchez, J.P.; Valtierra-Rodriguez, M. Short-circuited turn fault diagnosis in transformers by using vibration signals, statistical time features, and support vector machines on fpga. *Sensors* **2021**, *21*, 3598–3626. [CrossRef]
17. Bagheri, M.; Nezhivenko, S.; Naderi, M.S.; Zollanvari, A. A new vibration analysis approach for transformer fault prognosis over cloud environment. *Int. J. Electr. Power Energy Syst.* **2018**, *100*, 104–116. [CrossRef]
18. Cao, C.; Xu, B.; Li, X. Monitoring Method on Loosened State and Deformational Fault of Transformer Winding Based on Vibration and Reactance Information. *IEEE Access* **2020**, *8*, 215479–215492. [CrossRef]
19. Hong, K.; Wang, L.; Xu, S. A Variational Mode Decomposition Approach for Degradation Assessment of Power Transformer Windings. *IEEE Trans. Instrum. Meas.* **2019**, *68*, 1221–1229. [CrossRef]
20. Xie, T.; Huang, X.; Choi, S.K. Intelligent Mechanical Fault Diagnosis Using Multisensor Fusion and Convolution Neural Network. *IEEE Trans. Ind. Inform.* **2022**, *18*, 3213–3223. [CrossRef]
21. Saufi, S.R.; Ahmad, Z.A.B.; Leong, M.S.; Lim, M.H. Gearbox Fault Diagnosis Using a Deep Learning Model with Limited Data Sample. *IEEE Trans. Ind. Inform.* **2020**, *16*, 6263–6271. [CrossRef]
22. Zhang, Z.; Geiger, J.; Pohjalainen, J.; Mousa, A.E.D.; Jin, W.; Schuller, B. Deep learning for environmentally robust speech recognition: An overview of recent developments. *ACM Trans. Intell. Syst. Technol.* **2018**, *9*, 1–28.
23. Jiang, F.; Lu, Y.; Chen, Y.; Cai, D.; Li, G. Image recognition of four rice leaf diseases based on deep learning and support vector machine. *Comput. Electron. Agric.* **2020**, *179*, 105824–105832. [CrossRef]
24. Rastgoo, M.N.; Nakisa, B.; Maire, F.; Rakotonirainy, A.; Chandran, V. Automatic driver stress level classification using multimodal deep learning. *Expert Syst. Appl.* **2019**, *138*, 112793–112803. [CrossRef]
25. Hong, K.; Jin, M.; Huang, H. Transformer winding fault diagnosis using vibration image and deep learning. *IEEE Trans. Power Deliv.* **2021**, *36*, 676–685. [CrossRef]
26. Xiao, R.; Zhang, Z.; Wu, Y.; Jiang, P.; Deng, J. Multi-scale information fusion model for feature extraction of converter transformer vibration signal. *Meas. J. Int. Meas. Confed.* **2021**, *180*, 109555–109566. [CrossRef]
27. Arroyo, A.; Martinez, R.; Manana, M.; Pigazo, A.; Minguez, R. Detection of ferroresonance occurrence in inductive voltage transformers through vibration analysis. *Int. J. Electr. Power Energy Syst.* **2019**, *106*, 294–300. [CrossRef]
28. Chen, B.; Shen, B.; Chen, F.; Tian, H.; Xiao, W.; Zhang, F.; Zhao, C. Fault diagnosis method based on integration of RSSD and wavelet transform to rolling bearing. *Meas. J. Int. Meas. Confed.* **2019**, *131*, 400–411. [CrossRef]
29. Gao, J.; Wang, B.; Wang, Z.; Wang, Y.; Kong, F. A wavelet transform-based image segmentation method. *Optik* **2020**, *208*, 164123–164130. [CrossRef]
30. Mojahed, A.; Bergman, L.A.; Vakakis, A.F. New inverse wavelet transform method with broad application in dynamics. *Mech. Syst. Signal Process.* **2021**, *156*, 107691–107712. [CrossRef]
31. Chen, R.; Huang, X.; Yang, L.; Xu, X.; Zhang, X.; Zhang, Y. Intelligent fault diagnosis method of planetary gearboxes based on convolution neural network and discrete wavelet transform. *Comput. Ind.* **2019**, *106*, 48–59. [CrossRef]
32. Guo, M.F.; Yang, N.C.; You, L.X. Wavelet-transform based early detection method for short-circuit faults in power distribution networks. *Int. J. Electr. Power Energy Syst.* **2018**, *99*, 706–721. [CrossRef]
33. Li, Z.; Liu, F.; Yang, W.; Peng, S.; Zhou, J. A Survey of Convolutional Neural Networks: Analysis, Applications, and Prospects. *IEEE Trans. Neural Netw. Learn. Syst.* **2021**, *33*, 6999–7019.
34. Yang, R.; Singh, S.K.; Tavakkoli, M.; Amiri, N.; Yang, Y.; Karami, M.A.; Rai, R. CNN-LSTM deep learning architecture for computer vision-based modal frequency detection. *Mech. Syst. Signal Process.* **2020**, *144*, 106885–106902. [CrossRef]
35. Liu, J.; Yang, Y.; Lv, S.; Wang, J.; Chen, H. Attention-based BiGRU-CNN for Chinese question classification. *J. Ambient Intell. Humaniz. Comput.* **2019**, 1–12. [CrossRef]
36. LeCun, Y. LeNet-5, Convolutional Neural Networks. 2015; Volume 20, p. 14. Available online: http://yann.lecun.com/exdb/lenet (accessed on 17 April 2023).
37. He, K.; Zhang, X.; Ren, S.; Sun, J. Identity mappings in deep residual networks. In Proceedings of the Computer Vision–ECCV 2016: 14th European Conference, Amsterdam, The Netherlands, 11–14 October 2016; Proceedings, Part IV 14; Springer: Berlin/Heidelberg, Germany, 2016; pp. 630–645.
38. Tan, M.; Le, Q. Efficientnet: Rethinking model scaling for convolutional neural networks. In Proceedings of the 36th International Conference on Machine Learning (ICML), Long Beach, CA, USA, 10–15 June 2019; pp. 6105–6114.

39. Zhao, B.; Zhang, X.; Li, H.; Yang, Z. Intelligent fault diagnosis of rolling bearings based on normalized CNN considering data imbalance and variable working conditions. *Knowl.-Based Syst.* **2020**, *199*, 105971–105986. [CrossRef]
40. Ben Ali, J.; Fnaiech, N.; Saidi, L.; Chebel-Morello, B.; Fnaiech, F. Application of empirical mode decomposition and artificial neural network for automatic bearing fault diagnosis based on vibration signals. *Appl. Acoust.* **2015**, *89*, 16–27. [CrossRef]
41. Shao, H.; Jiang, H.; Zhang, X.; Niu, M. Rolling bearing fault diagnosis using an optimization deep belief network. *Meas. Sci. Technol.* **2015**, *26*, 115002. [CrossRef]

Review

Progress in Active Infrared Imaging for Defect Detection in the Renewable and Electronic Industries

Xinfeng Zhao [1], Yangjing Zhao [2], Shunchang Hu [2,3], Hongyan Wang [2,3], Yuyan Zhang [2,*] and Wuyi Ming [2,3,*]

[1] College of Water Conservancy Engineering, Yellow River Conservancy Technical Institute, Kaifeng 475000, China; zhaoxinfeng@yrcti.edu.cn
[2] Henan Key Laboratory of Intelligent Manufacturing of Mechanical Equipment, Zhengzhou University of Light Industry, Zhengzhou 450002, China; zhaoyj0529@163.com (Y.Z.); hushunchang2022@gmail.com (S.H.); hongyanwang923@163.com (H.W.)
[3] Guangdong Provincial Key Laboratory of Digital Manufacturing Equipment, Guangdong HUST Industrial Technology Research Institute, Dongguan 523808, China
* Correspondence: 2020022@zzuli.edu.cn (Y.Z.); mingwuyi@zzuli.edu.cn (W.M.)

Abstract: In recent years, infrared thermographic (IRT) technology has experienced notable advancements and found widespread applications in various fields, such as renewable industry, electronic industry, construction, aviation, and healthcare. IRT technology is used for defect detection due to its non-contact, efficient, and high-resolution methods, which enhance product quality and reliability. This review offers an overview of active IRT principles. It comprehensively examines four categories based on the type of heat sources employed: pulsed thermography (PT), lock-in thermography (LT), ultrasonically stimulated vibration thermography (UVT), and eddy current thermography (ECT). Furthermore, the review explores the application of IRT imaging in the renewable energy sector, with a specific focus on the photovoltaic (PV) industry. The integration of IRT imaging and deep learning techniques presents an efficient and highly accurate solution for detecting defects in PV panels, playing a critical role in monitoring and maintaining PV energy systems. In addition, the application of infrared thermal imaging technology in electronic industry is reviewed. In the development and manufacturing of electronic products, IRT imaging is used to assess the performance and thermal characteristics of circuit boards. It aids in detecting potential material and manufacturing defects, ensuring product quality. Furthermore, the research discusses algorithmic detection for PV panels, the excitation sources used in electronic industry inspections, and infrared wavelengths. Finally, the review analyzes the advantages and challenges of IRT imaging concerning excitation sources, the PV industry, the electronics industry, and artificial intelligence (AI). It provides insights into critical issues requiring attention in future research endeavors.

Keywords: infrared thermographic; renewable industry; electronic industry; algorithms; artificial intelligence

Citation: Zhao, X.; Zhao, Y.; Hu, S.; Wang, H.; Zhang, Y.; Ming, W. Progress in Active Infrared Imaging for Defect Detection in the Renewable and Electronic Industries. *Sensors* **2023**, *23*, 8780. https://doi.org/10.3390/s23218780

Academic Editors: Dong Wang, Shilong Sun and Changqing Shen

Received: 20 September 2023
Revised: 20 October 2023
Accepted: 25 October 2023
Published: 27 October 2023

1. Introduction

Any object in nature that is above absolute temperature (-273 °C) radiates heat (electromagnetic waves) outward [1]. Electromagnetic waves with a wavelength range of 760 nm to 1 mm are called infrared and cannot be seen by the naked eye. The higher the temperature of an object, the greater the energy radiated. Infrared thermographic (IRT) technology involves sensing infrared waves through special materials, converting them into electrical signals, and then converting the electrical signals into digital images. Using thermal imaging technology, the detection device (an infrared thermal imager) receives varying degrees of infrared radiation from the surface of a sample, generating a temperature field map. This temperature field map characterizes the infrared radiation distribution and can be used to evaluate the differences in the external and internal structures of the sample. This is because the differences in the external and internal structures of the evaluated object will generate different heat conduction in the material, thereby affecting the heat flow [2]. This means that samples with defects, due to differences in internal structure, will cool or heat up at different ratios, resulting in different thermal contrasts in infrared thermal radiation imaging.

Therefore, IRT technology can be used in the field of defect detection [3], especially in the electronic [4,5] and renewable industries [6]. According to the structural characteristics and defect properties of different materials, different types of thermal excitation sources need to be designed to actively heat the surface or interior of the tested object. The thermal excitation source can be modulated or not. Common excitation sources include flash/halogens lamps, hot air, lasers, ultrasound, electromagnetics, etc. Due to the presence of defects on the surface or inside of the tested object, there will be certain differences in the ways in which the thermal waves generated propagate towards the surface of the object. The main advantages of IRT over other technologies are: (1) non-contact and non-invasive; (2) high-speed; (3) large-area; (4) simple operation; (5) intuitive and easy-to-understand results; and (6) a wide range of inspection objects such as metallic, non-metallic, and composite materials. For example, with IRT technology, it is possible to measure the temperature of extremely hot objects or dangerous products (e.g., strong acid, hot steel) at high speed in a non-contact, non-invasive, and large-area way so that their temperature distribution can be safely measured and users can be kept away from danger [5,7]. In addition, it is possible to perform high-speed scanning not only of stationary targets but also of fast-moving targets. In contrast to the harmful radiation effects of techniques such as X-ray imaging, IRT is radiation-free and suitable for long-term and repeated use. Figure 1 illustrates the search results for the citation frequency and publication count of IRT keywords in the Web of Science database. The chart clearly reflects a gradual increase in both citation frequency and publication count, underscoring the continuous growth of research interest and study in the field of IRT. This upward trend suggests the increasing significance of IRT across various academic disciplines, motivating researchers to delve deeper into the applications and advancements of IRT technology. This is distinctly demonstrated in the chart, providing robust support and impetus for current and future IRT research.

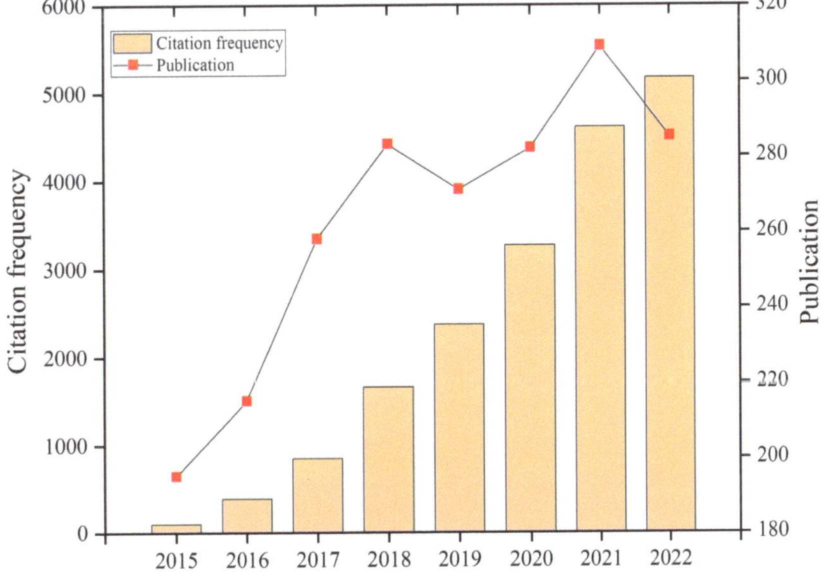

Figure 1. The citation frequency and number of publications for the keyword "IRT" were searched for in Web of Science.

In the past decade, the global photovoltaic (PV) market has grown almost exponentially in size. PV solar energy has strong competitiveness in the global energy market and has become a mainstream renewable energy technology [6]. The IRT imaging method is an efficient and potent tool for qualitative examination of PV modules when compared to conventional I–V characteristics. It can reliably pinpoint the specific position of defects in PV power plants in addition to detecting their presence in the system. For example, for a normal PV module, the incident irradiance causes a uniform temperature distribution on its surface. On the contrary, for most faulty PV modules, the thermal behavior of the PV module affects its surface temperature distribution, resulting in various inhomogeneities in the temperature distribution. This means that with minimal instrumentation, no direct contact, and no interruption of the functioning of the PV system in real-world conditions [8,9], details regarding the thermal characteristics and the precise physical location of the fault can be quickly obtained to quantitatively diagnose the presence of a faulty cell, cell bank, or module.

The electronics industry, stemming from the advancement and application of electronic science and technology, is not only one of the pillar industries of the national economy, but also an emerging science and technology development industry. In integrated circuits, for example, the electronic circuits of printed circuit boards (PCBs) are widely made [10,11], and these contain a high density of electronic components in the board power supply and many electronic connections, which are potential manufacturing defects. And the identification and localization of these defects are critical to the error-free performance of PCBs. Typically, defects produce abnormal temperature patterns that can be detected by the IRT. For example, transparent components are a key core component of smart terminals (one of the pillar industries of the electronics industry). Common transparent components mainly include the cover of the display, the light guide plate, etc., in this field [3]. 3D glass cover components are prone to defects (scratches, microcracks, microbubbles, water ripples, etc.) during the manufacturing process. According to statistics, the yield rate of 3D glass cover components is less than 75% [12–14], so high-performance detection of defects to improve the yield rate of the final smart terminal products is a technical challenge to be overcome. However, the defect detection process of transparent components has special characteristics with high light transmission and reflectivity, and the existing process is mainly manual. Relative to the traditional optical machine vision detection method, thermal spray infrared imaging will be controlled by a high-temperature gas through a moving nozzle, heating intelligent terminal transparent components due to the thermal resistance effect, component defects of the geometry, spatial location, etc. In the process of heat transfer, therefore, the defects in the vicinity of the spatial temperature evolution have a certain degree of variability compared to the normal, so the difference can be captured by thermal infrared imaging.

Due to the rapid expansion of the renewable energy sector, a dedicated section has been included in this paper to delve into the intricacies and developments within this industry. The fusion of IRT and advanced deep learning techniques represents a substantial leap forward in improving the accuracy and efficacy of detecting and diagnosing defects in PV panels [15]. For example, commonly used algorithms include convolutional neural networks (CNN), chaos synchronization detection method (CSDM), and genetic algorithm (GA) [16–19]. This integration harnesses the power of IRT's thermal imaging to capture nuanced temperature variations across PV panel surfaces, and when combined with deep learning algorithms [20], the system can not only identify defects but also offer enhanced predictive capabilities [21]. Through this interdisciplinary approach, the ability to precisely pinpoint and diagnose issues in PV panels is substantially elevated [22,23].

The detection of electronic components presents a unique set of challenges owing to their complex and intricate structures [24]. In this context, the process of detecting defects and anomalies typically necessitates external excitation to induce heating within these electronic components. This, in turn, enables the capture of thermal radiation emitted by the object under inspection, facilitating the creation of a thermal image. In the realm of infrared thermal imaging for the detection of electronic components, lasers have emerged as a common and preferred excitation source [25,26]. The utilization of lasers at the 808 nm wavelength has demonstrated several advantages in electronic component inspection [27]. Firstly, it ensures the accurate targeting of specific areas of interest on the component, facilitating a controlled heating process. Additionally, this wavelength is well-matched to the spectral response of many infrared cameras, enhancing the efficiency of data acquisition [28]. As a result, the thermal images captured exhibit clarity and detail, enabling the detection of defects or anomalies with high precision.

2. Principle and Key Techniques

Active IRT is a technique whereby the surface or interior of an object to be inspected is excited in a controlled manner by a controlled heat source, causing its temperature to change. In this way, the changing temperature field of the object to be detected in space and time can be recorded using an infrared thermal camera to obtain the dynamic response of the heat wave. Afterwards, the thermal series of images obtained by the camera are processed and analyzed by image processing algorithms to determine whether the object to be detected is defective. In addition to high-performance infrared cameras, active IRT needs to focus on stimulating sources, heat transfer mechanisms, and image processing algorithms.

2.1. Principle

Active IRT, a subset of infrared imaging-based machine vision (IRMV), refers to a computer's capability to produce images from infrared (IR) rays emitted or reflected by an object. A distinct demarcation exists between IRMV and traditional machine vision (MV). A typical IRMV system comprises an infrared camera with a lens, an infrared light source (stimulating heat source), a PC for image processing, a control module, and actuators. Traditional MV does not require the infrared camera and the stimulating heat source, but only a traditional camera and a common light source.

Infrared is an electromagnetic wave with a wavelength between microwaves and visible light. The infrared band is usually subdivided into several sub-bands based on their wavelengths, as shown in Figure 2. Typically, near-infrared (NIR) waves have wavelengths ranging from 0.76 to 1 µm, short-wave infrared (SWIR) has wavelengths ranging from 1 to 2.5 µm, mid-wave infrared (MWIR) has wavelengths ranging from 3 to 5 µm, long-wave infrared (LWIR) rays have wavelengths ranging from 7.5 to 14 µm, and far-infrared (FIR) rays ranges from 15 to 1000 µm. Terahertz (THz) rays are FIR rays with wavelengths between 0.1 and 1 mm. As a result, TeraSense or another THz camera could be used to define terahertz machine vision (THzMV) [29]. Infrared applications are divided into three main categories: short-wave infrared, mid-wave infrared, and long-wave infrared. Short-wave infrared utilizes the short-wave infrared radiation prevalent in the target's reflective environment and is similar in resolution and detail to visible light images. Long-wave and mid-wave infrared imaging utilizes thermal radiation emitted by the room-temperature target itself and is used in a variety of infrared thermal vision devices.

2.2. Excitation Sources

Various excitation sources (e.g., flash/halogens lamps, hot air, lasers, ultra-sound, electromagnetics, etc.) are employed to thermally stimulate either the object's surface or interior of the object according to the needs of different detection objects, to measure the temperature change after induction, as shown in Figure 3. Depending on the type of excitation heat source used, active IRT is mainly categorized into pulsed thermography (PT), locked-in thermography (LT), ultrasonically stimulated vibration thermography (UVT), and eddy current thermography (ECT).

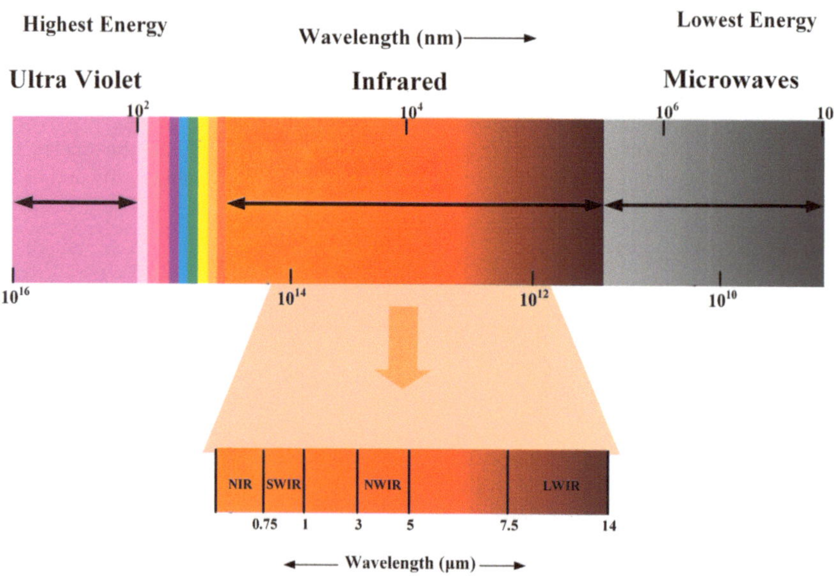

Figure 2. Infrared electromagnetic spectrum and its detection by infrared imaging.

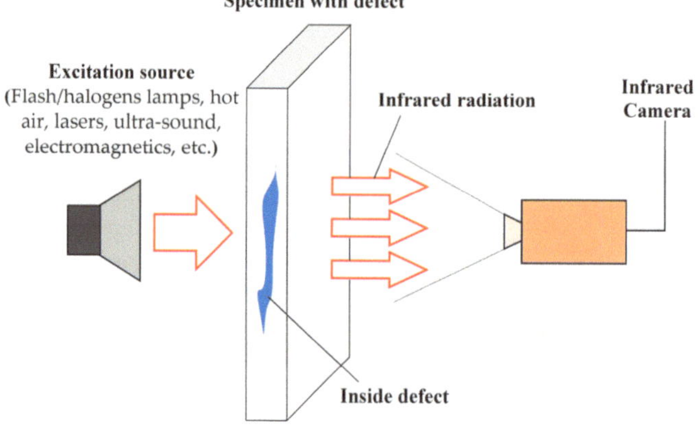

Figure 3. Active IRT method and its excitation source in the defect detection process.

PT This technique utilizes a pulsed heat source (e.g., flash lamps, lasers, etc.) to emit heat pulses to the specimen into be inspected and heat it, as shown in Figure 4. Due to the very concentrated energy of the pulsed heat source and the very short pulses, the thermal equilibrium of the specimen is disturbed, and heat is rapidly conducted inside the specimen. If a defect exists, this results in a temperature difference between the surface of the specimen above the defect and the rest of the area. At this time, the fast infrared camera can record continuous thermal infrared imaging images, and the captured images are analyzed by the computer through algorithms for real-time pixel analysis. After the pulsed heat source is injected into the specimen, the variation of temperature profiles in different areas provides information about the internal defect characteristics of the material. This pulsed IRT can be used to detect defects on the surface or inside the specimen. The pulsed method has the advantage of being independent of compound heating inhomogeneities and possible changes in surface properties [4,30].

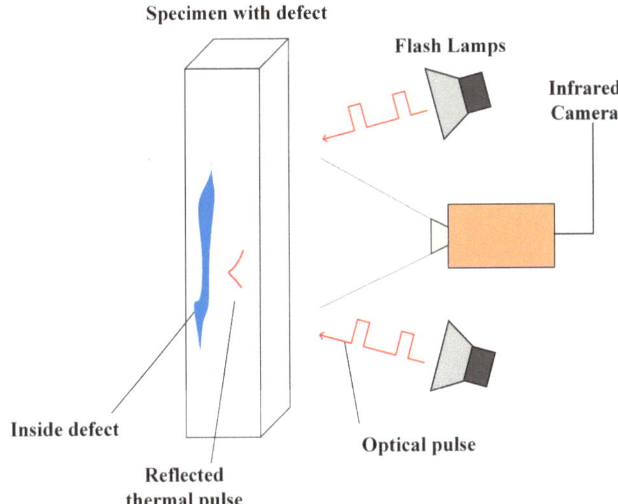

Figure 4. Schematic of PT tests in the defect detection process.

LT Lock-in thermography, also known as thermal wave imaging or modulated thermography, was proposed by Busse et al. [31], and has since been further developed by various researchers [32]. The method entails subjecting a specimen to a frequency-specified periodic (typically sinusoidal, with a given modulation frequency ω and amplitude I) thermal excitation in a steady state and then capturing the surface heating using an infrared camera. The thermal excitation time is at least one modulation cycle until the surface temperature of the specimen reaches a quiescent state. The detection method of LT is shown in Figure 5 and consists of a signal generator and an infrared camera. The signal generator provides the modulation frequency and intensity for the halogen lamp to generate thermal waves, and the IR camera has a high resolution to capture the thermal response of the specimen under thermal excitation. By analyzing the phase shift between the thermal excitation signal and the thermal surface response, it is possible to not only locate the presence of defects in the inspected specimen, but also to precisely locate the defects and determine the defect depth. The LT technique is used in much the same way as the PT technique, with the difference being the sampling frequency. In the former, the specimen to be tested is subjected to thermal excitation that lasts for several cycles, resulting in a longer testing time. However, this method is insensitive to external disturbances and works well even under difficult conditions. In addition, the LT technique allows defect detection on large surfaces, and the excitation frequency can determine the depth of test defects with a good signal-to-noise ratio.

UVT It is well known that PT and LT are the two primary forms of optical excitation. Like optical excitation, acoustic excitation can also be used for active IRT detection. Figure 6 illustrates the schematic diagram of UVT, primarily consisting of an ultrasonic transducer and an infrared camera [33]. The ultrasonic transducer excites the specimen to be detected, and the vibration propagates inside the material, leading to localized heating of cracks through internal friction, and then through the high-performance infrared camera to capture its temperature changes. The interaction between the thermal and mechanical waves can localize the presence of defects in the specimen. If there are defects, because they release more heat through friction, there will be some difference in the thermal image from other normal areas. Since ultrasonic waves propagate to deeper layers, UVT can detect deep defects within the specimen. Analyzing the phase shift between the ultrasonic excitation and the thermal response enables precise defect localization.

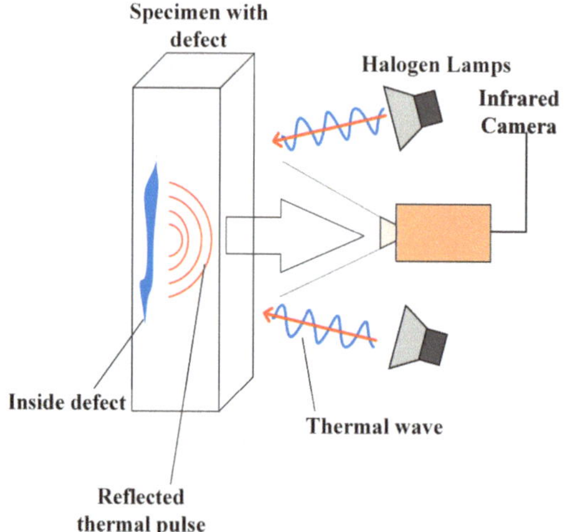

Figure 5. Schematic of LT tests in the defect detection process.

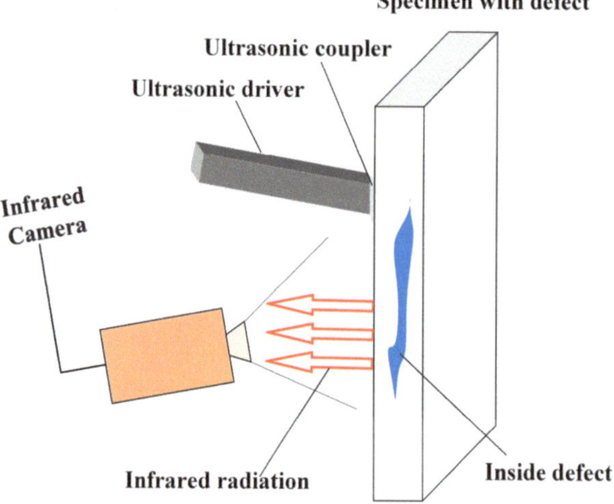

Figure 6. Schematic of UVT tests in in the defect detection process.

ECT Eddy-current-induced IRT is a technique that uses external excitation to induce eddy currents inside the specimen and an infrared camera to capture the heat flowing from the surface (shown in Figure 7) [34]. For instance, when a coil with pulsed excitation is brought near the test specimen, if the test specimen is free of defects and made of a uniform material, the induced eddy currents will be uniformly distributed across the test specimen. On the contrary, if the test piece surface or internal part has cracks and other defects or is mixed with other impurities due to material inhomogeneity, the induced eddy current will be around these defects or impurities, which will inevitably lead to the entire test specimen's defective local temperature rising faster than the test specimen, thus forming a temperature field distribution.

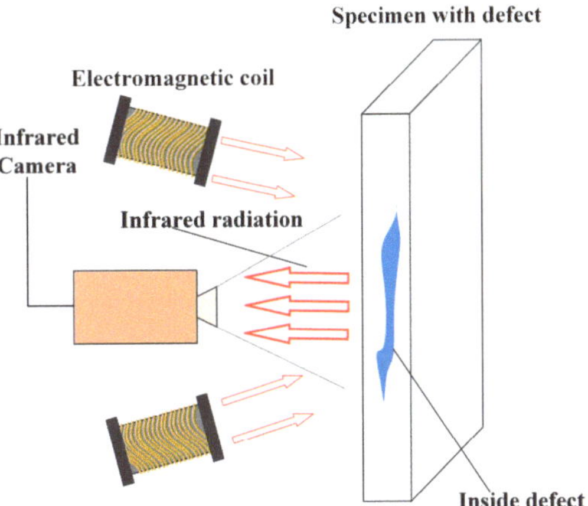

Figure 7. Schematic of ECT tests in in the defect detection process.

2.3. Heat Transfer Mechanisms

In addition to the type of excitation heat source used for defect detection, the form of the excitation wave of the heat source is also very important due to the heat transfer mechanism involved. Evidently, manipulating the amplitude (energy level), frequency, and duration of the excitation heat source has a great impact on the outcomes of active thermography [35,36]. Suitable process parameters for the excitation heat source can enhance the accuracy and robustness of the detection. This means that for a specific specimen, we need to select the appropriate thermal excitation waveform according to the nature of the defect to better utilize the heat transfer effect and improve the signal-to-noise ratio between the defective region and the normal region. Thermal imaging techniques can be broadly categorized into two types: transient and static. The former is the use of pulsed (given the stimulus time of the waveform) forms of energy waveforms to stimulate the specimen to produce a thermal response; infrared data acquisition is carried out in the transient mode before the specimen is heated to a steady state. The latter is the use of modulated (given the frequency of the stimulus waveform) energy waves to stimulate the specimen to produce a thermal response. The specimen is heated to reach a steady state after infrared imaging to obtain the modulated waveform of the thermal response. Table 1 summarizes the energy waveforms and their equivalent temperature response. Take a point on the surface of the sample and a point on the defect, which is Sound P#1 and Defect P#2, respectively. The color of Sound P#1 is green, and the color of Defect P#2 is red. The colors of the curves correspond to the colors of the two points, respectively. The green curve shows what happens at the surface, while the red curve shows what happens at the defect site. The comparison of the two curves describes the different results produced by different methods in these two sampling areas, and the curve changes at the defect can be seen. The contrast between the two curves describes the different results produced by different methods in the two sample areas.

For transient heat transfer in the defect detection experiments, the material specimen is subjected to relatively short energy pulses and the temperature rise and decay curves over time are recorded. The diffusion of the thermal front under the surface of the specimen is calculated according to the Fourier diffusion equation [37], as shown in Equation (1):

$$\frac{\partial T}{\partial t} = \alpha \nabla^2 T \tag{1}$$

where α is the thermal diffusivity (m^2/s).

Table 1. Summary of the energy waveforms and their equivalent temperature response for sound (P#1) and defective (P#2) points at the surface.

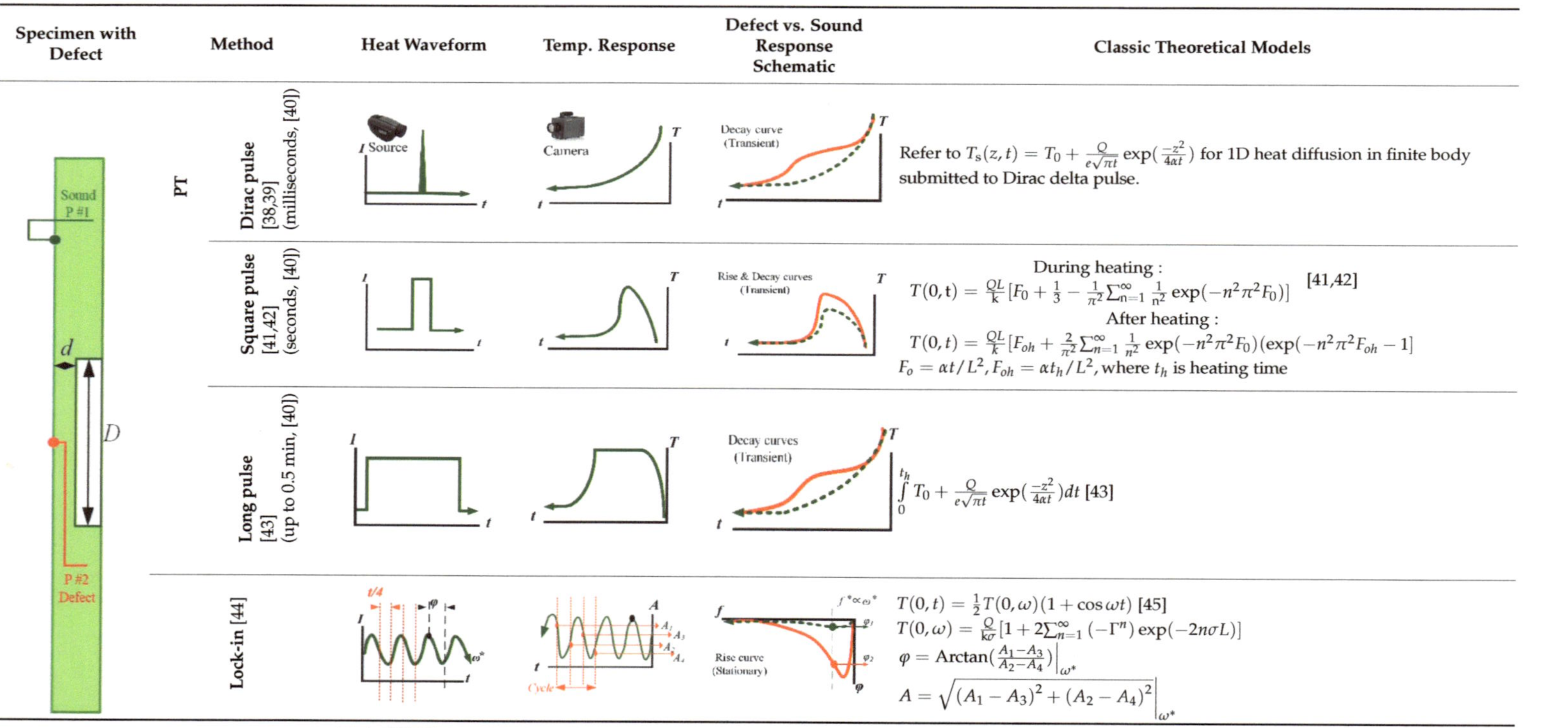

Specimen with Defect	Method	Heat Waveform	Temp. Response	Defect vs. Sound Response Schematic	Classic Theoretical Models
	Dirac pulse [38,39] (milliseconds, [40])	Source	Camera	Decay curve (Transient)	Refer to $T_s(z,t) = T_0 + \frac{Q}{e\sqrt{\pi t}}\exp(\frac{-z^2}{4\alpha t})$ for 1D heat diffusion in finite body submitted to Dirac delta pulse.
	Square pulse [41,42] (seconds, [40])			Rise & Decay curves (Transient)	During heating : $T(0,t) = \frac{QL}{k}[F_0 + \frac{1}{3} - \frac{1}{\pi^2}\sum_{n=1}^{\infty}\frac{1}{n^2}\exp(-n^2\pi^2 F_0)]$ [41,42] After heating : $T(0,t) = \frac{QL}{k}[F_{oh} + \frac{2}{\pi^2}\sum_{n=1}^{\infty}\frac{1}{n^2}\exp(-n^2\pi^2 F_0)(\exp(-n^2\pi^2 F_{oh} - 1]$ $F_o = \alpha t/L^2, F_{oh} = \alpha t_h/L^2$, where t_h is heating time
	Long pulse [43] (up to 0.5 min, [40])			Decay curves (Transient)	$\int_0^{t_h} T_0 + \frac{Q}{e\sqrt{\pi t}}\exp(\frac{-z^2}{4\alpha t})dt$ [43]
	Lock-in [44]			Rise curve (Stationary)	$T(0,t) = \frac{1}{2}T(0,\omega)(1 + \cos\omega t)$ [45] $T(0,\omega) = \frac{Q}{k\sigma}[1 + 2\sum_{n=1}^{\infty}(-\Gamma^n)\exp(-2n\sigma L)]$ $\varphi = \text{Arctan}(\frac{A_1 - A_3}{A_2 - A_4})\big|_{\omega^*}$ $A = \sqrt{(A_1 - A_3)^2 + (A_2 - A_4)^2}\big|_{\omega^*}$

$-\sigma = (1+)\sqrt{\omega/2\alpha}$: Complex thermal wave number; k: Thermal conductivity; L: Specimen thickness (it will be replaced with "*d*" for thickness over defect); α: Thermal diffusivity; e: Thermal effusivity.

Considering that the surface of the specimen is uniformly heated, the heat propagation into its interior can be regarded as a one-dimensional heat flow process [38,46]. Therefore, the one-dimensional heat flow of Equation (1) can be expressed as follows [37]:

$$\frac{\partial T}{\partial t} = \alpha \frac{\partial^2 T}{\partial z^2} \tag{2}$$

where z corresponds to the coordinate parallel to specimen thickness.

For a Dirac delta pulse plane source of strength $Q/\rho C$ released into a semi-infinite medium ($z \gg 0$) from its surface ($z = 0$), Equation (2)'s solution is as follows [37]:

$$T_s(z, t) = T_0 + \frac{Q}{e\sqrt{\pi t}} \exp\left(\frac{-z^2}{4\alpha t}\right) \tag{3}$$

where T_s is the transient temperature in the semi-infinite body, T_0 is the initial temperature, and e is the thermal effusivity.

Static heat transfer In the lock-in detection process, the specimen is subjected to periodic thermal waves and the one-dimensional solution for an isotropic semi-infinite specimen is as follows [47]:

$$T_s(z, t) = T_0 \exp\left(-\frac{z}{\mu}\right) \cos\left(\frac{2\pi z}{\lambda} - \omega t\right) \tag{4}$$

where ω is the modulated frequency, λ corresponds to thermal wavelength, T_0 is the initial temperature, and μ is expressed as thermal diffusion length, which is equivalent to the rate of decay of the thermal wave as it penetrates through the material [47].

In contrast to the PT technique (which records the temperature decay), the LT technique records the changes during the temperature rise period in a stationary state by means of a thermal imaging camera [35]. In the case of static heat transfer, the LT technique makes it easy to analyze the time dependence of the response waveform over a complete modulation period by using a sinusoidal waveform thermal excitation, which allows the reference waveform to maintain good shape and frequency, thus determining the type and location of defects.

2.4. IR Image Processing Algorithms

Compared to visible light imaging, infrared imaging characterizes the temperature distribution of the specimen and is a grayscale image with no color or shading, low resolution, and poor resolution potential. Therefore, the clarity of infrared imaging is lower than that of visible light images. Additionally, the infrared imaging process is susceptible to random external interference and imperfections in the thermal imaging system, resulting in a very low signal-to-noise ratio for the infrared image. This means that after IR imaging, when the acquired image does not provide satisfactory information about the condition of the detected object, it also needs to be preprocessed using appropriate algorithms [48,49]. Non-uniformity correction algorithms and image enhancement algorithms are typical representatives of IR image preprocessing algorithms.

It is well known that non-uniformity correction algorithms are mainly divided into two categories: calibration-based nonuniformity correction (CBNUC) and scene-based non-uniformity correction (SBNUC) [50]. CBNUC encompasses a range of algorithms used to mitigate non-uniformity in thermal infrared imaging devices. Representative algorithms in this category include two-point correction (TPC) [51], multi-point correction (MPC) [52], radiometric correction [53], and scene-based non-uniformity correction (SBNUC) [50]. Certain CBNUC methods can provide highly accurate non-uniformity correction, ensuring that thermal images accurately represent temperature differences. In some cases, the correction process may be time-consuming, especially when using SBNUC methods that require substantial computational effort [54,55]. As an example, the scene-based non-uniformity correction algorithm can adapt to the non-uniformity change caused by the ambient temperature change, in which the representative algorithms are the temporal high-pass filtering (THF) method [56], constant statistics (CS) method [57], Kalman filtering (KF) method [58], neural network (NN) method [59], and registration-based (RB) [60]. Infrared image enhancement algorithms mainly include traditional frequency domains, space domain, and new image enhancement methods. The traditional enhancement method is to adjust the histogram of the image through grayscale mapping so that its distribution is balanced to achieve the enhancement of the whole image contrast, which is fast and effective, suitable for the scene depth, and does not change much. At the same time, the image distribution is relatively uniform. Most of the traditional algorithms are based on the histogram equalization (HE) algorithm for infrared image enhancement, which can be classified into two categories according to the area of action of the mapping function: the global contrast enhancement (GCE) algorithm and the local contrast enhancement (LCE) algorithm [61,62]. Among the new image enhancement methods, Edwin Land proposed the Retinex theory, an image enhancement algorithm that removes the effect of irradiated light in the original image and obtains the reflective properties possessed by the object itself [63] to analyze the intrinsic nature of the image. This algorithm has the advantages of local contrast enhancement, high dynamic range compression, and image color constancy that can be maintained.

Recently, with the development of MV and artificial intelligence (AI) image processing techniques, the level of a computer's ability to process and comprehend images has increased [64]. Machine learning (ML) is a branch of AI that learns from data through computer programs and automatically improves and adapts its performance [65,66]. Thus, ML aims to help computers learn and adapt, without having to perform explicitly extensive manual programming, to automated analytical algorithms that deal with multivariate and multiparameter problems [3]. For example, Saintey and Almond [67] utilized an artificial neural network (ANN) as an expert system to obtain detailed information on defect size and depth from transient thermographic data. This type of method [67,68] is generally based on pre-training the ANN on normal, defect-characterized experimental datasets to obtain the thermal contrast, phase contrast, etc., after infrared imaging as a function of the presence or absence of defects, the defect shape and size categories, and the range of defect depths. Notably, clustering algorithms (including various improved versions) have also been widely used in defect detection in IR imaging [8,69,70].

3. Renewable Industry

PV solar power generation has become an indispensable component of the global energy landscape [71,72]. The long-term performance and overall reliability of PV modules are significantly influenced by faults occurring both in real-world operational conditions and during transportation and installation [73,74]. These faults lead to specific abnormal operations, primarily characterized by reduced power output, abnormal module surface temperature distribution, excessive thermal/mechanical stress, and even safety risks [75,76]. Traditional electrical performance testing of PV modules is a mature testing method, but it has limited fault-detection capabilities [77]. With the advent of digital cameras, charge-coupled devices (CCDs), and uncooled focal plane array (UFPA) detectors, optical-based infrared thermal imaging detection has gained popularity [78,79]. Specifically, electroluminescence (EL) and IR imaging prove to be potent tools for the qualitative assessment of PV modules, enabling the detection of faults in PV installations and precise identification of their exact locations [80]. In conducting this research, a total of 94 literature reviews published between 2000 and 2023 were identified on the Web of Science. These reviews covered various domains, including energy fuels, engineering, and computer science, and were obtained by limiting the search to reviews related to IRT detection in PV. Table 2 presents the top five most-cited reviews in this domain.

Table 2. The most cited review on the application of IRT in PV.

Authors	Year	Citations	Title
Tsanakas et al. [6]	2016	200	Faults and infrared thermographic diagnosis in operating c-Si photovoltaic modules: A review of research and future challenges.
Aghaei et al. [81]	2015	102	Innovative Automated Control System for PV Fields Inspection and Remote Control.
Herraiz et al. [22]	2020	75	Photovoltaic plant condition monitoring using thermal images analysis by convolutional neural network-based structure.
Gallardo-Saavedra et al. [82]	2018	71	Technological review of the instrumentation used in aerial thermographic inspection of photovoltaic plants.
Niazi et al. [83]	2019	68	Hotspot diagnosis for solar photovoltaic modules using a Naive Bayes classifier.

Regarding the number of literature searches on the infrared detection of photovoltaic panels in Web of Science, Table 3 provides an overview of key annual performance indicators. The table demonstrates a noticeable upward trajectory in the volume of the literature, reflecting the growing interest in the research field. However, it is worth noting that both the average citation count and H-Index for each publication exhibit a declining trend, as indicated in Table 3. This trend can be attributed to the tendency for older literature to accumulate more citations. Of particular interest is the anomaly in 2019, where, despite a substantial increase in publications, there was a sharp decrease in the average citation count per publication and the average yearly citation rate per publication.

Table 3. Annual performance metrics of renewable industry.

Year	Documents	Citations	Average Citations per Document	H-Index
2013	65	1954	30.06	25
2014	96	1782	18.56	23
2015	81	1547	19.1	25
2016	99	1485	15	21
2017	140	3820	27.16	29
2018	99	1482	14.97	22
2019	136	1890	13.9	24
2020	136	2436	17.15	26
2021	124	1400	11.29	19
2022	150	641	4.27	11

Numerous investigations have been carried out, and there has been a recent surge in publications focusing on assessing the suitability of IRT for the detection of PV anomalies [84]. Kandeal et al. [21] accomplished this by meticulously analyzing the available data from the Scopus database and using the VOSviewer tool [85] to create a bibliometric network to illustrate the literature. These networks were presented in the schematic representation of keyword relationships (Figure 8). In this illustration, the size of each circle signifies the occurrence frequency of keywords, the thickness of the connecting lines represents how frequently these keywords co-occur, and the color-coding denotes the year of publication. As depicted, the IRT method has been widely employed across various imaging applications and has found substantial utility in the monitoring of PV conditions, particularly from 2016 onward.

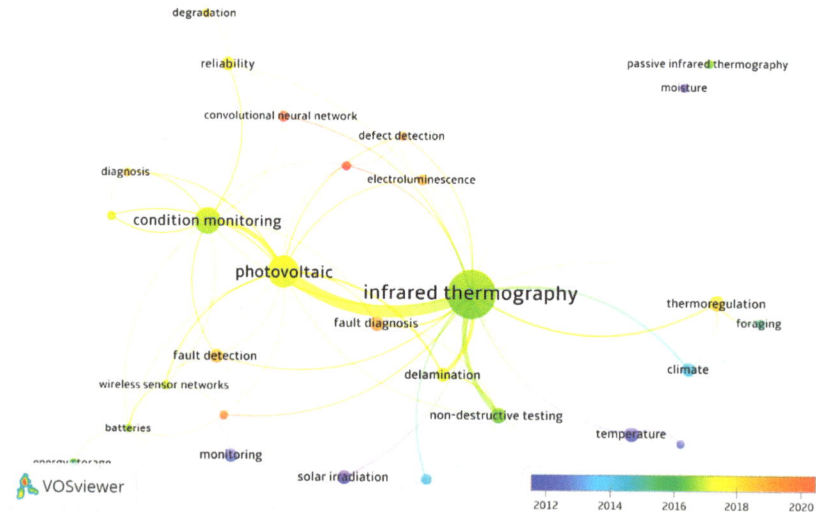

Figure 8. Map of keywords co-occurrence in IRT-PV context [21].

3.1. Optical Degradation

One of the pivotal attributes of high-quality PV front encapsulation materials is achieving optimal optical transmission efficiency [86,87]. However, when deployed in real-world conditions, PV modules encounter an array of environmental challenges, including elevated temperatures, humidity, exposure to ultraviolet (UV) radiation, wind, and snow pressure [88,89]. Among these environmental stressors, moisture can infiltrate the interior of the solar panel through various pathways, including its edges, rear section, or any voids like cracks in the panel structure [90,91]. The pathway leading to optical deterioration in PV modules as a consequence of moisture infiltration is depicted in Figure 9a [92]. As time progresses, the concern regarding optical degradation intensifies and, in the most severe scenarios, can result in a reduction of over 50% in the rated power output of the PV module [93]. Therefore, it is crucial to understand the attributes of imperfections and the fault mechanisms responsible for the optical deterioration of PV devices. This understanding is essential for preventing further degradation and the development of additional failure mechanisms [94].

IR imaging offers insights into the temperature distribution across the surface of the PV module and the location of defects or fault modes [95]. Faulty cells result in mismatch losses, thereby leading to an uneven distribution of cell temperature (Tc) across the PV module. The malfunctioning cells operate at elevated Tc levels, creating hotspots that subsequently affect the module temperature (Tm) [96]. Figure 9b displays the IR image of PV Module X, while Figure 9c–e present magnified EL images of the highlighted regions in Figure 9b [97]. These highlighted areas in Figure 9b are in proximity to the module's frame and represent the most critical hotspots, indicating the presence of significant leakage current during operation. It's worth noting that hotspots are distributed throughout the module. The positioning of hotspot cells near the PV module's frame aligns with the findings from the electroluminescence (EL) images. The abundance of hotspot cells implies that a substantial portion of the cells in field-aged PV Module X are experiencing various stages of degradation. In Figure 9c, no evident cracks are detected, but the highlighted region in Figure 9b shows hotspots. These hotspots in Figure 9b may result from metal grid corrosion and/or solar cell degradation. Moving to Figure 9d, it reveals the existence of microcracks, with the warmest cells identified in this area on the IR image (as seen in Figure 9b). In contrast, Figure 9e displays some cracks, but the hotspots in its corresponding area on the IRT image are not as pronounced as those in Figure 9d. The significance of cracks in facilitating current flow underscores the occurrence and severity of the hotspots observed in Figure 9d. The ΔT of PV Module X was approximately ~8.2 $\pm$ 2 °C [98].

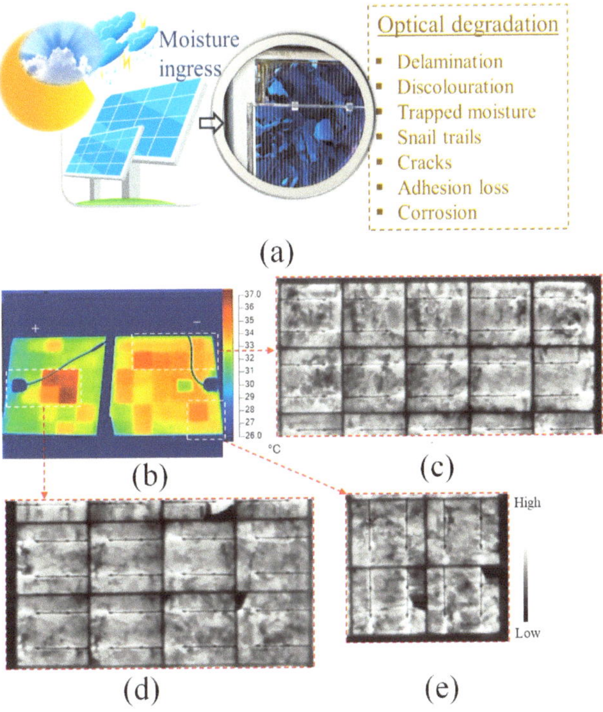

Figure 9. PV module in the field: (**a**) under environmental stressors e.g., high humidity, temperature, and UV radiation, moisture can enter the PV module [92]; (**b**) IRT characteristics of PV Module X acquired under clear sky outdoor conditions; (**c**–**e**) EL characteristics acquired under I_{sc} bias conditions of the corresponding marked areas in (**b**) [97].

Addressing the current drawbacks in industrial production lines, such as low defect detection efficiency, limited data, and high error rates, is crucial due to the significant impact of defects in the silicon photovoltaic (Si-PV) cell manufacturing process on the normal power generation of PV systems. Hence, defect detection is of utmost importance. Du et al. [99] introduced a defect detection and classification method for Si-PV cells based on IRT and CNN. The method involved fine-tuning the LeNet-5, VGG-16, and GoogleNet models after generating the dataset. After 71 training iterations, the GoogleNet model consistently achieved 100% defect classification accuracy with a verification accuracy of 100% and a loss of 0.002. However, training was halted at this point since no significant improvements were observed, and the model reached its peak stability at the highest accuracy. The VGG-16 model attained its highest defect classification accuracy after 121 training iterations, achieving a verification accuracy of 97.67% and a loss of 0.15. While the LeNet-5 model could also achieve a 100% precision value, it exhibited instability and significant fluctuations during the training process. Balasubramani et al. [100] proposed a method for detecting ethylene vinyl acetate (EVA) discoloration and delamination defects based on the thermal pixel counting (TPC) algorithm. Temperature indicators, namely T_{15} and T_{20}, were introduced to highlight the temperature pixel distribution at ΔT °C = 15 °C and 20 °C, respectively. These indicators were compared with healthy panels to validate the algorithm's effectiveness. The classification was automated using a fuzzy classifier, adjusting classification boundaries by modifying fuzzy IF-THEN rule certainty levels while keeping membership function parameter values constant. This approach, particularly the use of the certainty factor (CF) in the fuzzy classifier, significantly improved classification accuracy, surpassing other methods by an average of 10%.

3.2. Electrical Mismatches and Degradation

The term "electrical mismatches" encompasses a range of fault types, including cell cracks, snail trails, broken interconnecting ribbons and busbars, shunts, and poor soldering [100]. These faults are not always discernible through straightforward visual inspection, especially when it comes to optical degradation. Typically, power loss and thermal degradation in faulty modules can lead to an increased risk of safety issues in the entire PV system [6]. Mismatched voltage characteristics can lead to uneven current distribution, thereby affecting the overall performance of the system. Current imbalances between different components may result in electrical mismatch issues among the modules [101].

The commonly employed method for diagnosing faults in solar PV panels is the measurement of Current–Voltage (I–V) characteristics. However, this approach is time-consuming and lacks the ability to categorize defects like delamination, discoloration of EVA, and isolation of cell parts resulting from cell cracks [102]. Pei and Hao [103] presented fault indicators based on current and voltage to detect faults in PV systems. According to the experimental results by Tsanakas et al. [6], cracks in PV modules were actually diagnosed through the I–V characteristics. These interconnection material issues in a single cell or within a cell string occur due to physical strains during transport or installation, thermal cycling leading to thermomechanical stresses, subpar soldering, and potential hotspots arising from extended PV system operation in real-world conditions [104]. Detecting broken interconnections is straightforward using optical techniques such as EL, IRT, ultraviolet (UV) imaging, or through basic I–V characterization (see Figure 10). Figure 10a illustrates the typical I–V characteristic output. Figure 10b,c display the thermal images of PV modules with electrical mismatches, attributed to interconnection ribbon fractures (Figure 10b) and soldering/busbar defects (Figure 10c) as observed through IRT. Figure 10a shows the typical I–V characteristic output, while Figure 10b,c display thermal images of PV modules. These thermal images reveal electrical mismatches due to interconnection ribbon fractures (Figure 10b) and soldering/busbar defects (Figure 10c), as observed through IRT. Belhaouas et al. [105] employed thermal imaging to investigate the performance of solar PV modules after outdoor exposure. The thermographic inspection revealed that the temperature of PV cells inside the PV modules ranges from 32 °C to 68.2 °C, as given in Figure 10d. This temperature variation occurs while the average ambient temperature during the thermal inspection is 23 °C. The thermal inspection found that the deployed PV modules, regardless of their glass types, primarily experience minor temperature mismatch (ΔT) at 90.27%, followed by major ΔT mismatch at 9.58%, and a critical ΔT mismatch case at 0.13%. Nonetheless, PV modules with textured glass exhibit slightly lower thermal stress levels compared to those with float glass. Tsanakas et al. [106] assessed the suitability of thermal image processing and edge detection for defect detection in PV modules. The approach combined image segmentation with Canny edge detection and has yielded favorable results through on-site thermal imaging measurements of two PV arrays: PV-1 and PV-2. It successfully identified 13 out of 14 faulty cells in PV-1 and 27 out of 29 faulty cells in PV-2 by detecting hotspots within the edge maps. These identified hotspots were validated against the standard electrical tests conducted on each module before the experiments, revealing a performance decline of 9.5% for PV-1 and 9.7% for PV-2, respectively. Aziz et al. [107] exploited continuous wavelet transform to generate two-dimensional (2D) images from PV system data and utilized CNN for PV system fault classification, achieving a circuit fault detection accuracy of 73.53%.

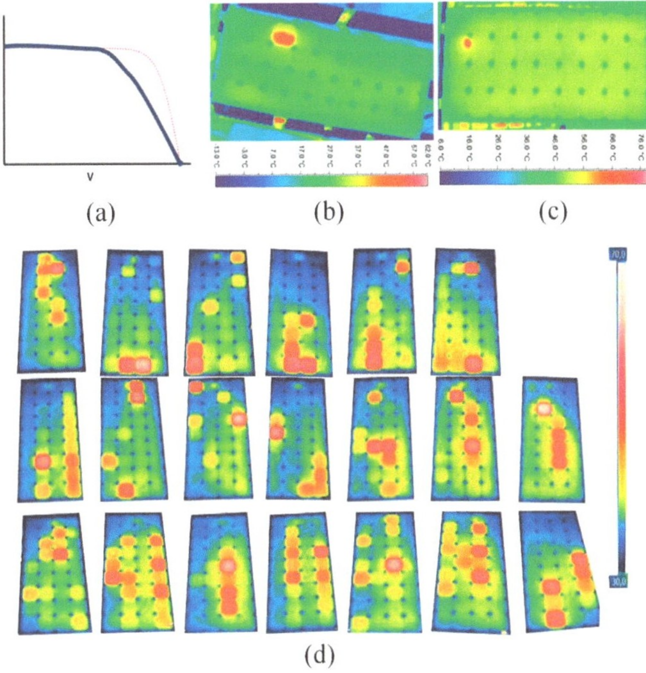

Figure 10. PV module in the field: (**a**) typical *I–V* characteristic output [6]. IRT of an electrically mismatched PV module, due to broken interconnection ribbons; (**b**) defective soldering/busbar (**c**) [6]; (**d**) thermal image of PV module [105].

3.3. Non-Classified Faults

In addition to optical degradation and electrical mismatches and degradation, faults such as potential-induced degradation (PID) and defective bypass diodes (short circuits) are informally referred to as "non-classified" faults. PID is a relatively newly identified fault mechanism in operational PV modules and remains an area with limited comprehensive research and understanding. It involves a crucial externally induced factor, typically accelerated in hot and humid conditions, resulting in significant degradation and power loss within the affected PV modules [108,109].

Researchers have conducted various algorithm and laboratory tests to detect "non-classified" faults. For instance, Bouaichi et al. [110] assessed the PID recovery process in affected PV modules using IR evaluation. PID can be considered a factor affecting the durability and power output of crystalline silicon modules. Lu et al. [16] employed a hybrid algorithm combining chaos synchronization detection method (CSDM) and CNN for the investigation of fault detection in PV modules. The discussion encompassed four prevalent states observed in PV modules: the normal state, module damage state, module contact defect state, and module bypass diode failure state. The research findings showcased the proposed method's remarkable recognition accuracy of 99.5% when 400 sets of randomly generated fault data (with 100 data points for each fault) were inputted, surpassing the traditional edited nearest neighbor (ENN) algorithm's recognition rate of 86.75%. Tao et al. [17] introduced a genetic algorithm-optimized deep belief network (GA-DBN) for diagnosing PV faults, covering normal operation, grounded short circuit, open-circuit in series, partial shadow, and abnormal aging. Although achieving an impressive overall diagnostic accuracy of 95.73%, it's important to note that the average training time was relatively long at 316.34 s, primarily due to the intricate optimization process involving the initial weight and bias of the DBN through GA. Manno et al. [18] achieved optimal performance with CNN using thresholding as a preprocessing method, achieving a 99% accuracy on mid-range CPUs in less than 30 min. Additionally, simplification of thermal imaging images, representing various operational states of PV modules, can achieve high precision. Considering a dataset consisting of 200 sliced images, the same configuration resulted in 90% accuracy for the MLP network and 100% accuracy for CNN. Figure 11 displays various thermographic images utilized for CNN training, the thermographic image in Figure 11a was taken by an operator using a standard lens. Figure 11b shows a non-perpendicular thermographic image angle, and Figure 11c, captured with a standard lens, includes multiple PV modules. In Figure 11d, the thermographic image was acquired using a wide-angle lens and encompasses several PV modules. Mellit [111] adopted an embedded system for fault detection and diagnosis in PV modules, utilizing IRT and deep convolutional neural networks (DCNNs). Two DCNN-based models were developed, one for fault detection and the other for fault diagnosis. Despite the limited dataset size, simulation results indicate a remarkable accuracy of 99% for fault detection and a quite impressive 95.55% accuracy for fault diagnosis. As shown in Figure 11e, the classifier accurately identifies instances of dust deposition on the PV surface, with a recognition accuracy of only 95.5%. In fact, this is due to the similarity in contours between partial shading effects and dust accumulation, as well as PV modules with short circuits and damaged bypass diodes. Dhimish et al. [112] imported a novel PV hotspot fault detection algorithm based on cumulative density function (CDF) modeling technique, achieving an accuracy of 80%.

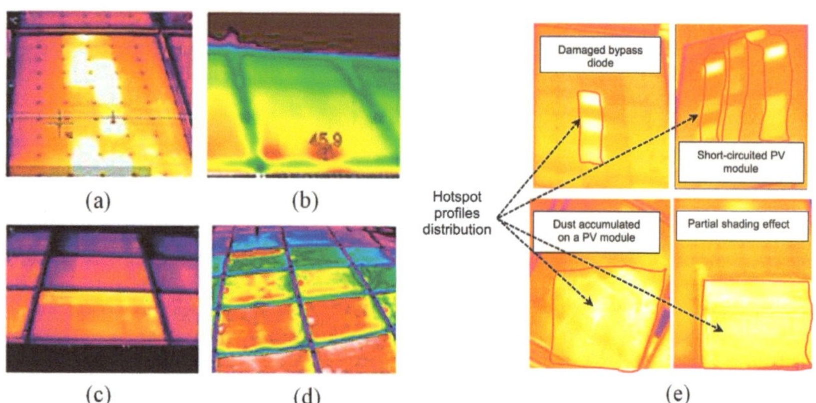

Figure 11. Different thermographic images used: (**a**) thermal images taken by operators using standard lenses [18]; (**b**) the thermographic image was captured at an angle that is not perpendicular to the module [18]; (**c**) thermal image obtained by standard lens [18]; (**d**) thermal image obtained by wide-angle lens [18]; (**e**) host spot profiles variation for different examined PV module defects [111].

3.4. Summary

The advantages of the machine-learning-based method over traditional methods are manifold. Machine learning algorithms can adapt and learn from data, allowing them to improve their performance over time without the need for manual adjustments. This adaptability is a significant advantage when dealing with complex and dynamic systems [113]. Machine-learning-based methods undoubtedly offer numerous advantages for IRT applications. However, like any approach, they do come with certain disadvantages that need to be considered in the context of IRT. Machine learning models, especially deep learning models, require large amounts of data for effective training. In the case of IRT, acquiring a substantial dataset, particularly for rare or specific defects, can be challenging [114]. Furthermore, efforts must be made to make machine learning models more interpretable and transparent in the context of IRT to establish trust and confidence in their results.

Machine learning is a widely-used technology that relies on algorithms and models to enable computers to learn from data and make decisions. Deep learning, on the other hand, is a branch of machine learning that involves artificial neural networks, which can simulate the workings of the human brain to process vast amounts of complex data. The integration of IRT with deep learning plays a pivotal role in detecting and diagnosing defects in PV panels [115,116]. Initially, the technique of IRT is employed to capture thermal images of the PV panels. These thermal images depict the temperature distribution across the surface of the PV panels, where defects typically manifest as anomalous temperature patterns. Preprocessing of the thermal images may be necessary to eliminate noise, enhance contrast, or adjust image dimensions to ensure compatibility with deep learning models [117]. Deep learning models [118], such as CNN [99] or GA-DBN [17], are then utilized to learn and extract features pertaining to defects from the thermal images [119]. These models possess the capability to autonomously acquire knowledge and recognize patterns within the thermal images, including potential defects. The deep learning models excel in automatically discerning complex patterns and temperature distributions within the IRT, thereby enhancing the accuracy of fault detection and diagnosis [120]. This amalgamation enables the automation of the detection and diagnosis processes, reducing the reliance on manual intervention and significantly enhancing overall efficiency.

For instance, despite the relatively limited scale of the dataset employed in Mellit's study [111], simulation results demonstrated a fault detection accuracy of 99% and a fault diagnosis accuracy of 95.55%, as shown in the Figure 12. In most cases, this method can identify various types of defects in PV panels, including but not limited to hotspots, cracks, dirt, and cell damage. Performance metrics for detection may encompass accuracy, recall, and precision, among others, and these metrics are generally contingent on the specific problem and model configurations. In summary, the fusion of IRT and deep learning offers an efficient and highly accurate solution for detecting defects in PV panels. It holds the potential to play a crucial role in the monitoring and maintenance of PV energy systems. Table 4 summarizes the application of the combination of IRT and deep learning techniques for defect detection and diagnosis of PV panels.

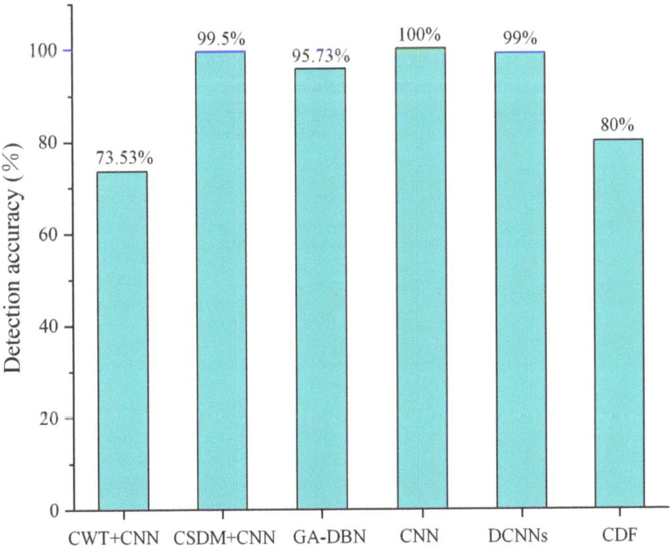

Figure 12. PV panel defect detection accuracy.

Table 4. Summary of the combination of IRT and deep learning techniques for defect detection and diagnosis of PV panels.

Algorithms	Authors, Year	Purpose	Findings	Remarks
CNN	Du et al. [99] 2020	To enhance the detection efficiency of Si-PV cells and achieved extensive defect detection and classification of Si-PV cells.	The classification results of traditional classification methods were significantly lower than those of CNN models.	IRT and CNN have significant potential for applications in defect detection and automatic recognition in Si-PV cells.
TPC algorithm	Balasubramani et al. [100] 2020	The TPC algorithm detects discoloration and layering defects on PV panels.	The CF's fuzzy classifier exhibited superior classification accuracy, resulting in an average classification accuracy improvement of 10%.	The TPC algorithm demonstrates a high level of effectiveness in detecting EVA discoloration and layering defects.
Canny edge detection	Tsanakas et al. [106] 2013	Rapid detection and diagnosis of hotspots in PV modules.	The diagnosed hotspots have been validated against the standard electrical tests for each module, indicating a performance decrease of 9.5% for PV-1 and 9.7% for PV-2, respectively.	This method utilizes qualitative and quantitative data from processed thermal images of two PV arrays, providing easily interpretable results.
CSDM and CNN	Lu et al. [16] 2021	A hybrid algorithm combining CSDM and CNN is employed to study fault detection in PV modules.	The proposed method achieved an impressive recognition accuracy of 99.5%.	The algorithm simplifies a substantial amount of raw measurement data through CSDM and subsequently employs CNN to accurately identify the fault states of PV modules.
GA-DBN	Tao et al. [17] 2020	GA are utilized for diagnosing faults in PV arrays to optimize the DBN.	The GA-DBN method effectively enables diagnostic detection of five operational states in PV arrays, achieving an overall diagnostic accuracy of 95.73%.	Compared to the DBN, SVM, and GA-BP models, this model exhibits higher accuracy in both overall diagnosis and individual fault type diagnosis.
CNN	Manno et al. [18] 2021	Utilizing CNN for the automatic classification of thermal images to identify faults in PV panels.	A dataset consisting of 200 segmented images achieved a 100% accuracy rate used CNN.	The CNN method proves to be an effective tool, enhancing the image classification resolution for remote fault detection issues.
DCNNs	Mellit [111] 2022	Embedded PV module fault detection and diagnosis using IRT and DCNNs.	Two DCNN-based models, namely the fault detection and diagnosis models, achieved an accuracy rate of 99% for fault detection and an accuracy rate of 95.55% for fault diagnosis.	Embedded solutions can detect and diagnose faulty PV modules with acceptable accuracy.

Table 5 presents a comprehensive comparative analysis between research conducted by scholars in the past and the current state of research. Historically, the majority of studies were primarily focused on the conventional methods for PV panel inspection. In contrast, contemporary scholars are placing significant emphasis on the integration of deep learning with IRT techniques. This shift in focus reflects the evolving landscape of research in this field and the recognition of the potential of advanced methods for more precise and efficient PV panel defect detection. The utilization of deep learning in conjunction with IRT is emerging as a promising avenue for achieving higher accuracy and reliability in the inspection of PV panels.

Table 5. Early and current authors are conducting research on the infrared detection of PV panels.

Authors	Year	Citations	Title	Remarks
Dincer et al. [121]	2014	29	Polarization Angle Independent Perfect Metamaterial Absorbers for Solar Cell Applications in the Microwave, Infrared, and Visible Regime.	The proposed metamaterial-based solar cell demonstrates high absorption in both the infrared and visible spectra, enhancing the potential for more efficient next-gen solar cells.
Chandel et al. [122]	2015	33	Degradation analysis of 28 year field exposed mono-c-Si photovoltaic modules of a direct coupled solar water pumping system in western Himalayan region of India.	Utilizing thermal imaging technology to identify hotspots and quantifying degradation by measuring PV parameters under indoor and outdoor conditions.
Adams et al. [123]	2015	42	Water Ingress in Encapsulated Inverted Organic Solar Cells: Correlating Infrared Imaging and Photovoltaic Performance.	Utilizing infrared imaging for local, in-situ tracking of humidity-induced performance degradation to predict the lifespan of organic solar cells and modules.
Du et al. [124]	2017	38	Nondestructive inspection, testing and evaluation for Si-based, thin film and multi junction solar cells: An overview.	Non-destructive inspection, testing, and assessment of solar cells and modules.
Addabbo et al. [125]	2017	55	A UAV Infrared Measurement Approach for Defect Detection in Photovoltaic Plants.	Drones can swiftly inspect solar farms, employing this positioning technology for detecting, labeling anomalies, and identifying faulty panels.
He et al. [126]	2018	36	Noncontact Electromagnetic Induction Excited Infrared Thermography for Photovoltaic Cells and Modules Inspection.	The active electromagnetic induction infrared thermal imaging defect detection method has enabled the visual detection of defects in PV cells and modules.
Zefri et al. [127]	2018	48	Thermal Infrared and Visual Inspection of Photovoltaic Installations by UAV Photogrammetry-Application Case: Morocco.	Visual defects, such as cracks, contamination, and hotspots, have been identified in both visual RGB and thermographic inspections.
Akram et al. [128]	2020	80	Automatic detection of photovoltaic module defects in infrared images with isolated and develop-model transfer deep learning.	CNN are used to train an isolation learning model, achieving an average accuracy of 98.67%. Fine-tuning the pre-trained base model through transfer learning on an infrared image dataset increased accuracy to 99.23%.
Du et al. [99]	2020	43	Intelligent Classification of Silicon Photovoltaic Cell Defects Based on Eddy Current Thermography and Convolution Neural Network.	IRT and CNN demonstrate significant potential for defect detection and automatic recognition in Si-PV cells, providing a reliable approach for the research, testing, manufacturing, servicing, and maintenance of Si-PV cells.
Alves et al. [129]	2021	40	Automatic fault classification in photovoltaic modules using Convolutional Neural Networks.	Using cross-validation methods, CNN achieve an estimated accuracy of 92.5% in detecting anomalies in PV modules.

Automatic photovoltaic inspection has garnered significant interest from researchers in recent years. Numerous studies have explored automatic photovoltaic inspection using various imaging methods. Demant et al. [130] employed a support vector machine algorithm for the automatic classification of cracks in photoluminescence (PL) images. Stromer et al. [131] proposed an enhanced EL image crack segmentation framework. Li et al. [132] adopted image processing algorithms for the automatic detection of snail trails and dust in visible light images. Su et al. [133] utilized newly proposed feature descriptors to classify manufacturing defects in solar cell EL images. However, there has been limited research on the application of deep learning for defect detection in photovoltaic component images. These studies, including those by Chen et al. [134], Ding et al. [135], and Li et al. [136], have leveraged deep learning techniques to detect defects in visible light (red, green, blue) RGB images. Demant et al. [137] used CNN for automatic quality assessment and control during the production of solar cells in PL images. Deitsch et al. [138] and Akram et al. [139] employed deep learning methods for the automatic detection of faults in solar cell EL images. This represents a notable shift toward utilizing deep learning approaches for photovoltaic inspection.

4. Electronic Industry

With the progression of information electronic devices towards high reliability, miniaturization, light weight, and multifunctionality, high-density integrated circuits with numerous functional components have found extensive applications [140,141]. PCBs, serving as critical structures for electrical and pneumatic interconnection, signal transmission, mechanical linkage, and electronic system support, are also the primary failure-prone areas of components. The long-term reliability of PCBs has become a focal research topic, resulting in challenges associated with effectively and reliably detecting PCBs' defects. Traditional PCB defect detection methods have limitations, but active IRT, including techniques like pulsed thermography and lock-in thermography, has found extensive use in non-destructive testing for PCBs. The development of very large-scale integration (VLSI) technology, increasing silicon wafer diameters, and decreasing integrated circuit linewidths have imposed higher demands on silicon wafer manufacturing processes and surface quality [142]. During semiconductor silicon wafer production, the formation of microcrack defects is common, ultimately affecting the quality of silicon-based microelectronic products. Ensuring the quality and performance of products necessitates non-destructive testing of silicon wafers. Surface mount components achieve interconnection between chips/packages and substrates or PCBs using solder bumps. However, common manufacturing defects, including opens, cracks, or missing solder bumps, persist. As solder bumps are concealed within packages after assembly, the increasing trend towards high-density and ultra-fine pitch has made defect detection progressively more challenging, severely impeding the advancement of surface mount technology. Detecting defects in solder bump protrusions has become a critical issue in integrated circuit manufacturing technology. Concealed solder bump protrusions impede the entry of light beams, and infrared imaging proves to be an effective detection technique capable of identifying nearly all solder bump defects.

Refer to the number of literature searches on Web of Science on the application of infrared thermal imaging technology in electronic industry defect detection, and the results are shown in Table 6. Table 6 provides an overview of the key annual performance indicators. As can be seen from the table, the number of literatures is in a slightly fluctuating state each year, indicating that people's interest in this field has not changed much. It is worth noting that the average number of citations and the H-Index of each publication are almost horizontal, but suddenly decline in 2022. This trend can be attributed to the lack of in-depth research in the field. Of particular interest is the anomaly of 2017, in which the average number of citations per publication rose sharply, even though the number of publications was not as high as before.

Table 6. Annual performance metrics of electronic industry.

Year	Documents	Citations	Average Citations per Document	H-Index
2013	45	680	15.11	13
2014	52	483	9.29	10
2015	89	856	9.62	17
2016	47	493	10.49	11
2017	29	483	16.66	11
2018	58	650	11.21	13
2019	49	514	10.49	12
2020	53	531	10.02	14
2021	58	587	10.12	10
2022	55	287	5.22	9

Table 7 makes a comprehensive comparative analysis of the research of past scholars and the current research status. Historically, the feasibility of infrared non-destructive testing technology has brought a lot of convenience to the electronics industry, and accumulated experience for the subsequent research. With the improvement of technology and the deepening of research, it can be seen that contemporary scholars have added the cost factor to the concern of non-destructive testing in the electronics industry. Future cost reductions will also make infrared non-destructive testing technology have a better market.

Table 7. Early and current authors are conducting research on the infrared detection of electronics industry.

Authors	Year	Citations	Title	Remarks
Jadin et al. [143]	2012	19	Infrared Image Enhancement and Segmentation for Extracting the Thermal Anomalies in Electrical Equipment	The segmentation performance of infrared images is improved by image enhancement method which adjusts the image intensity.
Rogalski et al. [144]	2013	28	Semiconductor detectors and focal plane arrays for far-infrared imaging	The progress of far infrared and submillimeter wave semiconductor detector technology in focal plane array in recent 20 years is introduced.
Xu et al. [145]	2014	18	Using active thermography for defects inspection of flip chip	The feasibility of the flip chip defect detection method based on active thermal imaging is proved.
Daimon et al. [146]	2016	98	Thermal imaging of spin Peltier effect	The combination of spin Peltier effect and lock-in thermography technology provides a new direction for spintronics applications.
Christensen et al. [147]	2018	75	The OSIRIS-REx Thermal Emission Spectrometer (OTES) Instrument	It provides precise moving mirror control and infrared sampling at 772 Hz and minimizes surface reflection.
Aragon et al. [148]	2020	36	A Calibration Procedure for Field and UAV-Based Uncooled Thermal Infrared Instruments	A new calibration method of ambient temperature correlation for a variety of uncooled thermal infrared radiometers is proposed, which significantly improves the measurement accuracy.
Yu et al. [149]	2020	29	Low-Cost Microbolometer Type Infrared Detectors	The advantages of pixel size reduction are significant.

4.1. Chip

Since the 1960s, advancements in semiconductor technology have profoundly transformed our lives and facilitated the development of high-performance electronic devices. The emergence of smartphones, for instance, would not have been possible without the progress in miniaturized and high-performance semiconductors. The demand for lighter, more compact smartphones necessitates the production of smaller, thinner, and higher-performing semiconductor chips. With the growing trend of using thinner wafers for semiconductor chips, various issues have emerged, including a significant concern related to microcracks that can be found on the surface and sub-surface, varying in size from a few micrometers to several tens of micrometers. Semiconductor chip materials are inherently brittle, making them susceptible to stress-induced cracks during chip manufacturing and assembly. These cracks manifest primarily as scratches, fractures, orange peel effects, and pits [150]. Surface cracks can adversely affect the performance and reliability of the final electronic device, thus escalating the demand for inspecting surface cracks in semiconductor chips during the manufacturing process. Efficient and high-precision non-destructive testing is crucial for semiconductor chip inspection. Optical visual methods, while offering non-contact and non-destructive three-dimensional chip characterization, have limitations in detecting concealed defects. Active IRT bestows the following advantages for semiconductor chip inspection: complete non-contact, non-destructive, and non-invasive testing, along with the capability to examine large areas in a single test. IRT has emerged as one of the most promising techniques in non-destructive testing and evaluation [145].

Introducing non-contact active IRT technology into chip defect detection involves the use of an external heat source, such as a flash lamp or laser, for active thermal imaging. When subjected to external heating, the presence of defects within the chip leads to abnormal thermal resistance, enabling the capture of thermal distributions using an infrared imaging device. Analyzing thermal images aids in defect identification, with laser excitation being the most frequently used method for semiconductor chip defect detection among various external excitation techniques. Bu et al. [151] investigated a method utilizing Barker code-modulated pulse compression waveforms for detecting microcrack defects in semiconductor silicon wafers. This technique employed an optical infrared thermal imaging device for transmission, where an infrared camera captured the thermal wave signal response to the laser-modulated Barker code waveform. The acquired images were stored as sequences and analyzed for detectability using a full-harmonic distortion algorithm, resulting in improved defect detectability. An et al. [26] introduced the line laser lock-in thermal imaging technique for semiconductor chip inspection. This technique integrated a line-scanning laser source, an infrared camera with a dedicated lens, and a control computer, assembling a novel line laser lock-in thermal imaging system as shown in Figure 13a. The continuous wave laser beam was modulated into a pulsed laser beam by the excitation unit, and the cylindrical lens transformed the pulsed laser beam shape from point-like to linear. The control unit then issued control signals to the galvo scanner, directing the line laser beam onto the target surface. Subsequently, the line laser beam generated a thermal wave along the desired excitation line, performing horizontal and vertical scans on the target surface, effectively detecting randomly oriented cracks, as shown in Figure 13b. Yang et al. [152] proposed a multi-point laser lock-in thermal imaging system for real-time imaging of semiconductor chip cracks, as shown in Figure 13c. This system employed multi-point pulsed laser beams to simultaneously generate thermal waves at multiple points on the target semiconductor chip surface. The corresponding thermal response was measured using a high-speed infrared camera, enabling real-time detection during the semiconductor chip manufacturing process. Figure 13b,d illustrates a comparative diagram of semiconductor chip defect detection using the same excitation source—laser—in different modes. The integration of infrared sensing technology with the lock-in method significantly improved the sensitivity and resolution of thermal imaging. The sensitivity of thermal imaging was increased by two orders of magnitude, reaching approximately 100 μK, while the resolution for surface defects was lowered to 5 μm [153].

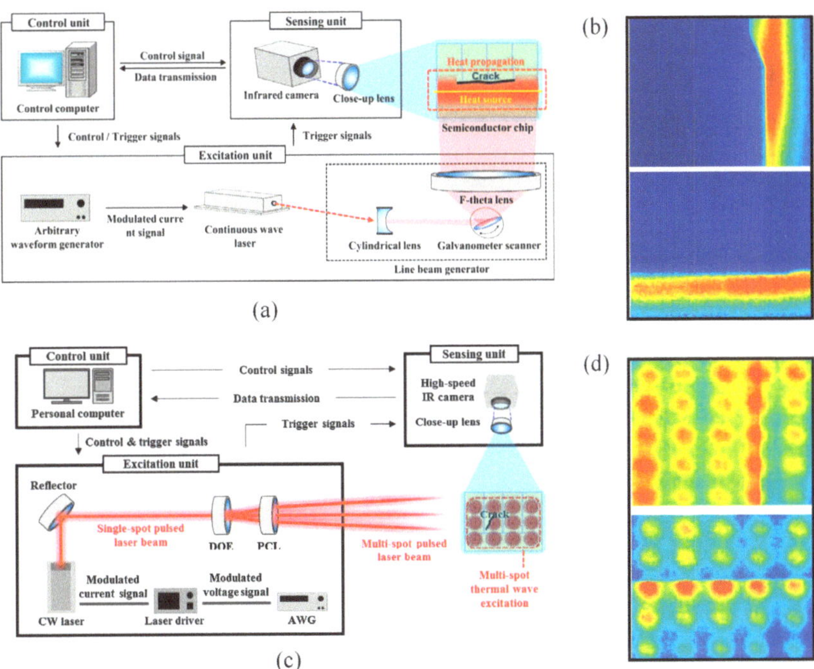

Figure 13. Schematic representations of two distinct laser-based thermographic inspection methods: (**a**) line laser lock-in thermography system and (**b**) corresponding thermograms of vertically oriented crack chips and horizontally oriented crack chips within this system [26]; (**c**) multi-point laser lock-in thermography system and (**d**) corresponding thermograms of vertically oriented crack chips and horizontally oriented crack chips within this system [152].

4.2. PCBs

PCBs serve as crucial structures for achieving electrical and pneumatic interconnection, signal transmission, mechanical linkage, and support for electronic systems. They also represent the primary failure-prone areas for components, especially in high-frequency and high-voltage circuits. Hence, the detection and maintenance of faults in PCBs are critical due to their complex multi-layered structures, leading to various defects such as layer separation, delamination, breakdown damage, and micro-holes during processing and usage. Conventional defect detection techniques for PCBs encompass visual inspection by human operators and automated optical inspection, X-ray, CT imaging, ultrasound, laser ultrasonics, and terahertz imaging. While manual visual inspection and automated optical inspection are the most common methods, they are limited to detecting visible surface-level defects and cannot guarantee the absence of internal flaws. IRT inspection, as a non-contact measurement method, has gradually found application in the field of PCB fault detection. PCB fault detection methods based on IRT mainly involve three steps: thermal source identification, feature extraction, and thermal pattern recognition [154]. Figure 14a shows 2D and 3D views of the PCBs transient amplitude images. Wang et al. [155] employed laser-induced lock-in thermography to detect various real defects in rigid or flexible PCBs. Phase characteristic images enabled effective detection of delamination defects with a depth of 1.2 mm and micro-hole defects with a depth of 400 μm. The reference regions for both defective and non-defective areas are illustrated in Figure 14c. Experimental results demonstrated that laser-induced thermography is suitable for detecting multiple types of PCB defects. Avdelidis et al. [156] utilized two different integrated pulse thermography systems: thermoscope and echotherm. In both cases, mid-wave infrared cameras were used; a merlin 3–5 μm thermoscope system and a phoenix 3–5 μm echotherm system. Both systems were state-of-the-art portable non-destructive testing and electronic inspection systems with integrated flash heating capability. The results showed that pulse thermography can be used for defect detection in circuit boards (i.e., delamination and/or soldering defects). Cong et al. [157] proposed and utilized optical/thermal fusion imaging technology to inspect PCBs. A semiconductor laser diode with a wavelength of 808 nm was employed as the radiation source. Sample data and images were acquired using a mid-infrared camera. Phase-locked thermal imaging was employed for the study of layered defects in PCBs, as illustrated in Figure 14b. Six different fusion algorithms were applied in the experimental study of image fusion, and four metrics were introduced to evaluate the fusion performance. The experimental results indicate that this fusion technology maintains a high level of accuracy and precision under diverse imaging conditions.

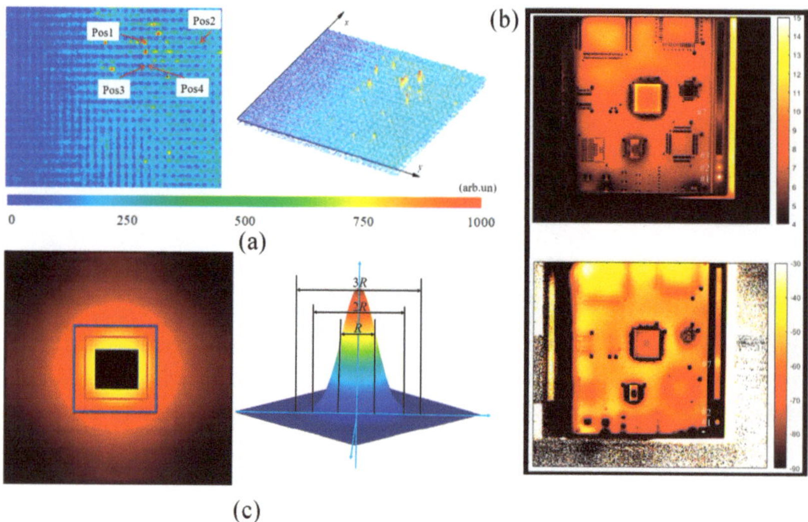

Figure 14. IRT in PCBs: (**a**) 2D and 3D views of transient amplitude images [158]; (**b**) amplitude and phase images in lock-in thermography [157]; (**c**) defective and non-defective reference regions [155].

4.3. Weld

Solder joints constitute crucial components on PCBs. Apart from serving as electrical conduits, they also provide mechanical connections between electronic components and the substrate. Solder joints are more susceptible to defects such as cracks, voids, and missing balls, as depicted in Figure 15a [159]. These flaws can adversely affect the performance and lifespan of flip-chip packages, leading to erratic circuit behavior and intermittent instability. This poses significant risks for debugging, operation, and maintenance of circuits. Therefore, the assessment of solder joint integrity holds paramount importance. Presently, conventional non-destructive testing methods such as X-ray, optical inspection, and flying probe testing struggle to effectively detect such welding defects. In contrast, infrared non-destructive testing offers a wide applicability, non-contact measurement, rapid detection, high precision, ease of qualitative and quantitative analysis, as well as convenient observability, presenting a comprehensive set.

Chai et al. [160] proposed an active transient thermography technique for detecting inverted solder balls. When a solder ball is defective, its resistance is significantly higher than that of a normal solder ball, leading to an abnormal temperature. Hence, using thermal image contrast from an infrared sensor, this method detects the presence and location of defective solder balls, primarily void defects and localized cracks. Lu et al. [161] investigated a pulse-phase thermography-based method for identifying solder joint defects. In this approach, the test chip is stimulated with a thermal pulse, and the subsequent transient response is captured using a commercial thermal imaging camera. Thethermal imager was employed to measure the transient response of the test chip under infrared photothermal excitation. The thermal imager, equipped with a micro-lens with a pixel resolution of 25 µm, enhances spatial resolution. The temperature resolution of the thermal imager, utilizing a microbolometer detector, is superior to 80 mK, with a spectral response range of 7.5 to 14 µm, and a frame size of 640 × 480 pixels. Wei et al. [162] developed an intelligent system for detecting solder joint defects using active thermography. Figure 15b illustrates the experimental setup, employing a fiber-coupled semiconductor laser with a central wavelength of 808 nm as the heat source, monitored by the thermal imager. Statistical features were extracted and classified using the M-SVM algorithm. All missing protrusions were identified, achieving the highest recognition accuracy. The results demonstrate that the combination of active thermography and M-SVM is an effective method for intelligent diagnosis of microelectronic packaging solder material defects. He et al. [163] utilized a pulsed laser with a central wavelength of 808 nm to heat the substrate of the test sample SFA1. The sample consisted of 25 solder balls arranged in a 5 × 5 pattern, with protrusion diameters and pitch distances of 500 µm and 1000 µm, respectively. Thermal images of the SFA1 package were acquired using the VH680 infrared imager. The experimental setup is depicted in Figure 15c, while Figure 15d shows the thermal image of the experimental sample SFA1. The matrix was used as the desired output vector, and a transformation function was applied to convert the desired output vector from an index to a vector. A PNN was then established with input vectors, output vectors, and propagation speed as parameters. The results indicate that the infrared detection system based on PNN is effective for defect detection in high-density packaging.

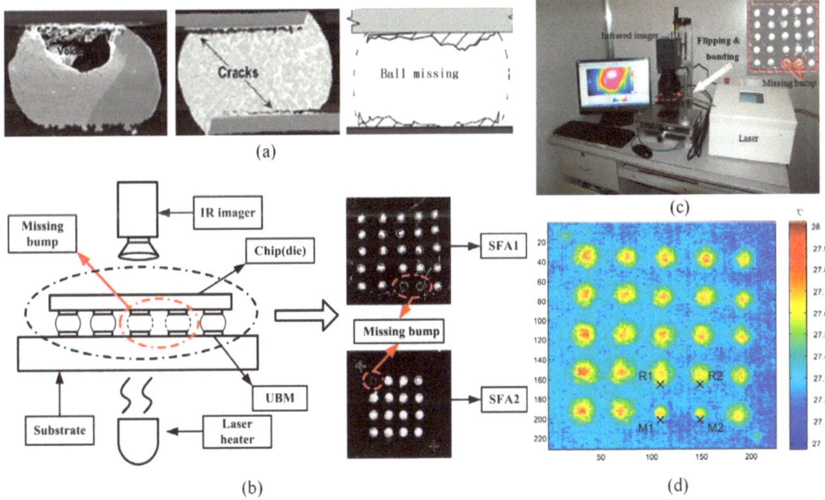

Figure 15. Schematic diagram of welding defect detection: types of weld defects (**a**) [159]; (**b**) schematic of experimental setup and distribution of welds in test samples [162]; (**c**) experimental setup and distribution diagram [161]; (**d**) infrared thermal images of weld defects [163].

4.4. Others

Glass fibers are extensively utilized as reinforcement materials, with glass-fiber-reinforced polymers (GFRP) commonly found in electrical and electronic devices, as well as in numerous components used in our daily lives [12]. Glass fibers present a competitive edge due to their lightweight nature and lower cost compared to other reinforcement materials like carbon fibers [164], showcasing superior properties within composite materials [165]. However, the manufacturing process may incur defects, especially the formation of voids. Fuel cells are essential components in emission-free energy conversion, directly converting chemical energy into electricity. The critical aspect of fuel cell functionality lies in the necessity for all distinct sealing layers to be both electrically insulating and hermetic. The material connecting the two steel interconnect sections of the cells is the glass solder layer, which incorporates artificially induced defects in the form of missing solder of varying diameters.

Meola et al. [166] conducted an assessment of GFRP under low-energy, low-velocity impact using IRT. They employed a equipped with a quantum well infrared photodetector (QWIP) operating in the 8–9 μm range, with a spatial resolution of 640 × 512 pixels at full frame. For the purpose of comparison, thermal imaging and visible light images of the same sample are presented in Figure 16a. The results demonstrated that non-destructive testing utilizing lock-in thermography could detect manufacturing defects such as uneven resin distribution, porosity, fiber misalignment, and impact damage. Dua et al. [167] introduced a high-depth resolution frequency-modulated thermal wave imaging technique for infrared characterization of GFRP laminates. Each GFRP sample comprised five patches, with a thickness of 2 mm. The selected samples were subjected to experiments using two 1 kW halogen lamps. Thermal distributions of the samples were recorded by an infrared camera at a frame rate of 25 Hz. The results indicated that the layer-wise detection capability of time-correlated coefficient images significantly outperformed the widely used phase-based post-processing methods. Muzaffar et al. [168] proposed a rapid and straightforward method for detecting faults in antenna arrays using infrared thermal imaging. The thermal imager employed was a 14-bit, 320 × 240 resolution mid-wave infrared (MWIR) camera from FLIR. The study demonstrated that IRT could be applied for detecting faulty elements in antenna arrays, with the variation of temperature rise on the absorptive screen being crucial for identifying the faulty components. Figure 16d,e respectively present the sample image and the corresponding thermal imaging of defects. Wei et al. [169] advocated the application of artificial intelligence techniques for automatic processing of infrared images to detect defects within the glass seal layer of solid oxide fuel cells. Three methods were investigated: (1) support vector machine, (2) adaptive enhancement, and (3) U-Net. The results indicated that features extracted from individual thermal profiles might be insufficient for defect identification, while U-Net displayed significant potential in thermal image segmentation. Wang et al. [170] conducted experimental studies on the detection of impact damage in GFRP using pulse radar thermal wave imaging technology. They utilized a high-performance, cooled focal plane infrared imager with a response wavelength of 3.6–5.2 μm and pixel dimensions of 640 × 512. An 808 nm semiconductor laser was used, and various time/frequency domain analysis algorithms were applied to extract features from the thermal image sequences. The thermal image sequence was acquired using an IRT camera, The results showed that the dual-channel orthogonal demodulation algorithm exhibited excellent recognition capabilities for delamination defects in GFRP. Within the specified defect diameter and depth range, it could identify delamination defects with a depth $\geq$1.70 mm and a diameter-depth ratio (D/H) $\geq$2.35. By analyzing the signal-to-noise ratio (SNR) of feature images, gong et al. [171] quantitatively evaluated the detection ability of laser bidirectional thermal wave radar imaging (BTWRI) to detect defects of carbon/glass fiber reinforced polymer (C/GFRP). Figure 16b is the sample used in the experiment. By comparing the signal-to-noise ratio of feature images on a frame-by-frame basis, the optimal ACC detection image was obtained. Figure 16c shows the defect phase diagram and amplitude diagram of the sample.

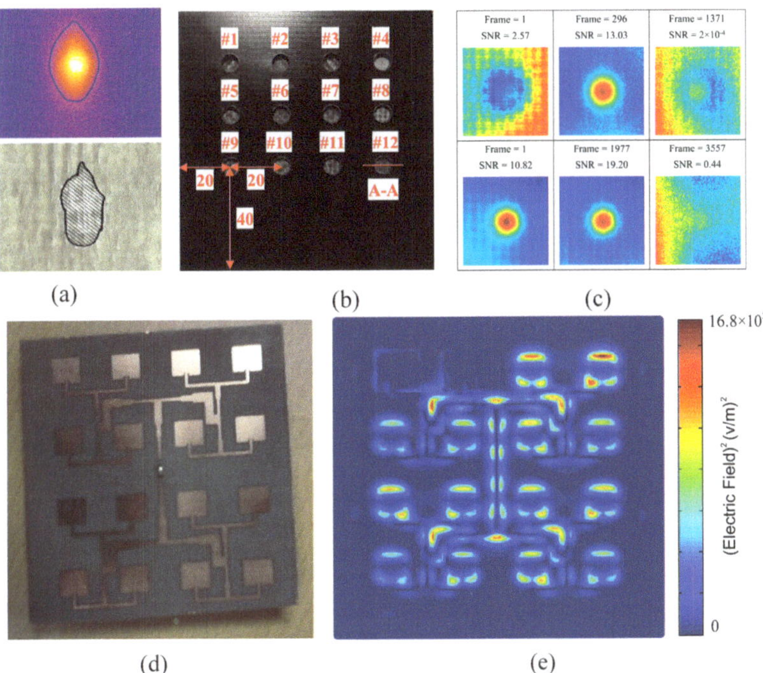

Figure 16. Other applications of IRT: (**a**) thermal and visible images of the sample [166]; (**b**) C/GFRP specimens with artificial flat-bottom holes [171]; (**c**) patch antenna array em-ployed in the experiment [171]; (**d**) phase images of defect #2 in S2 and amplitude images of defect #2 in S2 [168]; (**e**) thermal imaging of defects [168].

4.5. Summary

The advancement of technology has led to increasingly stringent requirements for the quality of electronic components [11]. This chapter provides an overview of the application of IRT in electronic component defect detection from four aspects. Firstly, it introduces the application of IRT in semiconductor chip defect detection. Laser is commonly used as the excitation source, but not all thermal imaging techniques are suitable for detecting defects within semiconductor chip encapsulation. To address this, phase-locked thermography has been developed, which can overcome two limitations of IRT: the inability to differentiate surface and sub-surface features, and the lack of sensitivity. Next, it discusses the application of IRT in PCBs. The structure of PCBs and their relative positions on components are generally fixed. Defect detection in PCBs involves feature matching, and the accuracy of results varies with different parameters. Establishing a neural network in infrared non-destructive defect detection during soldering can enhance the feasibility of defect detection in soldering. In summary, IRT technology, by observing thermal distribution, can identify and address potential thermal issues, faults, or deficiencies in electronic components such as PCBs, chips, soldering, and GFRP. Table 8 provides a summary of the applications of IRT excitation sources in electronic component defect detection and diagnosis.

Table 8. Applications of IRT excitation sources in electronic component defect detection and diagnosis.

Exciting Source	Authors, Year	Purpose	Fundings	Remarks
Linear frequency modulation (LFM) laser	Tang et al. [150] 2020	To perform non-destructive testing on surface/sub-surface damage during the production process of semiconductor silicon wafers.	It could effectively identified microcracks of 10 µm, with a theoretical minimum detectable temperature difference less than 0.401 K.	In theory, microcracks with a width of 10 µm can be detected.
Barker-code laser	Bu et al. [151] 2022	Conducting non-destructive testing on semiconductor silicon wafers.	The BCLIT technology enhanced the signal-to-noise ratio and defect detectability.	Providing theoretical basis and operational reference for BCLIT technology in detecting microcrack defects in semiconductor silicon wafers.
Multi-spot laser	Yang et al. [152] 2016	Real-time inspection during the process of semiconductor chip manufacturing.	Successfully detected cracks within a range of 20 µm.	The MLLT system can be further developed into a standalone system for semiconductor manufacturing facilities.
Line laser	An et al. [26] 2015	To conduct instantaneous detection of surface cracks in semiconductor chips during actual manufacturing.	Successfully conducted visual inspection of cracks in semiconductor chips with widths ranging from 28–54 µm.	Expanding from chip-level to wafer-level for more efficient and faster detection.
Semiconductor laser diode (808 nm)	Wang et al. [155] 2023	To conduct research on the multi-type defect detection of multi-layered complex structured PCBs.	Effective detection of PCBs delamination defected with a depth of 1.2 mm and micro-hole defected with a depth of 400 µm.	Laser-induced lock-in thermography is suitable for detecting defects in the complex, multilayered structure of PCBs.
Laser (808 nm)	Xu et al. [145] 2014	To investigate a thermography-based active method for solder joint inspection.	The detection method based on active thermal imaging was effective for identifying missing protrusions in flip-chip packages.	Further research is needed to differentiate subtle defects in flip-chip packaging.
IR lamp	Lu et al. [161] 2011	To investigate the defect identification method for solder joints based on pulse-phase thermography.	The phase profilometry technique employed phase difference can characterize missing solder bumps defects in high-density packaging.	The detection method based on PPT is effective in identifying missing protrusions in high-density packaging.
Fiber-coupled semiconductor laser	Wei et al. [162] 2015	To develop an intelligent system utilizing active thermal imaging technology for detecting solder joint defects.	Resolved the issue of small sample sizes in solder defect detection, achieving the highest level of identification accuracy.	The combination of active thermography with M-SVM is an effective approach for intelligent diagnosis of solder defects in microelectronic packaging.

5. Discussion

5.1. Algorithmic Detection of PV Panels

The integration of IRT and deep learning techniques significantly enhances the precision of detecting and diagnosing defects in PV panels [29,172]. This approach typically demonstrates a high level of accuracy in its detection performance, with specific metrics depending on the chosen deep learning model and the quality and scale of the dataset used [173]. The Figure 17 provides a comprehensive overview of the various research studies related to defect detection in PV panels. An analysis of the data highlights several key trends and significant findings in this area. First and foremost, it is evident that research efforts employing CNN as the primary algorithm for PV panel defect detection have been the most prolific. This dominance underscores the efficiency and high defect recognition rates achieved through CNN-based approaches. These neural networks have demonstrated remarkable capabilities in pattern recognition and have significantly advanced the field of PV panel inspection. Furthermore, the integration of cutting-edge deep learning techniques with unmanned aerial vehicles (UAVs) has resulted in a substantial boost in the efficiency of IRT for PV panel inspection [132,174,175].

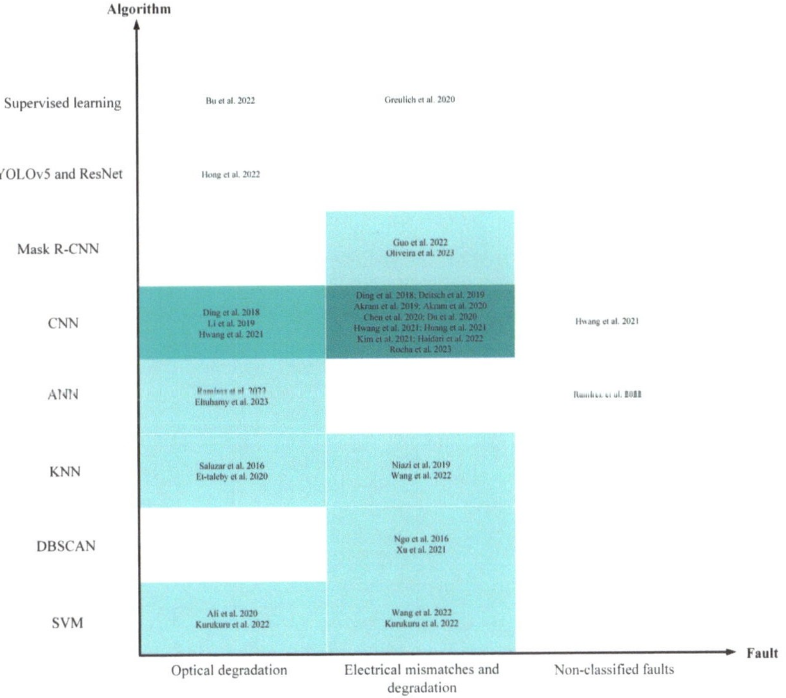

Figure 17. A summary of fault detection in PV panels based on various algorithms and techniques (K-Nearest Neighbours (KNN), You Only Look Once (YOLOv5), Deep Residual Network (ResNet), Adaptive neuro-fuzzy inference system (ANFIS), Naive Bayes (NB), Density-Based Spatial Clustering of Applications with Noise (DBSCAN), Support vector machine (SVM), Artificial neural network (ANN) [83,99,128,134–136,138,139,176–194].

Additionally, it is essential to acknowledge that earlier studies in this area have been relatively scarce, which emphasizes the rapid advancement and evolving landscape of deep learning's role in photovoltaic research. In summary, the research landscape in PV panel defect detection is marked by a strong reliance on CNN algorithms for their efficiency and high recognition rates. Additionally, the synergistic use of deep learning and UAVs with

IRT has greatly enhanced the speed and effectiveness of PV panel inspection, promising a brighter future for this field.

5.2. Excitation Sources of Electronic Industry

In the IRT inspection of PV panels, it is common practice to utilize external natural light sources or indoor lighting, such as sunlight or thermal radiation from the PV cells, as the thermal excitation source [195]. These light sources illuminate the surface of the PV panel, resulting in the absorption of energy by the panel and subsequent temperature elevation. Subsequently, an infrared thermal imaging camera captures the thermal radiation emitted by the PV panel, generating thermal images for the purpose of further analyzing and detecting anomalies or defects in the surface temperature distribution [196]. It is noteworthy that this approach typically obviates the need for additional artificial excitation sources, relying instead on naturally occurring or ambient light sources for the thermal imaging inspection. This inherent advantage enhances the convenience of the detection process and renders it suitable for the monitoring and maintenance of practical PV panels.

In contrast to infrared thermal imaging detection in PV panels, the detection of electronic components differs due to their complex and intricate structures. Often, external excitation is required to induce heating for these electronic components. This allows the thermal radiation of the object under inspection to be captured by the infrared camera, generating a thermal image. Subsequently, the obtained thermal image is subjected to further analysis to diagnose any defects in the specimen. In the infrared thermal imaging detection of electronic components, lasers are commonly used as the excitation source. This preference arises from the fact that lasers do not induce stress concentration or subsequent damage on the surface of brittle materials. The prevalent laser wavelength used for this purpose is 808 nm. Various factors influence the interaction between the laser and the sample surface during laser stimulation. The primary influencing factors encompass laser power, sampling frequency, convective heat transfer, laser beam diameter, spatial resolution, and thermal camera noise. Table 9 summarizes the characteristics of the excitation source for the detection object.

Table 9. Summary of characteristics of the excitation source for the detection object.

Detection Object	Excitation Source	Purpose	Characteristic	Remarks
PV panels	Thermal radiation [21]	The PV components exhibit abnormal temperature distribution at faulty and damaged areas.	IRT is characterized by its non-destructive testing technology for safety.	The use of machine learning methods based on IRT has been proven to have high accuracy (up to 99%) in PV detection and fault diagnosis.
Chip	Electromagnetic waves [145]	Active thermal imaging for solder joint inspection.	Inverted chips are heated by a non-contact heating source.	The active thermal imaging detection method can effectively identify missing bumps in inverted chips.
Weld	Thermal radiation [197]	Thermal imaging testing is used for the detection of sub-surface cracks in welding.	Thermal data is used to study the cooling trends in both defective and non-defective areas.	After detecting defects, they can be differentiated based on their morphology.
PCBs	Thermal radiation, electromagnetic waves [155]	Laser-induced phase-locked thermography technology is used to detect various defects in PCBs.	It can accurately identify defects with flat-bottom holes at depths of 0.2 mm and 0.6 mm.	Laser-induced lock-in thermography is suitable for detecting various types of defects in multi-layer and complex structured PCBs.

In the last few years, the field of non-destructive testing in the electronics industry has made remarkable progress. Figure 18 shows the types of excitation sources commonly used in the electronics industry in recent years, especially in the field of chips, PCBS, and welding. These excitation sources include lasers, heaters, ultrasound, electricity, and flashlights. As can be clearly seen from the figure, a wide variety of laser sources have been used in past research, and these laser sources show different advantages in different application scenarios. However, recent studies have shown that laser is increasingly used as an excitation source in non-destructive testing in the electronics industry. There are several reasons behind this trend. First, the relatively low cost of the laser makes it the preferred incentive source for many researchers and engineers. Second, the laser is able to cover a large, heated area, which is important when dealing with complex electronic components. Compared with other excitation sources, the laser has a wider heating range and can detect the properties of the target material more comprehensively. In addition, in different practical applications, the researchers found that the wavelength of 808 nm laser is the most commonly chosen laser in the electronics industry in non-destructive testing performance. In general, past and present studies have shown that lasers, as the main excitation source in non-destructive testing in the electronics industry, have the advantages of lower cost, wide heating range, and wavelength. This trend not only reflects the importance of laser technology in the electronics industry, but also provides useful enlightenment for future research and application.

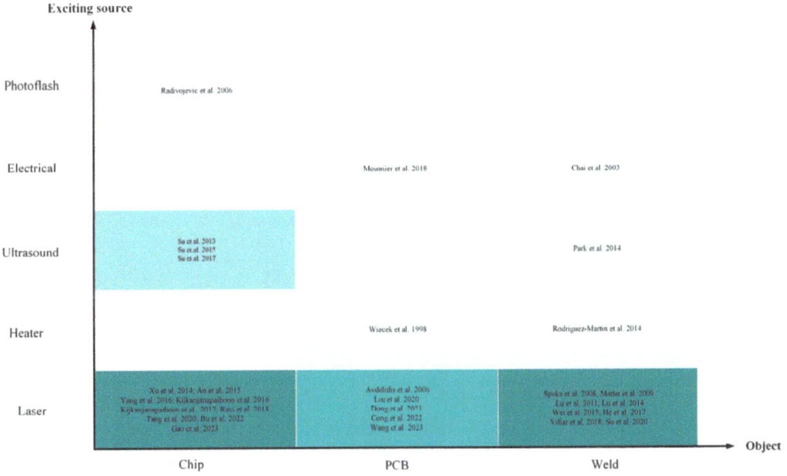

Figure 18. Type of excitation source commonly used in the electronics industry [25,26,145,150–153,155–163,197–212].

5.3. Wavelengths

The application of IRT technology in electronic components not only enables effective detection of defects at the micron level but also facilitates real-time monitoring during the manufacturing process of electronic components such as semiconductors. For instance, in the study conducted by Yang et al. [152], which encompasses data acquisition and processing, the total inspection time for each semiconductor chip is less than 1 s, successfully detecting cracks within a 20 µm range. In summary, IRT technology provides an efficient, non-destructive, and highly accurate method for defect detection in electronic components, enhancing detectability while also serving as a reference for non-destructive testing of similar materials.

Figure 19 shows the proportion of different bands in our selected references. It can be seen from the figure that the utilization rate of long-wave infrared and medium-wave infrared in the electronic industry is relatively high. In the non-destructive testing of PCB, only a single mid-infrared wave is used, and its wavelength range is 3 to 5 μm. In the non-destructive testing of chips, long infrared waves with a wavelength range from 7.5 to 14 μm are commonly used, and in the non-destructive testing of welding, both mid-infrared waves and long infrared waves appear. The absorption of infrared radiation by different materials is different, and it can be seen that the choice of wavelength is related to the size, material, structure, and other aspects of the product to be tested.

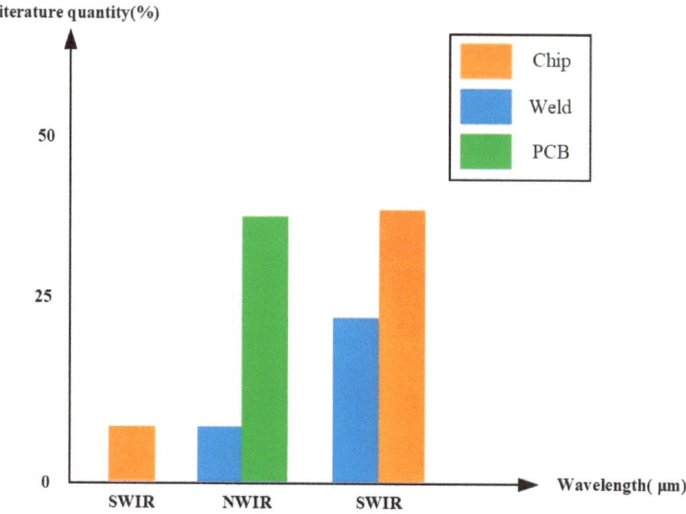

Figure 19. The proportion of the number of references at different wavelengths in the electronics industry.

6. Outlooks

The future of active infrared imaging for defect detection in the renewable and electronic industries will be characterized by advancements in excitation sources, improvements in PV panels, widespread adoption in electronics manufacturing, and seamless integration with AI, leading to more efficient, accurate, and cost-effective defect detection solutions. The outlook is given in the following areas:

(1) The future of active infrared imaging for defect detection holds promising developments in excitation sources. Research is expected to focus on more efficient, compact, and versatile excitation methods. Emerging technologies such as advanced lasers and LED arrays may provide more controlled and tailored excitation, enhancing defect visibility [213]. Future research may also delve into multi-modal excitation sources that combine various energy types, such as ultrasound and electromagnetic radiation, with infrared illumination. This fusion of excitation modalities could unlock new possibilities in defect detection by exploiting complementary interactions between materials and different energy sources.

(2) Future research endeavors should prioritize the development and refinement of an expanded array of algorithms tailored for the detection of PV panels irregularities and defects. This emphasis on algorithmic innovation is essential to further enhance the precision and efficiency of PV panel inspection, making it an exciting and crucial avenue for future research. These advanced algorithms should encompass a wide range of imaging techniques and modalities, including IRT, EL, and optical imaging, among others. By diversifying the algorithmic approaches, researchers can effectively address the multifaceted challenges associated with PV panel inspection.

(3) The electronics industry will increasingly adopt active infrared imaging for quality control and defect detection during manufacturing. Active infrared imaging will

provide real-time quality control during manufacturing processes. As electronic components are assembled, the integrated infrared sensors will continuously monitor for defects, irregularities, and variations in temperature or electrical performance. This real-time feedback loop allows for immediate adjustments and corrections, reducing the likelihood of defects propagating downstream [214]. Compact and cost-effective infrared imaging equipment will be incorporated into manufacturing lines, facilitating swift and accurate examination of electronic components. This integration will contribute to the reduction of defects, improvement of product dependability, and reduction of production expenditures.

(4) The integration of active infrared imaging with AI will revolutionize defect detection. Machine learning algorithms, particularly deep learning techniques like CNN, will become more adept at recognizing complex defect patterns and distinguishing anomalies from normal operation. AI-driven defect detection systems will be capable of real-time analysis, reducing false positives and improving overall accuracy. Beyond detecting defects, AI can predict when components or systems are likely to fail based on their thermal behavior captured through infrared imaging. This enables predictive maintenance, where machinery and equipment are serviced or replaced before they break down, reducing downtime and costly repairs.

Author Contributions: X.Z. and Y.Z. (Yangjing Zhao) wrote the first draft. S.H. and H.W. carried out formal analysis. H.W. and W.M. mainly carried out the visitation of the literature review. Y.Z. (Yuyan Zhang) and W.M. proposed the concept, methodology, and undertook fund acquisition. X.Z. and Y.Z. (Yangjing Zhao) contributed equally to the review article. All authors have read and agreed to the published version of the manuscript.

Funding: This work was supported by the Guangdong Basic and Applied Basic Research Foundation (No. 2022A1515140066) and by the Guangdong Provincial Key Laboratory of Manufacturing Equipment Digitization (No. 2023B1212060012). In addition, this work was also supported by the National Natural Science Foundation of China (No. 52105536).

Institutional Review Board Statement: Not applicable.

Informed Consent Statement: Not applicable.

Conflicts of Interest: The authors declare no conflict of interest.

References

1. Hu, Z.; Mu, E. Fundamental of Radiative Cooling. In *Infrared Radiative Cooling and Its Applications*; Energy and Environment Research in China; Springer Nature: Singapore, 2022; pp. 33–74. ISBN 978-981-19660-8-8.
2. Conley-Tyler, M. A Fundamental Choice: Internal or External Evaluation? *Eval. J. Australas.* **2005**, *4*, 3–11. [CrossRef]
3. Ming, W.; Shen, F.; Zhang, H.; Li, X.; Ma, J.; Du, J.; Lu, Y. Defect Detection of LGP Based on Combined Classifier with Dynamic Weights. *Measurement* **2019**, *143*, 211–225. [CrossRef]
4. Jorge Aldave, I.; Venegas Bosom, P.; Vega González, L.; López De Santiago, I.; Vollheim, B.; Krausz, L.; Georges, M. Review of Thermal Imaging Systems in Composite Defect Detection. *Infrared Phys. Technol.* **2013**, *61*, 167–175. [CrossRef]
5. Usamentiaga, R.; Venegas, P.; Guerediaga, J.; Vega, L.; Molleda, J.; Bulnes, F. Infrared Thermography for Temperature Measurement and Non-Destructive Testing. *Sensors* **2014**, *14*, 12305–12348. [CrossRef] [PubMed]
6. Tsanakas, J.A.; Ha, L.; Buerhop, C. Faults and Infrared Thermographic Diagnosis in Operating C-Si Photovoltaic Modules: A Review of Research and Future Challenges. *Renew. Sustain. Energy Rev.* **2016**, *62*, 695–709. [CrossRef]
7. Cabello, G.; Cuesta, A.; Gutiérrez, C. A Method for Obtaining in Situ External Reflectance Infrared Spectra in Strongly Acidic Solutions Using Fluorite Windows. *Electrochem. Commun.* **2009**, *11*, 616–618. [CrossRef]
8. Mansfield, J.R.; Sowa, M.G.; Scarth, G.B.; Somorjai, R.L.; Mantsch, H.H. Fuzzy C-Means Clustering and Principal Component Analysis of Time Series from near-Infrared Imaging of Forearm Ischemia. *Comput. Med. Imaging Graph.* **1997**, *21*, 299–308. [CrossRef] [PubMed]
9. Tankisi, H.; Burke, D.; Cui, L.; De Carvalho, M.; Kuwabara, S.; Nandedkar, S.D.; Rutkove, S.; Stålberg, E.; Van Putten, M.J.A.M.; Fuglsang-Frederiksen, A. Standards of Instrumentation of EMG. *Clin. Neurophysiol.* **2020**, *131*, 243–258. [CrossRef]
10. Ming, W.; Cao, C.; Zhang, G.; Zhang, H.; Zhang, F.; Jiang, Z.; Yuan, J. Review: Application of Convolutional Neural Network in Defect Detection of 3C Products. *IEEE Access* **2021**, *9*, 135657–135674. [CrossRef]
11. Ming, W.; Xie, Z.; Jiang, Z.; Chen, Y.; Zhang, G.; Xu, Y.; He, W. Progress in Optical Adhesive and Lamination Process of Touch Screen in 3C Products. *J. Soc. Inf. Disp.* **2022**, *30*, 851–876. [CrossRef]

12. Ming, W.; Jiang, Z.; Chen, Z.; He, W.; Li, X. Modelling and Analysis of Energy Consumption in Glass Molding Process for Smartphone Covers Using Different Heating Strategies. *Int. J. Adv. Manuf. Technol.* **2023**, *124*, 1491–1512. [CrossRef]
13. Yang, W.; Zhang, Z.; Ming, W.; Yin, L.; Zhang, G. Study on Shape Deviation and Crack of Ultra-Thin Glass Molding Process for Curved Surface. *Ceram. Int.* **2022**, *48*, 6767–6779. [CrossRef]
14. Ming, W.; Shen, F.; Li, X.; Zhang, Z.; Du, J.; Chen, Z.; Cao, Y. A Comprehensive Review of Defect Detection in 3C Glass Components. *Measurement* **2020**, *158*, 107722. [CrossRef]
15. Sun, C.; Zhang, C.; Xiong, N. Infrared and Visible Image Fusion Techniques Based on Deep Learning: A Review. *Electronics* **2020**, *9*, 2162. [CrossRef]
16. Lu, S.-D.; Wang, M.-H.; Wei, S.-E.; Liu, H.-D.; Wu, C.-C. Photovoltaic Module Fault Detection Based on a Convolutional Neural Network. *Processes* **2021**, *9*, 1635. [CrossRef]
17. Tao, C.; Wang, X.; Gao, F.; Wang, M. Fault Diagnosis of Photovoltaic Array Based on Deep Belief Network Optimized by Genetic Algorithm. *Chin. J. Electr. Eng.* **2020**, *6*, 106–114. [CrossRef]
18. Manno, D.; Cipriani, G.; Ciulla, G.; Di Dio, V.; Guarino, S.; Lo Brano, V. Deep Learning Strategies for Automatic Fault Diagnosis in Photovoltaic Systems by Thermographic Images. *Energy Convers. Manag.* **2021**, *241*, 114315. [CrossRef]
19. Shen, D.; Ming, W.; Ren, X.; Xie, Z.; Zhang, Y.; Liu, X. A Cuckoo Search Algorithm Using Improved Beta Distributing and Its Application in the Process of EDM. *Crystals* **2021**, *11*, 916. [CrossRef]
20. Du, W.; Yang, L.; Wang, H.; Gong, X.; Zhang, L.; Li, C.; Ji, L. LN-MRSCAE: A Novel Deep Learning Based Denoising Method for Mechanical Vibration Signals. *J. Vib. Control* **2023**, 107754632311517. [CrossRef]
21. Kandeal, A.W.; Elkadeem, M.R.; Kumar Thakur, A.; Abdelaziz, G.B.; Sathyamurthy, R.; Kabeel, A.E.; Yang, N.; Sharshir, S.W. Infrared Thermography-Based Condition Monitoring of Solar Photovoltaic Systems: A Mini Review of Recent Advances. *Sol. Energy* **2021**, *223*, 33–43. [CrossRef]
22. Huerta Herraiz, Á.; Pliego Marugán, A.; García Márquez, F.P. Photovoltaic Plant Condition Monitoring Using Thermal Images Analysis by Convolutional Neural Network-Based Structure. *Renew. Energy* **2020**, *153*, 334–348. [CrossRef]
23. Du, W.; Hu, P.; Wang, H.; Gong, X.; Wang, S. Fault Diagnosis of Rotating Machinery Based on 1D–2D Joint Convolution Neural Network. *IEEE Trans. Ind. Electron.* **2023**, *70*, 5277–5285. [CrossRef]
24. Liu, Z.; Qu, B. Machine Vision Based Online Detection of PCB Defect. *Microprocess. Microsyst.* **2021**, *82*, 103807. [CrossRef]
25. Kölzer, J.; Oesterschulze, E.; Deboy, G. Thermal Imaging and Measurement Techniques for Electronic Materials and Devices. *Microelectron. Eng.* **1996**, *31*, 251–270. [CrossRef]
26. An, Y.-K.; Yang, J.; Hwang, S.; Sohn, H. Line Laser Lock-in Thermography for Instantaneous Imaging of Cracks in Semiconductor Chips. *Opt. Lasers Eng.* **2015**, *73*, 128–136. [CrossRef]
27. Kozlowska, A.; Latoszek, M.; Tomm, J.W.; Weik, F.; Elsaesser, T.; Zbroszczyk, M.; Bugajski, M.; Spellenberg, B.; Bassler, M. Analysis of Thermal Images from Diode Lasers: Temperature Profiling and Reliability Screening. *Appl. Phys. Lett.* **2005**, *86*, 203503. [CrossRef]
28. Wang, C.; Zhang, X.; Hu, W. Organic Photodiodes and Phototransistors toward Infrared Detection: Materials, Devices, and Applications. *Chem. Soc. Rev.* **2020**, *49*, 653–670. [CrossRef] [PubMed]
29. Khodayar, F.; Sojasi, S.; Maldague, X. Infrared Thermography and NDT: 2050 Horizon. *Quant. InfraRed Thermogr. J.* **2016**, *13*, 210–231. [CrossRef]
30. Mouahid, A. Infrared Thermography Used for Composite Materials. *MATEC Web Conf.* **2018**, *191*, 00011. [CrossRef]
31. Busse, G.; Wu, D.; Karpen, W. Thermal Wave Imaging with Phase Sensitive Modulated Thermography. *J. Appl. Phys.* **1992**, *71*, 3962–3965. [CrossRef]
32. Wu, D.; Busse, G. Lock-in Thermography for Nondestructive Evaluation of Materials. *Rev. Générale Therm.* **1998**, *37*, 693–703. [CrossRef]
33. Montinaro, N.; Cerniglia, D.; Pitarresi, G. Detection and Characterisation of Disbonds on Fibre Metal Laminate Hybrid Composites by Flying Laser Spot Thermography. *Compos. Part B Eng.* **2017**, *108*, 164–173. [CrossRef]
34. Weekes, B.; Almond, D.P.; Cawley, P.; Barden, T. Eddy-Current Induced Thermography—Probability of Detection Study of Small Fatigue Cracks in Steel, Titanium and Nickel-Based Superalloy. *NDT E Int.* **2012**, *49*, 47–56. [CrossRef]
35. Doshvarpassand, S.; Wu, C.; Wang, X. An Overview of Corrosion Defect Characterization Using Active Infrared Thermography. *Infrared Phys. Technol.* **2019**, *96*, 366–389. [CrossRef]
36. Milovanović, B.; Banjad Pečur, I. Review of Active IR Thermography for Detection and Characterization of Defects in Reinforced Concrete. *J. Imaging* **2016**, *2*, 11. [CrossRef]
37. Smithies, F. Conduction of Heat in Solids. By H. S. Carslaw and J. C. Jaeger Pp. Viii 386. 30s. 1947. (Oxford, at the Clarendon Press). *Math. Gaz.* **1952**, *36*, 142–143. [CrossRef]
38. Parker, W.J.; Jenkins, R.J.; Butler, C.P.; Abbott, G.L. Flash Method of Determining Thermal Diffusivity, Heat Capacity, and Thermal Conductivity. *J. Appl. Phys.* **1961**, *32*, 1679–1684. [CrossRef]
39. Balageas, D.L.; Krapez, J.C.; Cielo, P. Pulsed Photothermal Modeling of Layered Materials. *J. Appl. Phys.* **1986**, *59*, 348–357. [CrossRef]
40. Vavilov, V.; Maldague, X. Optimization of Heating Protocol in Thermal NDT, Short and Long Heating Pulses: A Discussion. *Res. Nondestruct. Eval.* **1994**, *6*, 1–18. [CrossRef]

41. Vavilov, V.; Grinzato, E.; Bison, P.G.; Marinetti, S.; Bales, M.J. Surface Transient Temperature Inversion for Hidden Corrosion Characterisation: Theory and Applications. *Int. J. Heat Mass Transf.* **1996**, *39*, 355–371. [CrossRef]
42. Vavilov, V.P. Pulsed Thermal NDT of Materials: Back to the Basics. *Nondestruct. Test. Eval.* **2007**, *22*, 177–197. [CrossRef]
43. Almond, D.P.; Angioni, S.L.; Pickering, S.G. Long Pulse Excitation Thermographic Non-Destructive Evaluation. *NDT E Int.* **2017**, *87*, 7–14. [CrossRef]
44. Maclachlan Spicer, J.W.; Kerns, W.D.; Aamodt, L.C.; Murphy, J.C. Measurement of Coating Physical Properties and Detection of Coating Disbonds by Time-Resolved Infrared Radiometry. *J. Nondestruct. Eval.* **1989**, *8*, 107–120. [CrossRef]
45. Osiander, R.; Spicer, J.W.M. Time-Resolved Infrared Radiometry with Step Heating. A Review. *Rev. Générale Therm.* **1998**, *37*, 680–692. [CrossRef]
46. Lau, S.K.; Almond, D.P.; Milne, J.M. A Quantitative Analysis of Pulsed Video Thermography. *NDT E Int.* **1991**, *24*, 195–202. [CrossRef]
47. Ibarra-Castanedo, C.; Genest, M.; Piau, J.-M.; Guibert, S.; Bendada, A.; Maldague, X.P.V. Active infrared thermography techniques for the nondestructive testing of materials. In *Ultrasonic and Advanced Methods for Nondestructive Testing and Material Characterization*; World Scientific: Singapore, 2007; pp. 325–348. ISBN 978-981-270-409-2.
48. Ibarra-Castanedo, C.; González, D.; Klein, M.; Pilla, M.; Vallerand, S.; Maldague, X. Infrared Image Processing and Data Analysis. *Infrared Phys. Technol.* **2004**, *46*, 75–83. [CrossRef]
49. Vergura, S.; Falcone, O. Filtering and Processing IR Images of PV Modules. *Renew. Energy Power Qual. J.* **2011**, *1*, 1209–1214. [CrossRef]
50. Njuguna, J.C.; Alabay, E.; Celebi, A. Field Programmable Logic Arrays Implementation of Scene-Based Nonuniformity Correction Algorithm. In Proceedings of the 2021 8th International Conference on Electrical and Electronics Engineering (ICEEE), Antalya, Turkey, 9 April 2021; IEEE: New York, NY, USA, 2021; pp. 6–9.
51. Zhang, D.; Sun, H.; Wang, D.; Liu, J.; Chen, C. Modified Two-Point Correction Method for Wide-Spectrum LWIR Detection System. *Sensors* **2023**, *23*, 2054. [CrossRef]
52. Sun, J.; Zhao, K.; Jiang, T. A Multipoint Correction Method for Environmental Temperature Changes in Airborne Double-Antenna Microwave Radiometers. *Sensors* **2014**, *14*, 7820–7830. [CrossRef]
53. Liu, Y.; Long, T.; Jiao, W.; He, G.; Chen, B.; Huang, P. A General Relative Radiometric Correction Method for Vignetting and Chromatic Aberration of Multiple CCDs: Take the Chinese Series of Gaofen Satellite Level-0 Images for Example. *IEEE Trans. Geosci. Remote Sens.* **2022**, *60*, 5616725. [CrossRef]
54. Liu, C.; Sui, X.; Liu, Y.; Kuang, X.; Gu, G.; Chen, Q. FPN Estimation Based Nonuniformity Correction for Infrared Imaging System. *Infrared Phys. Technol.* **2019**, *96*, 22–29. [CrossRef]
55. Simonetto, A.; Mokhtari, A.; Koppel, A.; Leus, G.; Ribeiro, A. A Class of Prediction-Correction Methods for Time-Varying Convex Optimization. *IEEE Trans. Signal Process.* **2016**, *64*, 4576–4591. [CrossRef]
56. Qian, W.; Chen, Q.; Gu, G. Space Low-Pass and Temporal High-Pass Nonuniformity Correction Algorithm. *Opt. Rev.* **2010**, *17*, 24–29. [CrossRef]
57. Harris, J.G.; Chiang, Y.M. Nonuniformity Correction of Infrared Image Sequences Using the Constant-Statistics Constraint. *IEEE Trans. Image Process.* **1999**, *8*, 1148–1151. [CrossRef] [PubMed]
58. Wang, S.; Zhang, S.; Liu, N. Kalman Filter for Stripe Non-Uniformity Correction in Infrared Focal Plane Arrays. In Proceedings of the 2016 International Symposium on Computer, Consumer and Control (IS3C), Xi'an, China, 4–6 July 2016; pp. 124–127.
59. Scribner, D.A.; Sarkady, K.A.; Kruer, M.R.; Caulfield, J.T.; Hunt, J.D.; Herman, C. *Adaptive Nonuniformity Correction for IR Focal-Plane Arrays Using Neural Networks*; Jayadev, T.S.J., Ed.; SPIE: San Diego, CA, USA, 1991; pp. 100–109.
60. Liu, N.; Xie, J. Interframe Phase-Correlated Registration Scene-Based Nonuniformity Correction Technology. *Infrared Phys. Technol.* **2015**, *69*, 198–205. [CrossRef]
61. Song, K.S.; Kang, M.G. Optimized Tone Mapping Function for Contrast Enhancement Considering Human Visual Perception System. *IEEE Trans. Circuits Syst. Video Technol.* **2019**, *29*, 3199–3210. [CrossRef]
62. Liu, C.; Sui, X.; Kuang, X.; Liu, Y.; Gu, G.; Chen, Q. Optimized Contrast Enhancement for Infrared Images Based on Global and Local Histogram Specification. *Remote Sens.* **2019**, *11*, 849. [CrossRef]
63. Land, E.H. Recent Advances in Retinex Theory and Some Implications for Cortical Computations: Color Vision and the Natural Image. *Proc. Natl. Acad. Sci. USA* **1983**, *80*, 5163–5169. [CrossRef]
64. He, W.; Sui, Z.; Liu, Y.; Liu, T.; Du, J.; Ma, J.; Cao, Y.; Ming, W. Feature Fusion Classifier With Dynamic Weights for Abnormality Detection of Amniotic Fluid Cell Chromosome. *IEEE Access* **2023**, *11*, 31755–31766. [CrossRef]
65. He, W.; Jiang, Z.; Ming, W.; Zhang, G.; Yuan, J.; Yin, L. A Critical Review for Machining Positioning Based on Computer Vision. *Measurement* **2021**, *184*, 109973. [CrossRef]
66. Zhang, Z.; Ming, W.; Zhang, G.; Huang, Y.; Wen, X.; Huang, H. A New Method for On-Line Monitoring Discharge Pulse in WEDM-MS Process. *Int. J. Adv. Manuf. Technol.* **2015**, *81*, 1403–1418. [CrossRef]
67. Saintey, M.B.; Almond, D.P. An Artificial Neural Network Interpreter for Transient Thermography Image Data. *NDT E Int.* **1997**, *30*, 291–295. [CrossRef]
68. Li, K.; Ma, Z.; Fu, P.; Krishnaswamy, S. Quantitative Evaluation of Surface Crack Depth with a Scanning Laser Source Based on Particle Swarm Optimization-Neural Network. *NDT E Int.* **2018**, *98*, 208–214. [CrossRef]

69. Yu, X. Fuzzy Infrared Image Segmentation Based on Multilayer Immune Clustering Neural Network. *Optik* **2017**, *140*, 959–963. [CrossRef]

70. Qi, C.; Li, Q.; Liu, Y.; Ni, J.; Ma, R.; Xu, Z. Infrared Image Segmentation Based on Multi-Information Fused Fuzzy Clustering Method for Electrical Equipment. *Int. J. Adv. Robot. Syst.* **2020**, *17*, 172988142090960. [CrossRef]

71. Zheng, C.; Kammen, D.M. An Innovation-Focused Roadmap for a Sustainable Global Photovoltaic Industry. *Energy Policy* **2014**, *67*, 159–169. [CrossRef]

72. Wilson, G.M.; Al-Jassim, M.; Metzger, W.K.; Glunz, S.W.; Verlinden, P.; Xiong, G.; Mansfield, L.M.; Stanbery, B.J.; Zhu, K.; Yan, Y.; et al. The 2020 Photovoltaic Technologies Roadmap. *J. Phys. Appl. Phys.* **2020**, *53*, 493001. [CrossRef]

73. Atsu, D.; Seres, I.; Aghaei, M.; Farkas, I. Analysis of Long-Term Performance and Reliability of PV Modules under Tropical Climatic Conditions in Sub-Saharan. *Renew. Energy* **2020**, *162*, 285–295. [CrossRef]

74. Sharma, V.; Chandel, S.S. Performance and Degradation Analysis for Long Term Reliability of Solar Photovoltaic Systems: A Review. *Renew. Sustain. Energy Rev.* **2013**, *27*, 753–767. [CrossRef]

75. Aghaei, M.; Fairbrother, A.; Gok, A.; Ahmad, S.; Kazim, S.; Lobato, K.; Oreski, G.; Reinders, A.; Schmitz, J.; Theelen, M.; et al. Review of Degradation and Failure Phenomena in Photovoltaic Modules. *Renew. Sustain. Energy Rev.* **2022**, *159*, 112160. [CrossRef]

76. Dhimish, M.; Tyrrell, A.M. Power Loss and Hotspot Analysis for Photovoltaic Modules Affected by Potential Induced Degradation. *Npj Mater. Degrad.* **2022**, *6*, 11. [CrossRef]

77. Hong, Y.-Y.; Pula, R.A. Methods of Photovoltaic Fault Detection and Classification: A Review. *Energy Rep.* **2022**, *8*, 5898–5929. [CrossRef]

78. Towaranonte, B.; Gao, Y. Application of Charge-Coupled Device (CCD) Cameras in Electrochemiluminescence: A Minireview. *Anal. Lett.* **2022**, *55*, 186–202. [CrossRef]

79. Burke, B.; Jorden, P.; Vu, P. CCD Technology. *Exp. Astron.* **2006**, *19*, 69–102. [CrossRef]

80. Buerhop, C.; Bommes, L.; Schlipf, J.; Pickel, T.; Fladung, A.; Peters, I.M. Infrared Imaging of Photovoltaic Modules: A Review of the State of the Art and Future Challenges Facing Gigawatt Photovoltaic Power Stations. *Prog. Energy* **2022**, *4*, 042010. [CrossRef]

81. Aghaei, M.; Grimaccia, F.; Gonano, C.A.; Leva, S. Innovative Automated Control System for PV Fields Inspection and Remote Control. *IEEE Trans. Ind. Electron.* **2015**, *62*, 7287–7296. [CrossRef]

82. Gallardo-Saavedra, S.; Hernández-Callejo, L.; Duque-Perez, O. Technological Review of the Instrumentation Used in Aerial Thermographic Inspection of Photovoltaic Plants. *Renew. Sustain. Energy Rev.* **2018**, *93*, 566–579. [CrossRef]

83. Niazi, K.A.K.; Akhtar, W.; Khan, H.A.; Yang, Y.; Athar, S. Hotspot Diagnosis for Solar Photovoltaic Modules Using a Naive Bayes Classifier. *Sol. Energy* **2019**, *190*, 34–43. [CrossRef]

84. Herraiz, Á.H.; Marugán, A.P.; Márquez, F.P.G. A Review on Condition Monitoring System for Solar Plants Based on Thermography. In *Non-Destructive Testing and Condition Monitoring Techniques for Renewable Energy Industrial Assets*; Elsevier: Amsterdam, The Netherlands, 2020; pp. 103–118. ISBN 978-0-08-101094-5.

85. Van Eck, N.J.; Waltman, L. Software Survey: VOSviewer, a Computer Program for Bibliometric Mapping. *Scientometrics* **2010**, *84*, 523–538. [CrossRef]

86. Madurai Elavarasan, R.; Mudgal, V.; Selvamanohar, L.; Wang, K.; Huang, G.; Shafiullah, G.M.; Markides, C.N.; Reddy, K.S.; Nadarajah, M. Pathways toward High-Efficiency Solar Photovoltaic Thermal Management for Electrical, Thermal and Combined Generation Applications: A Critical Review. *Energy Convers. Manag.* **2022**, *255*, 115278. [CrossRef]

87. Fernández-Solas, Á.; Micheli, L.; Almonacid, F.; Fernández, E.F. Optical Degradation Impact on the Spectral Performance of Photovoltaic Technology. *Renew. Sustain. Energy Rev.* **2021**, *141*, 110782. [CrossRef]

88. Feron, S.; Cordero, R.R.; Damiani, A.; Jackson, R.B. Climate Change Extremes and Photovoltaic Power Output. *Nat. Sustain.* **2020**, *4*, 270–276. [CrossRef]

89. Dechthummarong, C.; Wiengmoon, B.; Chenvidhya, D.; Jivacate, C.; Kirtikara, K. Physical Deterioration of Encapsulation and Electrical Insulation Properties of PV Modules after Long-Term Operation in Thailand. *Sol. Energy Mater. Sol. Cells* **2010**, *94*, 1437–1440. [CrossRef]

90. Smestad, G.P.; Germer, T.A.; Alrashidi, H.; Fernández, E.F.; Dey, S.; Brahma, H.; Sarmah, N.; Ghosh, A.; Sellami, N.; Hassan, I.A.I.; et al. Modelling Photovoltaic Soiling Losses through Optical Characterization. *Sci. Rep.* **2020**, *10*, 58. [CrossRef] [PubMed]

91. Segbefia, O.K.; Akhtar, N.; Sætre, T.O. Defects and Fault Modes of Field-Aged Photovoltaic Modules in the Nordics. *Energy Rep.* **2023**, *9*, 3104–3119. [CrossRef]

92. Segbefia, O.K. Temperature Profiles of Field-Aged Photovoltaic Modules Affected by Optical Degradation. *Heliyon* **2023**, *9*, e19566. [CrossRef] [PubMed]

93. Mahdi, H.A.; Leahy, P.; Morrison, A. Predicting Early EVA Degradation in Photovoltaic Modules From Short Circuit Current Measurements. *IEEE J. Photovolt.* **2021**, *11*, 1188–1196. [CrossRef]

94. Zhang, Z.; Qiu, W.; Zhang, G.; Liu, D.; Wang, P. Progress in Applications of Shockwave Induced by Short Pulsed Laser on Surface Processing. *Opt. Laser Technol.* **2023**, *157*, 108760. [CrossRef]

95. Sinha, A.; Sastry, O.S.; Gupta, R. Nondestructive Characterization of Encapsulant Discoloration Effects in Crystalline-Silicon PV Modules. *Sol. Energy Mater. Sol. Cells* **2016**, *155*, 234–242. [CrossRef]

96. Buerhop, C.; Schlegel, D.; Niess, M.; Vodermayer, C.; Weißmann, R.; Brabec, C.J. Reliability of IR-Imaging of PV-Plants under Operating Conditions. *Sol. Energy Mater. Sol. Cells* **2012**, *107*, 154–164. [CrossRef]

97. Segbefia, O.K.; Akhtar, N.; Sætre, T.O. Moisture Induced Degradation in Field-Aged Multicrystalline Silicon Photovoltaic Modules. *Sol. Energy Mater. Sol. Cells* **2023**, *258*, 112407. [CrossRef]

98. Kumar, S.; Meena, R.; Gupta, R. Imaging and Micro-Structural Characterization of Moisture Induced Degradation in Crystalline Silicon Photovoltaic Modules. *Sol. Energy* **2019**, *194*, 903–912. [CrossRef]

99. Du, B.; He, Y.; He, Y.; Duan, J.; Zhang, Y. Intelligent Classification of Silicon Photovoltaic Cell Defects Based on Eddy Current Thermography and Convolution Neural Network. *IEEE Trans. Ind. Inform.* **2020**, *16*, 6242–6251. [CrossRef]

100. Balasubramani, G.; Thangavelu, V.; Chinnusamy, M.; Subramaniam, U.; Padmanaban, S.; Mihet-Popa, L. Infrared Thermography Based Defects Testing of Solar Photovoltaic Panel with Fuzzy Rule-Based Evaluation. *Energies* **2020**, *13*, 1343. [CrossRef]

101. Köntges, M.; Kunze, I.; Kajari-Schröder, S.; Breitenmoser, X.; Bjørneklett, B. The Risk of Power Loss in Crystalline Silicon Based Photovoltaic Modules Due to Micro-Cracks. *Sol. Energy Mater. Sol. Cells* **2011**, *95*, 1131–1137. [CrossRef]

102. Dhoke, A.; Sharma, R.; Saha, T.K. A Technique for Fault Detection, Identification and Location in Solar Photovoltaic Systems. *Sol. Energy* **2020**, *206*, 864–874. [CrossRef]

103. Pei, T.; Hao, X. A Fault Detection Method for Photovoltaic Systems Based on Voltage and Current Observation and Evaluation. *Energies* **2019**, *12*, 1712. [CrossRef]

104. Muñoz, J.; Lorenzo, E.; Martínez-Moreno, F.; Marroyo, L.; García, M. An Investigation into Hot-Spots in Two Large Grid-Connected PV Plants. *Prog. Photovolt. Res. Appl.* **2008**, *16*, 693–701. [CrossRef]

105. Belhaouas, N.; Mehareb, F.; Kouadri-Boudjelthia, E.; Assem, H.; Bensalem, S.; Hadjrioua, F.; Aissaoui, A.; Hafdaoui, H.; Chahtou, A.; Bakria, K. The Performance of Solar PV Modules with Two Glass Types after 11 Years of Outdoor Exposure under the Mediterranean Climatic Conditions. *Sustain. Energy Technol. Assess.* **2022**, *49*, 101771. [CrossRef]

106. Tsanakas, J.A.; Chrysostomou, D.; Botsaris, P.N.; Gasteratos, A. Fault Diagnosis of Photovoltaic Modules through Image Processing and Canny Edge Detection on Field Thermographic Measurements. *Int. J. Sustain. Energy* **2015**, *34*, 351–372. [CrossRef]

107. Aziz, F.; Ul Haq, A.; Ahmad, S.; Mahmoud, Y.; Jalal, M.; Ali, U. A Novel Convolutional Neural Network-Based Approach for Fault Classification in Photovoltaic Arrays. *IEEE Access* **2020**, *8*, 41889–41904. [CrossRef]

108. Phinikarides, A.; Kindyni, N.; Makrides, G.; Georghiou, G.E. Review of Photovoltaic Degradation Rate Methodologies. *Renew. Sustain. Energy Rev.* **2014**, *40*, 143–152. [CrossRef]

109. Dhere, N.; Kaul, A.; Schneller, E.; Shiradkar, N. High-Voltage Bias Testing of PV Modules in the Hot and Humid Climate without Inducing Irreversible Instantaneous Degradation. In Proceedings of the 2012 38th IEEE Photovoltaic Specialists Conference, Austin, TX, USA, 3–8 June 2012; pp. 002445–002448.

110. Bouaichi, A.; Merrouni, A.A.; El Amrani, A.; Jaeckel, B.; Hajjaj, C.; Naimi, Z.; Messaoudi, C. Long-Term Experiment on p-Type Crystalline PV Module with Potential Induced Degradation: Impact on Power Performance and Evaluation of Recovery Mode. *Renew. Energy* **2022**, *183*, 472–479. [CrossRef]

111. Mellit, A. An Embedded Solution for Fault Detection and Diagnosis of Photovoltaic Modules Using Thermographic Images and Deep Convolutional Neural Networks. *Eng. Appl. Artif. Intell.* **2022**, *116*, 105459. [CrossRef]

112. Dhimish, M.; Mather, P.; Holmes, V. Novel Photovoltaic Hot-Spotting Fault Detection Algorithm. *IEEE Trans. Device Mater. Reliab.* **2019**, *19*, 378–386. [CrossRef]

113. Shen, D.; Zhang, S.; Ming, W.; He, W.; Zhang, G.; Xie, Z. Development of a New Machine Vision Algorithm to Estimate Potato's Shape and Size Based on Support Vector Machine. *J. Food Process Eng.* **2022**, *45*, e13974. [CrossRef]

114. Wang, Y.; Li, Q.; Chu, M.; Kang, X.; Liu, G. Application of Infrared Thermography and Machine Learning Techniques in Cattle Health Assessments: A Review. *Biosyst. Eng.* **2023**, *230*, 361–387. [CrossRef]

115. He, W.; Li, Z.; Liu, T.; Liu, Z.; Guo, X.; Du, J.; Li, X.; Sun, P.; Ming, W. Research Progress and Application of Deep Learning in Remaining Useful Life, State of Health and Battery Thermal Management of Lithium Batteries. *J. Energy Storage* **2023**, *70*, 107868. [CrossRef]

116. Ming, W.; Sun, P.; Zhang, Z.; Qiu, W.; Du, J.; Li, X.; Zhang, Y.; Zhang, G.; Liu, K.; Wang, Y.; et al. A Systematic Review of Machine Learning Methods Applied to Fuel Cells in Performance Evaluation, Durability Prediction, and Application Monitoring. *Int. J. Hydrog. Energy* **2023**, *48*, 5197–5228. [CrossRef]

117. Wu, Y.; Pan, F.; An, Q.; Wang, J.; Feng, X.; Cao, J. Infrared Target Detection Based on Deep Learning. In Proceedings of the 2021 40th Chinese Control Conference (CCC), Shanghai, China, 26–28 July 2021; IEEE: New York, NY, USA, 2021; pp. 8175–8180.

118. He, W.; Liu, T.; Han, Y.; Ming, W.; Du, J.; Liu, Y.; Yang, Y.; Wang, L.; Jiang, Z.; Wang, Y.; et al. A Review: The Detection of Cancer Cells in Histopathology Based on Machine Vision. *Comput. Biol. Med.* **2022**, *146*, 105636. [CrossRef]

119. Ming, W.; Zhang, S.; Liu, X.; Liu, K.; Yuan, J.; Xie, Z.; Sun, P.; Guo, X. Survey of Mura Defect Detection in Liquid Crystal Displays Based on Machine Vision. *Crystals* **2021**, *11*, 1444. [CrossRef]

120. Brunet, R.; Aumeunier, M.-H.; Miorelli, R.; Juven, A.; Reboud, C.; Artusi, X.; Bohec, M.L. Infrared Measurement Synthetic Database for Inverse Thermography Model Based on Deep Learning. *Fusion Eng. Des.* **2023**, *192*, 113598. [CrossRef]

121. Dincer, F.; Akgol, O.; Karaaslan, M.; Unal, E.; Sabah, C. Polarization angle independent perfect metamaterial absorbers for solar cell applications in the microwave, infrared, and visible regime. *Prog. Electromagn. Res.* **2014**, *144*, 93–101. [CrossRef]

122. Chandel, S.S.; Nagaraju Naik, M.; Sharma, V.; Chandel, R. Degradation Analysis of 28 Year Field Exposed Mono-c-Si Photovoltaic Modules of a Direct Coupled Solar Water Pumping System in Western Himalayan Region of India. *Renew. Energy* **2015**, *78*, 193–202. [CrossRef]

123. Adams, J.; Salvador, M.; Lucera, L.; Langner, S.; Spyropoulos, G.D.; Fecher, F.W.; Voigt, M.M.; Dowland, S.A.; Osvet, A.; Egelhaaf, H.; et al. Water Ingress in Encapsulated Inverted Organic Solar Cells: Correlating Infrared Imaging and Photovoltaic Performance. *Adv. Energy Mater.* **2015**, *5*, 1501065. [CrossRef]

124. Du, B.; Yang, R.; He, Y.; Wang, F.; Huang, S. Nondestructive Inspection, Testing and Evaluation for Si-Based, Thin Film and Multi-Junction Solar Cells: An Overview. *Renew. Sustain. Energy Rev.* **2017**, *78*, 1117–1151. [CrossRef]

125. Addabbo, P.; Angrisano, A.; Bernardi, M.L.; Gagliarde, G.; Mennella, A.; Nisi, M.; Ullo, S. A UAV Infrared Measurement Approach for Defect Detection in Photovoltaic Plants. In Proceedings of the 2017 IEEE International Workshop on Metrology for AeroSpace (MetroAeroSpace), Padua, Italy, 21–23 June 2017; IEEE: New York, NY, USA, 2021; pp. 345–350.

126. He, Y.; Du, B.; Huang, S. Noncontact Electromagnetic Induction Excited Infrared Thermography for Photovoltaic Cells and Modules Inspection. *IEEE Trans. Ind. Inform.* **2018**, *14*, 5585–5593. [CrossRef]

127. Zefri, Y.; ElKettani, A.; Sebari, I.; Lamallam, S.A. Thermal Infrared and Visual Inspection of Photovoltaic Installations by UAV Photogrammetry—Application Case: Morocco. *Drones* **2018**, *2*, 41. [CrossRef]

128. Akram, M.W.; Li, G.; Jin, Y.; Chen, X.; Zhu, C.; Ahmad, A. Automatic Detection of Photovoltaic Module Defects in Infrared Images with Isolated and Develop-Model Transfer Deep Learning. *Sol. Energy* **2020**, *198*, 175–186. [CrossRef]

129. Fonseca Alves, R.H.; Deus Júnior, G.A.D.; Marra, E.G.; Lemos, R.P. Automatic Fault Classification in Photovoltaic Modules Using Convolutional Neural Networks. *Renew. Energy* **2021**, *179*, 502–516. [CrossRef]

130. Demant, M.; Oswald, M.; Welschehold, T.; Nold, S.; Bartsch, S.; Schoenfelder, S.; Rein, S. Micro-Cracks in Silicon Wafers and Solar Cells: Detection and Rating of Mechanical Strength and Electrical Quality. In Proceedings of the 29th Solar Energy Conference and Exhibition, Amsterdam, The Netherlands, 22–26 September 2014; pp. 390–396. [CrossRef]

131. Stromer, D.; Vetter, A.; Oezkan, H.C.; Probst, C.; Maier, A. Enhanced Crack Segmentation (eCS): A Reference Algorithm for Segmenting Cracks in Multicrystalline Silicon Solar Cells. *IEEE J. Photovolt.* **2019**, *9*, 752–758. [CrossRef]

132. Li, X.; Yang, Q.; Chen, Z.; Luo, X.; Yan, W. Visible Defects Detection Based on UAV-based Inspection in Large-scale Photovoltaic Systems. *IET Renew. Power Gener.* **2017**, *11*, 1234–1244. [CrossRef]

133. Su, B.; Chen, H.; Zhu, Y.; Liu, W.; Liu, K. Classification of Manufacturing Defects in Multicrystalline Solar Cells With Novel Feature Descriptor. *IEEE Trans. Instrum. Meas.* **2019**, *68*, 4675–4688. [CrossRef]

134. Chen, H.; Pang, Y.; Hu, Q.; Liu, K. Solar Cell Surface Defect Inspection Based on Multispectral Convolutional Neural Network. *J. Intell. Manuf.* **2020**, *31*, 453–468. [CrossRef]

135. Ding, S.; Yang, Q.; Li, X.; Yan, W.; Ruan, W. Transfer Learning Based Photovoltaic Module Defect Diagnosis Using Aerial Images. In Proceedings of the 2018 International Conference on Power System Technology (POWERCON), Guangzhou, China, 6–8 November 2018; IEEE: New York, NY, USA, 2018; pp. 4245–4250.

136. Li, X.; Yang, Q.; Lou, Z.; Yan, W. Deep Learning Based Module Defect Analysis for Large-Scale Photovoltaic Farms. *IEEE Trans. Energy Convers.* **2019**, *34*, 520–529. [CrossRef]

137. Demant, M.; Virtue, P.; Kovvali, A.S.; Yu, S.X.; Rein, S. Deep Learning Approach to Inline Quality Rating and Mapping of Multi-Crystalline Si-Wafers. In Proceedings of the 35th European Photovoltaic Solar Energy Conference and Exhibition 2018, Brussels, Belgium, 24–28 September 2018; pp. 814–818. [CrossRef]

138. Deitsch, S.; Christlein, V.; Berger, S.; Buerhop-Lutz, C.; Maier, A.; Gallwitz, F.; Riess, C. Automatic Classification of Defective Photovoltaic Module Cells in Electroluminescence Images. *Sol. Energy* **2019**, *185*, 455–468. [CrossRef]

139. Akram, M.W.; Li, G.; Jin, Y.; Chen, X.; Zhu, C.; Zhao, X.; Khaliq, A.; Faheem, M.; Ahmad, A. CNN Based Automatic Detection of Photovoltaic Cell Defects in Electroluminescence Images. *Energy* **2019**, *189*, 116319. [CrossRef]

140. Ming, W.; Xie, Z.; Ma, J.; Du, J.; Zhang, G.; Cao, C.; Zhang, Y. Critical Review on Sustainable Techniques in Electrical Discharge Machining. *J. Manuf. Process.* **2021**, *72*, 375–399. [CrossRef]

141. Ming, W.; Xie, Z.; Cao, C.; Liu, M.; Zhang, F.; Yang, Y.; Zhang, S.; Sun, P.; Guo, X. Research on EDM Performance of Renewable Dielectrics under Different Electrodes for Machining SKD11. *Crystals* **2022**, *12*, 291. [CrossRef]

142. Ming, W.; Zhang, Z.; Wang, S.; Zhang, Y.; Shen, F.; Zhang, G. Comparative Study of Energy Efficiency and Environmental Impact in Magnetic Field Assisted and Conventional Electrical Discharge Machining. *J. Clean. Prod.* **2019**, *214*, 12–28. [CrossRef]

143. Jadin, M.S.; Taib, S. Infrared Image Enhancement and Segmentation for Extracting the Thermal Anomalies in Electrical Equipment. *Electron. Electr. Eng.* **2012**, *120*, 107–112. [CrossRef]

144. Rogalski, A. Semiconductor Detectors and Focal Plane Arrays for Far-Infrared Imaging. *Opto-Electron. Rev.* **2013**, *21*, 406–426. [CrossRef]

145. Xu, Z.; Shi, T.; Lu, X.; Liao, G. Using Active Thermography for Defects Inspection of Flip Chip. *Microelectron. Reliab.* **2014**, *54*, 808–815. [CrossRef]

146. Daimon, S.; Iguchi, R.; Hioki, T.; Saitoh, E.; Uchida, K. Thermal Imaging of Spin Peltier Effect. *Nat. Commun.* **2016**, *7*, 13754. [CrossRef] [PubMed]

147. Christensen, P.R.; Hamilton, V.E.; Mehall, G.L.; Pelham, D.; O'Donnell, W.; Anwar, S.; Bowles, H.; Chase, S.; Fahlgren, J.; Farkas, Z.; et al. The OSIRIS-REx Thermal Emission Spectrometer (OTES) Instrument. *Space Sci. Rev.* **2018**, *214*, 87. [CrossRef]

148. Aragon, B.; Johansen, K.; Parkes, S.; Malbeteau, Y.; Al-Mashharawi, S.; Al-Amoudi, T.; Andrade, C.F.; Turner, D.; Lucieer, A.; McCabe, M.F. A Calibration Procedure for Field and UAV-Based Uncooled Thermal Infrared Instruments. *Sensors* **2020**, *20*, 3316. [CrossRef] [PubMed]

149. Yu, L.; Guo, Y.; Zhu, H.; Luo, M.; Han, P.; Ji, X. Low-Cost Microbolometer Type Infrared Detectors. *Micromachines* **2020**, *11*, 800. [CrossRef] [PubMed]
150. Tang, Q.-J.; Gao, S.-S.; Liu, Y.-J.; Wang, Y.-Z.; Dai, J.-M. Theoretical Study on Infrared Thermal Wave Imaging Detection of Semiconductor Silicon Wafers with Micro-Crack Defects. *Therm. Sci.* **2020**, *24*, 4011–4017. [CrossRef]
151. Bu, C.; Li, R.; Liu, T.; Shen, R.; Wang, J.; Tang, Q. Micro-Crack Defects Detection of Semiconductor Si-Wafers Based on Barker Code Laser Infrared Thermography. *Infrared Phys. Technol.* **2022**, *123*, 104160. [CrossRef]
152. Yang, J.; Hwang, S.; An, Y.-K.; Lee, K.; Sohn, H. Multi-Spot Laser Lock-in Thermography for Real-Time Imaging of Cracks in Semiconductor Chips during a Manufacturing Process. *J. Mater. Process. Technol.* **2016**, *229*, 94–101. [CrossRef]
153. Ravi, B.V.; Xie, M.; Goyal, D. Investigation of Multiple Heat Source Effects in Lock-In Thermography Applications in Semiconductor Packages. *IEEE Trans. Compon. Packag. Manuf. Technol.* **2018**, *8*, 725–734. [CrossRef]
154. Dong, Z.; Chen, L. Image Registration in PCB Fault Detection Based on Infrared Thermal Imaging. In Proceedings of the 2019 Chinese Control Conference (CCC), Guangzhou, China, 27–30 July 2019; IEEE: New York, NY, USA, 2019; pp. 4819–4823.
155. Wang, F.; Zhou, Y.; Zhang, X.; Li, Z.; Weng, J.; Qiang, G.; Chen, M.; Wang, Y.; Yue, H.; Liu, J. Laser-Induced Thermography: An Effective Detection Approach for Multiple-Type Defects of Printed Circuit Boards (PCBs) Multilayer Complex Structure. *Measurement* **2023**, *206*, 112307. [CrossRef]
156. Avdelidis, N.P.; Nicholson, P.I.; Wallace, P. Pulsed Thermography in the Investigation of PCBs for Defect Detection & Analysis. In Proceedings of the 2006 International Conference on Quantitative InfraRed Thermography, Padua, Italy, 28–30 June 2006.
157. Cong, S.; Shang, Z.; Huang, Q. Detection for Printed Circuit Boards (PCBs) Delamination Defects Using Optical/Thermal Fusion Imaging Technique. *Infrared Phys. Technol.* **2022**, *127*, 104399. [CrossRef]
158. Dong, B.; Wang, F.; Song, P.; Liu, Y.; Wang, Y.; Wei, J.; Liu, J.; Wang, Y.; Chen, M.; Liu, G. Detection on the Coating Peeling of SiC Coated C/C Composites Using Infrared Thermal-Wave Tomography. *Infrared Phys. Technol.* **2021**, *114*, 103642. [CrossRef]
159. Su, L.; Yu, X.; Li, K.; Pecht, M. Defect Inspection of Flip Chip Solder Joints Based on Non-Destructive Methods: A Review. *Microelectron. Reliab.* **2020**, *110*, 113657. [CrossRef]
160. Chai, T.C.; Wongi, B.S.; Bai, W.M.; Trigg, A.; Lain, Y.K. A Novel Defect Detection Technique Using Active Transient Thermography for High Density Package and Interconnections. In Proceedings of the 53rd Electronic Components and Technology Conference, New Orleans, LA, USA, 27–30 May 2003; pp. 920–925.
161. Lu, X.; Liao, G.; Zha, Z.; Xia, Q.; Shi, T. A Novel Approach for Flip Chip Solder Joint Inspection Based on Pulsed Phase Thermography. *NDT E Int.* **2011**, *44*, 484–489. [CrossRef]
162. Wei, W.; Wei, L.; Nie, L.; Su, L.; Lu, X. Using Active Thermography and Modified SVM for Intelligent Diagnosis of Solder Bumps. *Infrared Phys. Technol.* **2015**, *72*, 163–169. [CrossRef]
163. He, Z.; Wei, L.; Shao, M.; Lu, X. Detection of Micro Solder Balls Using Active Thermography and Probabilistic Neural Network. *Infrared Phys. Technol.* **2017**, *81*, 236–241. [CrossRef]
164. Wang, P.; Zhang, Z.; Hao, B.; Wei, S.; Huang, Y.; Zhang, G. Investigation on Heat Transfer and Ablation Mechanism of CFRP by Different Laser Scanning Directions. *Compos. Part B Eng.* **2023**, *262*, 110827. [CrossRef]
165. Huang, H.; Yang, W.; Ming, W.; Zhang, G.; Xu, Y.; Zhang, Z. Mechanism of Springback Behavior in Ultra-Thin Glass Molding Process: A Molecular Dynamics Study. *J. Non-Cryst. Solids* **2022**, *596*, 121841. [CrossRef]
166. Meola, C.; Carlomagno, G.M. Infrared Thermography to Evaluate Impact Damage in Glass/Epoxy with Manufacturing Defects. *Int. J. Impact Eng.* **2014**, *67*, 1–11. [CrossRef]
167. Dua, G.; Mulaveesala, R.; Mishra, P.; Kaur, J. InfraRed Image Correlation for Non-Destructive Testing and Evaluation of Delaminations in Glass Fibre Reinforced Polymer Materials. *Infrared Phys. Technol.* **2021**, *116*, 103803. [CrossRef]
168. Muzaffar, K.; Giri, L.I.; Chatterjee, K.; Tuli, S.; Koul, S. Fault Detection of Antenna Arrays Using Infrared Thermography. *Infrared Phys. Technol.* **2015**, *71*, 464–468. [CrossRef]
169. Wei, Z.; Osman, A.; Gross, D.; Netzelmann, U. Artificial Intelligence for Defect Detection in Infrared Images of Solid Oxide Fuel Cells. *Infrared Phys. Technol.* **2021**, *119*, 103815. [CrossRef]
170. Wang, F.; Shi, Q.; Gao, M. Research on infrared pulsed radar thermal wave imaging detection technology for impact damage of C/GFRP laminates. *Aviation Sci. Technol.* **2023**, *34*, 78–85. [CrossRef]
171. Gong, J.; Zheng, Y.; Liu, J. A Study on the SNR Performance Analysis of Laser-Generated Bidirectional Thermal Wave Radar Imaging Inspection for Hybrid C/GFRP Laminate Defects. *Infrared Phys. Technol.* **2020**, *111*, 103526. [CrossRef]
172. He, Y.; Deng, B.; Wang, H.; Cheng, L.; Zhou, K.; Cai, S.; Ciampa, F. Infrared Machine Vision and Infrared Thermography with Deep Learning: A Review. *Infrared Phys. Technol.* **2021**, *116*, 103754. [CrossRef]
173. Ming, W.; Du, J.; Shen, D.; Zhang, Z.; Li, X.; Ma, J.R.; Wang, F.; Ma, J. Visual Detection of Sprouting in Potatoes Using Ensemble-Based Classifier. *J. Food Process Eng.* **2018**, *41*, e12667. [CrossRef]
174. Kim, D.; Youn, J.; Kim, C. Automatic Detection of Malfunctioning Photovoltaic Modules Using Unmanned Aerial Vehicle Thermal Infrared Images. *J. Korean Soc. Surv. Geod. Photogramm. Cartogr.* **2016**, *34*, 619–627. [CrossRef]
175. Kim, D.; Youn, J.; Kim, C. Automatic fault recognition of photovoltaic modules based on statistical analysis of uav thermography. *Int. Arch. Photogramm. Remote Sens. Spat. Inf. Sci.* **2017**, *XLII-2/W6*, 179–182. [CrossRef]
176. Kurukuru, V.S.B.; Haque, A.; Tripathy, A.K.; Khan, M.A. Machine Learning Framework for Photovoltaic Module Defect Detection with Infrared Images. *Int. J. Syst. Assur. Eng. Manag.* **2022**, *13*, 1771–1787. [CrossRef]

177. Ali, M.U.; Khan, H.F.; Masud, M.; Kallu, K.D.; Zafar, A. A Machine Learning Framework to Identify the Hotspot in Photovoltaic Module Using Infrared Thermography. *Sol. Energy* **2020**, *208*, 643–651. [CrossRef]
178. Xu, Z.; Shen, Y.; Zhang, K.; Wei, H. A Segmentation Method for PV Modules in Infrared Thermography Images. In Proceedings of the 2021 13th IEEE PES Asia Pacific Power & Energy Engineering Conference (APPEEC), Thiruvananthapuram, India, 21 November 2021; IEEE: New York, NY, USA, 2021; pp. 1–5.
179. Bu, C.; Liu, T.; Li, R.; Shen, R.; Zhao, B.; Tang, Q. Electrical Pulsed Infrared Thermography and Supervised Learning for PV Cells Defects Detection. *Sol. Energy Mater. Sol. Cells* **2022**, *237*, 111561. [CrossRef]
180. Greulich, J.M.; Demant, M.; Kunze, P.; Dost, G.; Ramspeck, K.; Vetter, A.; Probst, C. Comparison of Inline Crack Detection Systems for Multicrystalline Silicon Solar Cells. *IEEE J. Photovolt.* **2020**, *10*, 1389–1395. [CrossRef]
181. Hong, F.; Song, J.; Meng, H.; Wang, R.; Fang, F.; Zhang, G. A Novel Framework on Intelligent Detection for Module Defects of PV Plant Combining the Visible and Infrared Images. *Sol. Energy* **2022**, *236*, 406–416. [CrossRef]
182. Shuqiang Guo, S.G.; Shuqiang Guo, Z.W.; Zhiheng Wang, Y.L.; Yue Lou, X.L.; Xianjin Li, H.L. Detection Method of Photovoltaic Panel Defect Based on Improved Mask R-CNN. *J. Internet Technol.* **2022**, *23*, 397–406. [CrossRef]
183. Oliveira, A.K.V.D.; Bracht, M.K.; Aghaei, M.; Rüther, R. Automatic Fault Detection of Utility-Scale Photovoltaic Solar Generators Applying Aerial Infrared Thermography and Orthomosaicking. *Sol. Energy* **2023**, *252*, 272–283. [CrossRef]
184. Hwang, H.P.-C.; Ku, C.C.-Y.; Chan, J.C.-C. Detection of Malfunctioning Photovoltaic Modules Based on Machine Learning Algorithms. *IEEE Access* **2021**, *9*, 37210–37219. [CrossRef]
185. Huang, Z.; Duan, S.; Long, F.; Li, Y.; Zhu, J.; Ling, Q. Fault Diagnosis for Solar Panels Using Convolutional Neural Network. In Proceedings of the 2021 33rd Chinese Control and Decision Conference (CCDC), Kunming, China, 22 May 2021; IEEE: New York, NY, USA, 2021; pp. 153–158.
186. Kim, B.; Serfa Juan, R.O.; Lee, D.-E.; Chen, Z. Importance of Image Enhancement and CDF for Fault Assessment of Photovoltaic Module Using IR Thermal Image. *Appl. Sci.* **2021**, *11*, 8388. [CrossRef]
187. Haidari, P.; Hajiahmad, A.; Jafari, A.; Nasiri, A. Deep Learning-Based Model for Fault Classification in Solar Modules Using Infrared Images. *Sustain. Energy Technol. Assess.* **2022**, *52*, 102110. [CrossRef]
188. Rocha, D.; Alves, J.; Lopes, V.; Teixeira, J.P.; Fernandes, P.A.; Costa, M.; Morais, M.; Salome, P.M.P. Multidefect Detection Tool for Large-Scale PV Plants: Segmentation and Classification. *IEEE J. Photovolt.* **2023**, *13*, 291–295. [CrossRef]
189. Segovia Ramírez, I.; Das, B.; García Márquez, F.P. Fault Detection and Diagnosis in Photovoltaic Panels by Radiometric Sensors Embedded in Unmanned Aerial Vehicles. *Prog. Photovolt. Res. Appl.* **2022**, *30*, 240–256. [CrossRef]
190. Eltuhamy, R.A.; Rady, M.; Almatrafi, E.; Mahmoud, H.A.; Ibrahim, K.H. Fault Detection and Classification of CIGS Thin-Film PV Modules Using an Adaptive Neuro-Fuzzy Inference Scheme. *Sensors* **2023**, *23*, 1280. [CrossRef] [PubMed]
191. Salazar, A.M.; Macabebe, E.Q.B. Hotspots Detection in Photovoltaic Modules Using Infrared Thermography. *MATEC Web Conf.* **2016**, *70*, 10015. [CrossRef]
192. Et-taleby, A.; Boussetta, M.; Benslimane, M. Faults Detection for Photovoltaic Field Based on K-Means, Elbow, and Average Silhouette Techniques through the Segmentation of a Thermal Image. *Int. J. Photoenergy* **2020**, *2020*, 6617597. [CrossRef]
193. Wang, X.; Yang, W.; Qin, B.; Wei, K.; Ma, Y.; Zhang, D. Intelligent Monitoring of Photovoltaic Panels Based on Infrared Detection. *Energy Rep.* **2022**, *8*, 5005–5015. [CrossRef]
194. Ngo, G.C.; Macabebe, E.Q.B. Image Segmentation Using K-Means Color Quantization and Density-Based Spatial Clustering of Applications with Noise (DBSCAN) for Hotspot Detection in Photovoltaic Modules. In Proceedings of the 2016 IEEE Region 10 Conference (TENCON), Singapore, 22–26 November 2016; IEEE: New York, NY, USA, 2016; pp. 1614–1618.
195. Glavas, H.; Vukobratovic, M.; Primorac, M.; Mustran, D. Infrared Thermography in Inspection of Photovoltaic Panels. In Proceedings of the 2017 International Conference on Smart Systems and Technologies (SST), Osijek, Croatia, 18–20 October 2017; IEEE: New York, NY, USA, 2017; pp. 63–68.
196. Sohani, A.; Sayyaadi, H.; Miremadi, S.R.; Samiezadeh, S.; Doranehgard, M.H. Thermo-Electro-Environmental Analysis of a Photovoltaic Solar Panel Using Machine Learning and Real-Time Data for Smart and Sustainable Energy Generation. *J. Clean. Prod.* **2022**, *353*, 131611. [CrossRef]
197. Rodríguez-Martín, M.; Lagüela, S.; González-Aguilera, D.; Arias, P. Cooling Analysis of Welded Materials for Crack Detection Using Infrared Thermography. *Infrared Phys. Technol.* **2014**, *67*, 547–554. [CrossRef]
198. Villar, M.; Garnier, C.; Chabert, F.; Nassiet, V.; Samélor, D.; Diez, J.C.; Sotelo, A.; Madre, M.A. In-Situ Infrared Thermography Measurements to Master Transmission Laser Welding Process Parameters of PEKK. *Opt. Lasers Eng.* **2018**, *106*, 94–104. [CrossRef]
199. Radivojevic, Z.; Kassamakov, I.; Oinonen, M.; Saarikko, H.; Seppanen, H.; Vihinen, P. Transient IR Imaging of Light and Flexible Microelectronic Devices. *Microelectron. Reliab.* **2006**, *46*, 116–123. [CrossRef]
200. Mousnier, M.; Sanchez, K.; Locatelli, E.; Lebey, T.; Bley, V. Lock-in Thermography for Defect Localization and Thermal Characterization for Space Application. *Microelectron. Reliab.* **2018**, *88–90*, 67–74. [CrossRef]
201. Su, L.; Zha, Z.; Lu, X.; Shi, T.; Liao, G. Using BP Network for Ultrasonic Inspection of Flip Chip Solder Joints. *Mech. Syst. Signal Process.* **2013**, *34*, 183–190. [CrossRef]
202. Su, L.; Shi, T.; Du, L.; Lu, X.; Liao, G. Genetic Algorithms for Defect Detection of Flip Chips. *Microelectron. Reliab.* **2015**, *55*, 213–220. [CrossRef]
203. Su, L.; Shi, T.; Liu, Z.; Zhou, H.; Du, L.; Liao, G. Nondestructive Diagnosis of Flip Chips Based on Vibration Analysis Using PCA-RBF. *Mech. Syst. Signal Process.* **2017**, *85*, 849–856. [CrossRef]

204. Park, H.; Choi, M.; Park, J.; Kim, W. A Study on Detection of Micro-Cracks in the Dissimilar Metal Weld through Ultrasound Infrared Thermography. *Infrared Phys. Technol.* **2014**, *62*, 124–131. [CrossRef]
205. Wiecek, B.; Baetselier, E.D.; Mey, G.D. Active Thermography Application for Solder Thickness Measurement in Surface Mounted Device Technology. *Microelectron. J.* **1998**, *29*, 223–228. [CrossRef]
206. Kijkanjanapaiboon, K.; Xie, M.; Fan, X. Investigation of Geometry, Frequency and Material's Effects in Lock-In Thermography Applications in Semiconductor Packages. In Proceedings of the 2016 IEEE 66th Electronic Components and Technology Conference (ECTC), Las Vegas, NV, USA, 31 May–3 June 2016; IEEE: New York, NY, USA, 2016; pp. 1423–1429.
207. Kijkanjanapaiboon, K.; Xie, M.; Zhou, J.; Fan, X. Investigation of Dimensional and Heat Source Effects in Lock-In Thermography Applications in Semiconductor Packages. *Appl. Therm. Eng.* **2017**, *113*, 673–683. [CrossRef]
208. Gao, Y.; Du, J.; Liu, F.; Dai, F.; Guo, W.; Liu, Y. A Novel Chip-Scale Heterogeneous Integration Intelligent Micro-System of Short Wave Infrared Imaging. *Infrared Phys. Technol.* **2023**, *132*, 104737. [CrossRef]
209. Liu, H.; Tinsley, L.; Addepalli, S.; Liu, X.; Starr, A.; Zhao, Y. Detectability Evaluation of Attributes Anomaly for Electronic Components Using Pulsed Thermography. *Infrared Phys. Technol.* **2020**, *111*, 103513. [CrossRef]
210. Speka, M.; Matteï, S.; Pilloz, M.; Ilie, M. The Infrared Thermography Control of the Laser Welding of Amorphous Polymers. *NDT E Int.* **2008**, *41*, 178–183. [CrossRef]
211. Matteï, S.; Grevey, D.; Mathieu, A.; Kirchner, L. Using Infrared Thermography in Order to Compare Laser and Hybrid (laser+MIG) Welding Processes. *Opt. Laser Technol.* **2009**, *41*, 665–670. [CrossRef]
212. Lu, X.; Shi, T.; Han, J.; Liao, G.; Su, L.; Wang, S. Defects Inspection of the Solder Bumps Using Self Reference Technology in Active Thermography. *Infrared Phys. Technol.* **2014**, *63*, 97–102. [CrossRef]
213. Hadis, M.A.; Cooper, P.R.; Milward, M.R.; Gorecki, P.C.; Tarte, E.; Churm, J.; Palin, W.M. Development and Application of LED Arrays for Use in Phototherapy Research. *J. Biophotonics* **2017**, *10*, 1514–1525. [CrossRef]
214. Israelsen, N.M.; Petersen, C.R.; Barh, A.; Jain, D.; Jensen, M.; Hannesschläger, G.; Tidemand-Lichtenberg, P.; Pedersen, C.; Podoleanu, A.; Bang, O. Real-Time High-Resolution Mid-Infrared Optical Coherence Tomography. *Light Sci. Appl.* **2019**, *8*, 11. [CrossRef]

MDPI AG

Grosspeteranlage 5

4052 Basel

Switzerland

Tel.: +41 61 683 77 34

Sensors Editorial Office

E-mail: sensors@mdpi.com

www.mdpi.com/journal/sensors